Vision Models for High Dynamic Range and Wide Colour Gamut Imaging

Techniques and Applications

Computer Vision and Pattern Recognition Series

Series Editors

Also in the Series:

Lin and Zhang, Low-Rank Models in Visual Analysis: Theories, Algorithms and Applications, 2017, ISBN: 9780128127315
Zheng et al., Statistical Shape and Deformation Analysis: Methods, Implementation and Applications, 2017, ISBN: 9780128104934
De Marsico et al., Human Recognition in Unconstrained Environments: Using Computer Vision, Pattern Recognition and Machine Learning Methods for Biometrics, 2017, ISBN: 9780081007051
Saha et al., Skeletonization: Theory, Methods and Applications, 2017, ISBN: 9780081012918
Zheng et al., Statistical Shape and Deformation Analysis: Methods, Implementation and Applications, 2017, ISBN: 9780128104934
Murino et al., Group and Crowd Behavior for Computer Vision, 2017, ISBN: 9780128092767
Leo and Farinella, Computer Vision for Assistive Healthcare, 2018, ISBN: 9780128134450
Alameda-Pineda et al., Multimodal Behavior Analysis in the Wild, 2019, Advances and Challenges, ISBN: 9780128146019
Wang et al., Deep Learning Through Sparse and Low-Rank Modeling, 2019, ISBN: 9780128136591
Ying Yang, Multimodal Scene Understanding: Algorithms, Applications and Deep Learning, 2019, ISBN: 9780128173589
Hua et al., Spectral Geometry of Shapes: Principles and Applications, 2019, ISBN: 9780128138427

Vision Models for High Dynamic Range and Wide Colour Gamut Imaging

Techniques and Applications

Marcelo Bertalmío

Academic Press is an imprint of Elsevier
125 London Wall, London EC2Y 5AS, United Kingdom
525 B Street, Suite 1650, San Diego, CA 92101, United States
50 Hampshire Street, 5th Floor, Cambridge, MA 02139, United States
The Boulevard, Langford Lane, Kidlington, Oxford OX5 1GB, United Kingdom

Library of Congress Cataloging-in-Publication Data
A catalog record for this book is available from the Library of Congress

British Library Cataloguing-in-Publication Data
A catalogue record for this book is available from the British Library

ISBN: 978-0-12-813894-6

For information on all Academic Press publications
visit our website at https://www.elsevier.com/books-and-journals

Publisher: Mara Conner
Acquisition Editor: Tim Pitts
Editorial Project Manager: Lindsay Lawrence
Production Project Manager: Nirmala Arumugam
Designer: Christian J. Bilbow

Typeset by VTeX

A Luis, Jose, Mónica y Milagros.

A Vera, Lucas, Serrana y Graciela.

En recuerdo de mi padre.

Advance Praise for "Vision Models for High Dynamic Range and Wide Colour Gamut Imaging: Techniques and Applications"

Dr. Bertalmío has written a very useful text that connects key perceptual phenomena to image reproduction. His text nicely explains the scientific and engineering foundations for many image computations, including gamut mapping, tone mapping, and image quality metrics. The material is a thoughtful reference for people who wish to understand how analyses of human visual perception have been transformed into applications that measure image quality. I am particularly enthusiastic about his call for new vision models to guide the future generation of displays and rendering!

Professor Brian A. Wandell, Director of the Center for Cognitive and Neurobiological Imaging, Stanford University, USA

A lot of our colour science relies on experiments done decades ago, in viewing conditions that have very little to do with how we consume dramatic content today. Marcelo Bertalmío's book manages to move the discussion forward into the 21st century – by describing in detail how we see and process visual content, and the huge role that adaptation and efficient encoding of the visual signal play within it. But also by directly relating this knowledge to modern high dynamic range display technology. This book is essential reading for anyone who is interested in colour science for visual media, and I will keep bugging my colleagues at FilmLight to read it and to apply the wealth of knowledge contained in it towards our advancement of visual storytelling technology.

Wolfgang Lempp, founder and director, FilmLight Ltd, UK

A comprehensive and rare synthesis of vision science and imaging technology. This book is the go-to source for those wishing to learn more about the neuroscience and psychophysics of vision and its implications for the design of imaging and display pipelines to achieve highest quality for the human observer. Such a book is long overdue.

Professor Bruno Olshausen, Helen Wills Neuroscience Institute and School of Optometry, UC Berkeley, USA

This book provides an in-depth background on vision science for engineers working on imaging applications. It clearly explains how HDR imaging must essentially utilize the adaptation characteristics of the human visual system, and covers practical solutions for HDR/SDR and WCG production based on a deep knowledge of vision models.

Dr. Kenichiro Masaoka, NHK (Japan Broadcasting Corporation) Science & Technology Research Laboratories, Japan

This is a great book for anyone interested in tone and gamut mapping. But this is also a great book for vision scientists in general, linking our basic understanding of human vision gained through neuroscience, psychophysics and the theory of efficient coding to applications in imaging and cinema — and it nicely highlights the fundamental limitations of our current understanding of the human visual system. I thoroughly enjoyed reading Marcelo Bertalmío's book and I hope it finds the wide readership it deserves.

Professor Felix Wichmann, Tübingen University and Max Planck Institute, Germany

Contents

About the author

Marcelo Bertalmío (Montevideo, 1972) is a full professor at Universitat Pompeu Fabra, Spain, in the Information and Communication Technologies Department. He received the B.Sc. and M.Sc. degrees in electrical engineering from the Universidad de la República, Uruguay, and the Ph.D. degree in electrical and computer engineering from the University of Minnesota in 2001. He was awarded the 2012 SIAG/IS Prize of the Society for Industrial and Applied Mathematics (SIAM) for co-authoring the most relevant image processing work published in the period 2008–2012. Has received the Femlab Prize, the Siemens Best Paper Award, the Ramón y Cajal Fellowship, and the ICREA Academia Award, among other honours. He was Associate Editor for SIAM-SIIMS and elected secretary of SIAM's activity group on imaging. Has obtained an ERC Starting Grant for his project "Image processing for enhanced cinematography" and two ERC Proof of Concept Grants to bring to market tone mapping and gamut mapping technologies. He's co-coordinator of two H2020 projects, HDR4EU and SAUCE, involving world-leading companies in the film industry. Has written a book titled "Image Processing for Cinema", published by CRC Press in 2014, and edited the book "Denoising of Photographic Images and Video" published by Springer in 2018. His current research interests are in developing image processing algorithms for cinema that mimic neural and perceptual processes in the visual system, and to investigate new vision models based on the efficient representation principle, with fine-tuning by movie professionals.

Acknowledgements

I'm extremely grateful to all the researchers that I've had the pleasure to collaborate with, and learn from, on the topics covered in this book: Thomas Batard, Luca Calatroni, Trevor Canham, Praveen Cyriac, Valentina Franceschi, Benedetta Franceschiello, Adrián Galdrán, Raquel Gil Rodríguez, Alexander Gómez Villa, Antoine Grimaldi, David Kane, Jesús Malo, Adrián Martín, Dario Prandi, Yasuko Sugito, Javier Vázquez Corral, Jihyun Yeonan-Kim, Itziar Zabaleta and Syed-Waqas Zamir.

Many thanks to Harald Brendel, Jay Cassidy, Stephane Cattan, Jack Cowan, Remi Cozot, Jan Fröhlich, Jean-Philippe Jacquemin, Pierre Jasmin, Dirk Maes, John McCann, Albert Pascual, Alessandro Rizzi, Daniele Siragusano, and Beau Watson, for their help and the opportunity of engaging in very fruitful discussions.

José María Rojano has been of invaluable help for my work. And the administrative and research support personnel at UPF are outstanding, I'm indebted to Neus Balaguer, Magda Castellnou, Judith Champion, Joana Clotet, Mònica Díaz, Tomás Escuin, Lydia García, Vanessa Jiménez, Jana Safrankova, Ana Sagardoy, Joana Soria, and Ruth Temporal, with special thanks to Carme Buisan and Aurelio Ruiz.

A heartfelt thank you to Josep Blat, for his unwavering support.

This book was written while on sabbatical leave at Universidad Politécnica de Madrid, at Narciso García's GTI Lab: all my gratitude to him and his wonderful team for making me feel at home.

Many thanks to artists Javier Baliosian and Federico Lecumberry for their help with the illustrations. And to Lindsay Lawrence, Indhumathi Mani and Tim Pitts from Elsevier, for their assistance and, especially, their patience.

Very special thanks to Wolfgang Lempp, Kenichiro Masaoka, Bruno Olshausen, Brian A. Wandell and Felix Wichmann.

Finally, I want to acknowledge the support of the public funds that have made our lab's research (and this book) possible. Our work has received funding from: the EU Horizon 2020 programme under grant agreements 761544 (project HDR4EU) and 780470 (project SAUCE); the European Research Council through Starting Grant 306337 (Image processing for enhanced cinematography), ERC Proof of Concept Grant 2018 (GAMMAVISION) and ERC Proof of Concept Grant 2016 (VIPERCON); the ICREA Academia Award; and the Spanish Government and FEDER Fund through grants TIN2012-38112, TIN2015-71537-P (MINECO/FEDER, UE) and PGC2018-099651-B-I00 (MCIU/AEI/FEDER, UE).

Marcelo Bertalmío
Madrid, November 2019

CHAPTER

Introduction

1

1.1 HDR and WCG

The media industry is continuously striving to improve image quality and to enhance the overall viewing experience, with higher frame rates, larger resolution, more vivid colours, and greater contrast. Currently there is significant emphasis on high dynamic range (HDR) and wide colour gamut (WCG) imaging, let's briefly see what do these concepts mean.

The contrast in a scene is measured by the ratio, called dynamic range, of light intensity values between its brightest and darkest points. While common natural scenes may have a contrast of 1,000,000:1 or more, our visual system allows us to perceive contrasts of roughly 10,000:1, much greater than the available simultaneous contrast range of traditional cameras and screens. For this reason, the capture and display of content with a dynamic range approaching that of real scenes has been a long-term challenge, limiting the ability to reproduce more realistic images. Currently it is not uncommon for state-of-the-art camera sensors to have a dynamic range of 3 or 4 orders of magnitude, therefore matching in theory the range of human visual perception; computer generated images can have an arbitrarily high dynamic range; and recent display technologies have made brighter TV sets available without increasing the luminance of the black level, which has enabled the appearance of HDR displays.

The colour gamut of a display is the set of colours its capable of reproducing, and the wider the gamut is the closer the displayed images get to represent the colours we naturally perceive. The colour gamut of a display depends mostly on the properties of the light primaries it uses, and new display technologies like laser RGB projectors have monochromatic primaries with such a high colour purity that their colour gamut is truly wide, covering virtually every colour found in nature.

HDR and WCG are independent image attributes [1], i.e. a picture may have a high dynamic range but a reduced colour gamut, and the other way round. But there is definitely a connection since HDR displays, just by being brighter, can reproduce more saturated colours than what a SDR display can achieve.

1.2 Improved image appearance with HDR/WCG

An HDR system is specified and designed for capturing, processing, and reproducing a scene, conveying the full range of perceptible shadow and highlight detail, with

Vision Models for High Dynamic Range and Wide Colour Gamut Imaging. https://doi.org/10.1016/B978-0-12-813894-6.00006-5

sufficient precision and acceptable artefacts, including sufficient separation of diffuse white and specular highlights and separation of colour details, far beyond current SDR (Standard Dynamic Range) video and cinema systems capabilities [2].

HDR/WCG technology can provide a never before seen increase in contrast, colour and luminance and is lauded in the creative sector as a truly transformative experience: for the cinema and TV industries, HDR represents *"the most exciting format to come along since colour TV"* [3]. Besides the enthusiasm about the unparallelled improvement in picture quality for new content, the industry expects, through repurposing for HDR, to monetise existing film and TV libraries once HDR screens are established, in theatres, at home and on mobile devices.

The ultra high-definition (UHD) TV community has realised that the transition from high-definition (HD) to the 4K or 8K resolutions represent an insufficient improvement in viewing experience for the audience, therefore UHD is being aligned with HDR since the visual experience of material displayed on an HDR screen is significantly richer, looks more lifelike, and produces a better sense of immersion [4]. As a content creator puts it: *"Unlike other emerging distribution technologies such as Stereo 3D, high-frame-rate exhibition, wide gamuts, and ever-higher resolutions (4K, 8K) which engender quite a bit of debate about whether or not they're worth it, HDR is something that nearly everyone I've spoken with, professional and layperson alike, agree looks fantastic once they've seen it [...] Furthermore, it's easy for almost anyone to see the improvement, no matter what your eyeglass prescription happens to be"* [5].

The contribution of HDR to the sense of immersion stems from the fact that HDR images appear much more faithful to an actual perceived scene, allowing the viewer to resolve details in quite dark or quite bright regions, to distinguish subtle colour gradations, to perceive highlights as much brighter than diffuse white surfaces, all things that we associate with everyday experiences and that cannot be reproduced using SDR systems. Content creators that have been exposed to HDR are extremely enthusiastic about it, because they have realised that it allows them to overcome the artistic limitations (in terms of colour and contrast) imposed by SDR systems, giving them *"the perceptual tools that fine artists working in the medium of painting have had for hundreds of years"* [5]. With HDR/WCG technologies, colourists have previously unavailable means for creating dramatically differentiated planes of highlights, for directing the viewer's gaze by sprinkling HDR highlights strategically across the image, for letting the brighter mid tones of an image breathe, for surprising the audience with sudden flares or cuts to bright frames and, in general, for creating compelling narrative opportunities in storytelling [5].

1.3 The need for an HDR ecosystem

At present many if not most of the new TV sets in the market are HDR, and they generally have a wider colour gamut as well. Some TV programming is broadcast in

HDR, there's HDR streaming of select movies and series, most major new-release discs come in HDR, and there's substantial re-mastering of older movies for HDR.

But we are still a long way from a fully functional HDR ecosystem, upon whose existence depends the realisation of all the possibilities that the HDR format could offer in terms of revenue and market growth for companies and in terms of improved user experience for viewers. Very recent reports from several international organisations and standardisation bodies concur in that there are a number of challenges that need to be addressed for a successful and complete adoption of HDR technology.

For instance, a September 2018 guideline [6] by the UltraHDForum, an industry organisation of technology manufacturers set up to promote next generation UHD TV technology that has embraced HDR as a key component in the UHD roadmap, states that *"no de facto standards emerged which define best practices in configuring colour grading systems for 2160p/HDR/WCG grading and rendering."*

A report [7] by the International Telecommunication Union-Radiocommunication (ITU-R) from April 2019 states that: *"As HDR-TV is at a formative stage of research and development as presented in this Report, a call for further studies is made, in particular on the characteristics and performance of the recommended HDR-TV image parameter values, for use in broadcasting [...] The introduction of HDR imagery poses a number of housekeeping challenges, associated with the increased number of picture formats that will be in use."*

And the "Study Group report on an HDR imaging ecosystem" [2] by the Society for Motion Picture and Television Engineers (SMPTE) states that: *"The HDR Ecosystem needs imaging performance requirements that must be met with sufficient precision [...]. There is a need for a better understanding of the set of elements, including standards, required to form a complete functional and interoperable ecosystem for the creation, delivery and playback of HDR image content. The parameters that make up HDR [...] are beyond the capabilities of existing standards to deliver."*

1.4 Problems to solve to enable an HDR ecosystem

1.4.1 New colour management and grading tools

There is a need for colour management and grading tools so that the mastering process has a simple workflow that can deliver consistent images across the whole spectrum of possible display devices and environments.

The desire is to make sure that the standard post-production pipeline can handle smoothly the latest generation of HDR camera-shot footage, and that the final delivered footage has the fullest range for the new array of HDR displays. It is also essential to guarantee that the look specified by the cinematographer is applied correctly, and kept consistent and as expected throughout the whole chain.

The situation is significantly complex as there is not just a single "HDR", but a whole family of device technologies, formats and viewing environments. Differences between implementations and for varying display luminances are subtle but can be very significant, making the task really difficult: what might be acceptable in a TV showroom as an HDR viewing experience is not enough in a mastering suite.

1.4.2 New projection systems for movie theatres

Traditional digital cinema projectors rely on an illumination system that delivers a flat-field illumination onto a two dimensional light valve. The light valve will send the light that is needed to form the image on-screen towards the projection lens and either absorbs or directs the remainder of light away from the projection lens. The average light level of conventional cinema images has been evaluated to be below 10% of its maximum value, and thus the majority of the light is being blocked inside the projector. When moving to HDR cinema content, the average light level relative to the peak white brightness is expected to drop below 2%, so the conventional projector illumination approach becomes even more inefficient.

From the above, light-valve technology is limited in peak brightness, so the only way to achieve HDR with it would be to lower the black level. The sequential contrast is defined as the luminance of white, measured on a fully white screen, divided by the luminance of black, measured when the screen is fully dark. With a single, standard DLP projector the sequential contrast is around 2,000:1, and can be extended to 6,000:1 with state-of-the-art technology. Dolby Vision cinemas use two modulators in series, so the contrast is multiplied and Dolby claims values of 1,000,000:1. Recent studies have assessed the perceivable dynamic range in a cinema environment, considering the influence on the final contrast of scattering in the projection optics and room reflections. It was found that only in a very limited amount of image frames with extremely low average picture level, an improved projector black level can effectively result in an extended dynamic range. As soon as the average picture level rises, reflections by the room will dominate the on-screen contrast ratio. In addition, as the eye adapts to a higher average luminance level, the black detection threshold rises.

As such, with the current light valve technology, it appears impossible to substantially raise the white level, even with setups consisting of multiple projectors, and the actual gain in black level in a movie theatre by using two light valves in sequence is on average much lower than the theoretical improvement.

The conclusion is that there is a need for new projection systems for bringing HDR to movie theatres, since traditional cinema projector technology is not capable of providing images with a dynamic range that is actually high. Also, the conventional projector illumination approach is highly inefficient in terms of energy consumption.

1.4.3 Tools for conversion between HDR and SDR

Tone mapping (TM) and inverse tone mapping (ITM) algorithms are expected to be used at several stages of the HDR production chain. For instance, real-time TM will be used during shoots for on-set monitoring of HDR material (cinema, broadcast) on SDR displays, for combining HDR and SDR sources (live broadcast), and in cinema projection systems for presenting HDR material on SDR projectors. Real-time ITM is to be used for broadcast of SDR material over an HDR channel and, in cinema projection systems, for presenting SDR material on HDR projectors. Other methods for TM and ITM, allowing for interaction but not necessarily real-time, are to be used in post-production and grading as tools for performing a single grade regardless of the dynamic range of source and output.

Current TV programs are exchanged among broadcasters without difficulty in terms of tone mapping, because all program producers have similar understanding on how to map the scene from 0 to 100% video level, which is the "reference white level": this practice will not work for HDR systems, and a new practical guideline is needed, which should also be effective for ensuring consistency between SDR and HDR viewing. Efficient and high quality conversion among HDR and SDR, in both directions, is essential for the consolidation of the HDR format given that for the foreseeable future a mix of both SDR and HDR devices and techniques will coexist: the need to merge SDR and HDR content is inevitable. To avoid the massive cost of a complete system overhaul, as HDR technologies enter broadcast production, a single-stream, merged path of HDR and SDR content is economically necessary.

1.4.4 Tools for colour gamut conversions

The colour gamut of a device is the set of colours that it can reproduce, and brighter displays (as those associated with emerging technologies for HDR TV sets) have a larger colour volume. Pointer analysed many samples of frequently occurring real surface colours and derived what is commonly known as "Pointer's gamut". Although both the standard colour gamuts for cinema (DCI-P3) and TV (BT.709) cover a reasonable amount of Pointer's gamut, many interesting real world colours fall outside them. In 2012 ITU-R recommended a new standard gamut for the next generation UHDTV, called BT.2020, that encompasses DCI-P3 and BT.709 and covers 99.9% of Pointer's gamut.

New laser projectors are able to cover the very wide BT.2020 gamut, but if the inputs are movies with DCI-P3 gamut, as virtually all professional movies currently are, the full colour rendering potential of these new projectors cannot be realised. There is a pressing need then to develop gamut extension techniques that enlarge the gamut of the movie content while ensuring that the gamut extended result preserves as much as possible the artistic intent of the content creator.

Likewise, there is need for automatic gamut reduction techniques, for display of wide-colour-gamut HDR content on current screens, whose vast majority has a colour gamut no larger than DCI-P3. Professional movies are colour-graded in DCI-P3 and thus the final result is ready for cinema, but its gamut must be modified to fit into BT.709 for TV, disc release or streaming: this is done through a gamut modification process based on the use of three-dimensional look-up tables (LUTs). These LUTs contain millions of entries and colourists only specify a few colours manually, while the rest are interpolated without taking care of their spatial or temporal context. Subsequently, the resulting video may have false colours that were not present in the original material and intensive manual correction is usually necessary, commonly performed in a shot-by-shot, object-by-object basis.

In TV production, gamut mapping is performed in real-time with the use of predetermined LUTs, and therefore issues are unavoidable. There is also the need for gamut mapping when the output is an AR/VR headset: this type of technology mostly uses OLED screens, where the gamut is quite wider than BT.709, but not nearly as wide as BT.2020, hence either gamut extension or gamut reduction are desirable, depending on the input content.

1.4.5 Production and editing guidelines for HDR material

The current take on HDR imaging is still rather conservative: for instance, the standard encoding curve ST2084 aims to *"enable the creation of video images with an increased luminance range, not for the creation of video images with overall higher luminance levels,"* that is, ST2084 HDR does not attempt to make the whole image brighter, only highlight detail such as chrome reflections or light sources [8]. The movie industry, on the other hand, would like to assess what is the complete range of possibilities that HDR allows, so that movie makers can exploit the full potential of the technology and bring the medium to a new level.

This would require the elaboration of a set of guidelines to assist content creators in shooting and processing HDR material, and these guidelines will have to address a number of issues that may arise due to the complex and not yet fully understood interactions of HDR images with the human visual system, including: bright light exposure reduces pupil size, a shift from rod-dominated vision to cone-dominated vision when light conditions brighten, bright lights can reduce retinal sensitivity and produce after-images, adaptation to changes of average luminance levels may take seconds, high luminance can affect the perception of flicker and judder artefacts, fast cuts with high brightness elements might cause eye strain, the appearance of cross dissolves and fades in-out will be different in HDR movies, there might be a need of different production procedures if the intended output is TV or large cinema screens because of expected differences in eye-tracking, etc.

1.4.6 Tools for personalisation

The cinema industry has a successful, proven record of ensuring that moving pictures have the intended, optimal contrast and colour in the controlled scenarios given by cinema theatres and home TV viewing, and post-production is always performed for these two types of viewing conditions. But we are currently living an explosive growth of video consumption on mobile devices. Some manufacturers have started to tackle this issue with partial, proprietary solutions, but since media convergence and movie watching "on the go" may happen under a variety of very much uncontrolled viewing conditions, we can expect that image appearance will vary as well, both fluctuating as the surroundings change and departing from the creator's intent, sometimes very noticeably as in the case of memory colours and skin tones, which colourists take pains to preserve during post-production but only for a home TV or cinema screen output. A related issue is that individual differences in colour perception become more prominent as the colour capabilities of displays are improved.

1.5 Overview of the book

This book is focused on solutions, based on vision science, for several of the issues just mentioned but especially for the problems of tone and gamut mapping.

We start with an overview of the biology of vision. In Chapter 2 we will describe the impact of the optics of the eye in the formation of the retinal image, the layered structure of the retina and its different types of cell, and the neural interactions and transmission channels by which information is represented and conveyed. Whenever possible we try to stress two ideas: first, that many characteristics of the retina and its processes can be explained in terms of maximising efficiency, because they simply are the optimal choices; this concept of efficient representation is key for our applications and is the underlying theme in most of the book. And second, that much is still unknown about how the retina works.

In Chapter 3 we will describe the layered structure, cell types and neural connections in the lateral geniculate nucleus and the visual cortex, with an emphasis on colour representation. We will also introduce linear+nonlinear (L+NL) models, which are arguably the most popular form of model not just for cell activity in the visual system but for visual perception as well. An important take-away message is that despite the enormous advances in the field, the most relevant questions about colour vision and its cortical representation remain open: which neurons encode colour, how does the cortex transform the cone signals, how shape and form are perceptually bound, and how do these neural signals correspond to colour perception. Another important message is that the parameters of L+NL models change with the image stimulus, and the effectiveness of these models decays considerably when they are tested on natural images. This has grave implications for our purposes, since in colour imaging many essential methodologies assume a L+NL form.

In Chapter 4 we discuss adaptation and efficient representation. Adaptation is an essential feature of the neural systems of all species, a change in the input–output relation of the system that is driven by the stimuli and that is intimately linked with the concept of efficient representation. Through adaptation the sensitivity of the visual system is constantly adjusted taking into account multiple aspects of the input stimulus, matching the gain to the local image statistics through processes that aren't fully understood and contribute to make human vision so hard to emulate with devices. Adaptation happens at all stages of the visual system, from the retina to the cortex, with its effects cascading downstream; it's a key strategy that allows the visual system to deal with the enormous dynamic range of the world around us while the dynamic range of neurons is really limited.

In Chapter 5 we discuss brightness perception, the relationship between the intensity of the light (a physical magnitude) and how bright it appears to us (a psychological magnitude). It has been known for a long time that this relationship is not linear, that brightness isn't simply proportional to light intensity. But we'll see that determining the brightness perception function is a challenging and controversial problem: results depend on how the experiment is conducted, what type of image stimulus are used and what tasks are the observers asked to perform. Furthermore, brightness perception depends on the viewing conditions, including image background, surround, peak luminance and dynamic range of the display, and, to make things even harder, it also depends on the distribution of values of the image itself. This is a very important topic for imaging technologies, which require a good brightness perception model in order to encode image information efficiently and without introducing visible artifacts.

Chapter 6 deals with colour. There are models that for simple stimuli in controlled environments can predict very accurately the colour appearance of objects, as well as the magnitude of their colour differences. These models were developed and validated for SDR images, and their extension to the HDR case is not straightforward. For the general case of natural images in arbitrary viewing conditions, there are many perceptual phenomena that come into play and no comprehensive vision model that is capable of handling them all in an effective way. As a result, the colour appearance problem remains very much open, and this affects all aspects of colour representation and processing.

In Chapter 7 we show how an image processing method for colour and contrast enhancement, that performs local histogram equalisation (LHE), is linked to neural activity models and the Retinex theory of colour vision. The common thread behind all these subjects is efficient representation. The LHE method can be extended to reproduce an important visual perception phenomenon which is assimilation, a type of visual induction effect. The traditional view has been that assimilation must be a cortical process, but we show that it can start already in the retina. The LHE method is based on minimising an energy functional through an iterative process. If the functional is regularised, the minimisation can be achieved in a single step by convolving the input with a kernel, which has important implications for the performance of algorithms based on LHE.

In Chapter 8 we show how the local histogram equalisation approach presented earlier can be used to develop gamut mapping algorithms that are of low computational complexity, produce results that are free from artifacts and outperform state-of-the-art methods according to psychophysical tests. Another contribution of our research is to highlight the limitations of existing image quality metrics when applied to the gamut mapping problem, as none of them, including two state-of-the-art deep learning metrics for image perception, are able to predict the preferences of the observers.

Based on recent findings and models from vision science, we present in Chapter 9 effective tone mapping and inverse tone mapping algorithms for production, post-production and exhibition. These methods are automatic and real-time, and they have been both fine-tuned and validated by cinema professionals, with psychophysical tests demonstrating that the proposed algorithms outperform both the academic and industrial state-of-the-art. We believe these methods bring the field closer to having fully automated solutions for important challenges for the cinema industry that are currently solved manually or sub-optimally. As in the case of gamut mapping, we also show that state-of-the-art image quality metrics are not capable of predicting observers' preferences for tone mapping and inverse tone mapping results.

Chapter 10 starts by discussing how to combine gamut mapping and tone mapping, so as to be able to transform images where input and output can both be of any type of dynamic range and colour gamut. Next we recount how to extend our tone mapping methodology to take into account surround luminance and display characteristics. This is followed by a number of applications where our tone mapping framework has proven useful: improved encoding of HDR video, photorealistic style transfer, image dehazing and enhancement of backlit images. Then we introduce two common types of colour problems that might be encountered with HDR image creation and editing, where the encoding curve plays a role, and we propose methods to solve these issues.

Finally, in Chapter 11 we present our conclusions:

- Imaging techniques based on vision models are the ones that perform best for tone and gamut mapping and a number of other applications.
- The performance of these methods is still far below what cinema professionals can achieve.
- Vision models are lacking, most key problems in visual perception remain open.
- Rather than be improved or revisited, a change of paradigm seems to be needed for vision models, moving away from a L+NL framework.

And we end by proposing that a way forward is to explore models based on local histogram equalisation, with fine-tuning by movie professionals. This approach yields very promising outcomes in tone and gamut mapping, but results can still be improved, and new vision models developed. Local histogram equalisation is intrinsically nonlinear, and it's closely related with theories that advocate that spatial summation by neurons is nonlinear, and with works using nonlinear time series analysis of oscillations in brain activity.

A change of paradigm in vision models, with intrinsically nonlinear frameworks developed by mimicking the techniques of cinema professionals, could clearly have a really wide impact, much wider than the HDR/WCG domain, given that the L+NL formulation is prevalent not only in vision science and imaging applications, it's the basis of artificial neural networks as well.

References

[1] François E, Fogg C, He Y, Li X, Luthra A, Segall A. High dynamic range and wide color gamut video coding in hevc: status and potential future enhancements. IEEE Transactions on Circuits and Systems for Video Technology 2016;26(1):63–75.

[2] Report SHSG. Tech rep., SMPTE. https://www.smpte.org/standards/reports, 2015.

[3] Goeller K. Building the HDR economy nit by nit. In: SMPTE 2015 annual technical conference and exhibition. SMPTE; 2015. p. 1–9.

[4] Reinhard E, François E, Boitard R, Chamaret C, Serre C, Pouli T. High dynamic range video production, delivery and rendering. SMPTE Motion Imaging Journal 2015;124(4):1–8.

[5] http://vanhurkman.com/wordpress/?p=3548.

[6] https://ultrahdforum.org/wp-content/uploads/Ultra-HD-Forum-Guidelines-Phase-A-v1.5-final.pdf.

[7] https://www.itu.int/dms_pub/itu-r/opb/rep/R-REP-BT.2390-6-2019-PDF-E.pdf.

[8] http://www.lightillusion.com/uhdtv.html.

CHAPTER

The biological basis of vision: the retina

2

The retina is a thin sheet of neural tissue that lines the back of the eye and transforms light into electrical signals.

The classical, textbook view of the retina has always been that of a remarkably effective but simple spatiotemporal prefilter, mainly performing adaptation to ambient light, wavelength discrimination (in order to see colours) and contrast enhancement, then sending the visual image to the brain through a bundle of fibers, the optic nerve, so that the cortex can perform the really complex processing required by vision [1].

It has now become apparent that the retina is doing much more than originally assumed, including tasks that were previously thought could only take place in the brain, like orientation detection, object motion detection, motion prediction and very fast contrast adaptation.

In this chapter we will describe the impact of the optics of the eye in the formation of the retinal image, the layered structure of the retina and its different types of cell, and the neural interactions and transmission channels by which information is represented and conveyed.

Whenever possible we try to stress two ideas: first, that many characteristics of the retina and its processes can be explained in terms of maximising efficiency, because they simply are the optimal choices; this concept of efficient representation is key for our applications and is the underlying theme in most of the book.

And second, that much is still unknown about how the retina works.

2.1 How the retinal image is created

In order to arrive at the retina, light that reaches the eye at the cornea has to go through the anterior chamber, then the iris through its opening the pupil, next the crystalline lens, and finally the vitreous humour [2], see Fig. 2.1. About 3% of the light is reflected at the cornea and 50% is absorbed by the pigment in ocular media, leaving some 47% of the light to reach the retina. But since photoreceptors are spaced, some of the quanta that get to the retinal surface fall between photoreceptors and are lost to the visual system [3]. All in all, the light that is finally absorbed by photoreceptors is less than 10% of the total light entering the eye [4].

The cornea is a thin transparent layer. The anterior chamber is filled with a liquid called the aqueous humour. The iris is a muscle with a central hole, the pupil, whose diameter is modified by the contraction (via a circular set of muscles) or dilation (via

Vision Models for High Dynamic Range and Wide Colour Gamut Imaging. https://doi.org/10.1016/B978-0-12-813894-6.00007-7

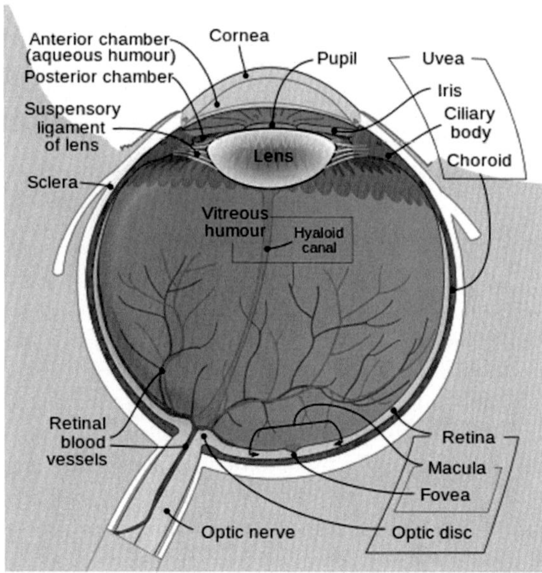

FIGURE 2.1 Eye diagram

Eye diagram. Image from [5].

a radial set of muscles) of the iris, thus regulating and limiting the amount of light entering the eye [6]. The difference in refractive index between the air and the cornea, combined with the spherical shape of the cornea, have the effect of making parallel light rays that come from a far-away point in the scene and reach our eye, be refracted at the air-cornea interface so that they cross again at the retina, at the back of the eye. When we look at a far-away object, our eye can be in a relaxed state and the image of the object is formed on the retina. On the other hand, if the object is close and the eye is relaxed, its image will form behind the retina so we would be seeing it blurry, see Fig. 2.2(A). It's only for closer points that the crystalline lens, an elastic bi-concave element of slightly different refraction index from that of the aqueous and vitreous humour that surround it, changes shape by being contracted by ciliary muscles that make the crystaline thicker and hence reduce its focal length so that the image forms on the retina, see Fig. 2.2(B). This is the phenomenon known as accommodation, and it assists our estimation of depth.

Given that the eye's focal length is ~ 17 mm, Fig. 2.3 shows how we can estimate the size S_i of the retinal image of an object of size S_o that is at a distance d_o from the eye:

$$S_i = S_o \frac{17 \text{ mm}}{d_o}.$$

As Cornsweet points out in [3], there's a common misconception regarding the intensity of the retinal image, coming from a misunderstanding of the inverse-square law, which states that the intensity of light falling on a surface (in units such as quanta per second per surface area) is inversely proportional to the square of the distance

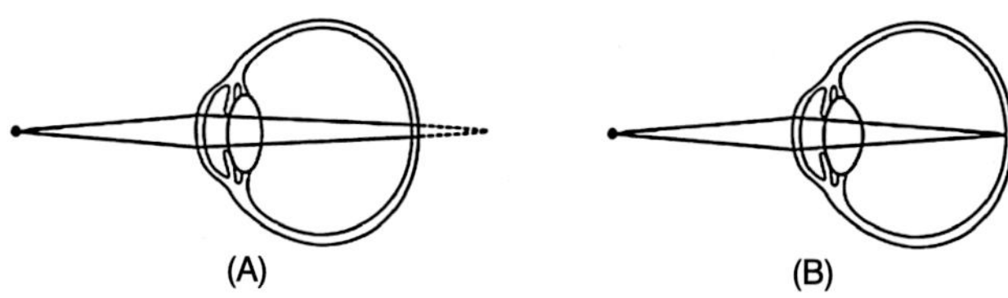

FIGURE 2.2 Accommodation

Left: close object is seen blurry. Right: accommodation makes its image form on the retina.

between the light source and the surface. Let's say an object at distance d produces a retinal image of area A where the light at each point of the retinal image is I_p; the retinal intensity is, by definition, proportional to the ratio I_p/A. By increasing the object's distance to, for instance, $2d$, the retinal intensity at each point is reduced, to $I_p/4$, so it would appear then that the retinal intensity must be reduced. But we must not forget that the retinal area of the image is reduced as well and by the same factor, to $A/4$, so therefore the retinal intensity does not change. In practice, so long as an object is away from our eyes a distance which is larger than 10 times the focal length of the eye (i.e. more than 17 cm away), then the retinal image will essentially have a constant intensity. Also, the intensity will be proportional to the pupil area, and the size of the retinal image will not depend on pupil size (in the same way that changing the camera aperture does not affect the size of the image formed in the camera sensor).

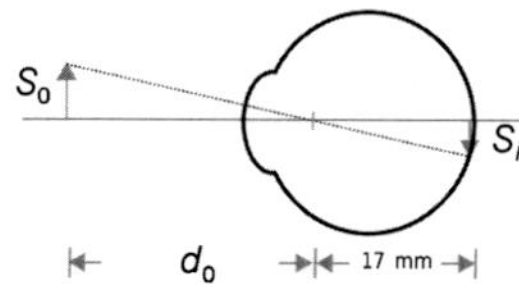

FIGURE 2.3

Retinal image size.

Different ocular media have different refractive indexes, e.g. for the aqueous humour it's 1.337, for the crystalline it's 1.42, for the vitreous humour it's 1.336 [2]. But these are only mean values, in each medium there are spatial irregularities in the refractive index that spread the light at large angles in a phenomenon called intraocular light scattering. The net result is that the retinal image contrast might be greatly reduced, especially in dark scenes with very bright light sources.

The diameter of the pupil is 2 mm in bright light conditions and more than 8 mm in a dark environment, corresponding, in photography terms, to aperture values or f-numbers of f/8 and f/2, respectively. When the aperture is really small, due to the wave nature of light there is diffraction at the pupil that distorts the image by introducing blur. When the aperture is wide, the optical imperfections of cornea and crystalline produce even more significant blur. The optical quality of the image decays noticeably with increasing eccentricity, as optical aberrations become more prominent under larger apertures, with more blur and less contrast. As a consequence, the density of photoreceptors and other retinal cells decreases with eccentricity, since a more sparse sampling of the retinal image is enough for off-center locations [2].

The eye as an optical system can be well described using Fourier optics, and the retinal image expressed as the convolution of the input image with a point spread function (PSF) that characterises the eye's response. Due to the factors mentioned above, the eye is not a perfect system and therefore its PSF isn't a δ function, but rather a blur kernel with very long tails, as seen in Fig. 2.4 The central part of the PSF is determined by the pupil aperture and cornea and lens aberrations, whereas the wide-angle part is due to the scattering. The central part of the PSF causes blur, while the tails reduce image contrast by allowing any bright point source to contribute to all points in the retinal image (remember that convolution with the PSF can be understood as placing a weighted PSF onto each point of the image and adding up all contributions). Fig. 2.5 shows a simulation of a retinal image, obtained by convolving a "radiance map" image (whose pixel values are proportional to light intensity) with the PSF of the eye.

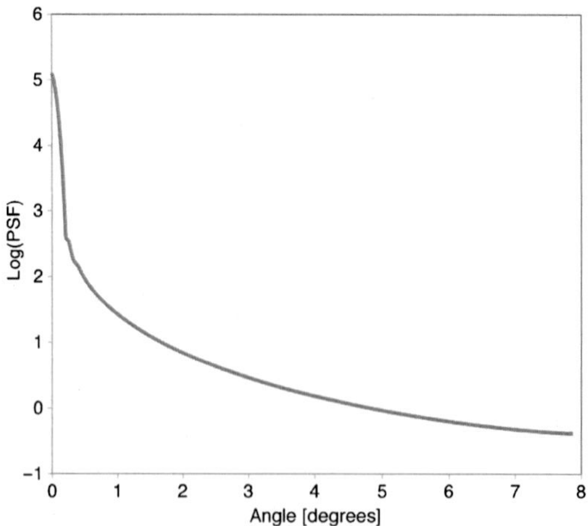

FIGURE 2.4 PSF of the eye

Plot shows the measurement of the PSF of the eye. Adapted from [2].

The aberrations discussed above are monochromatic, but there are chromatic aberrations as well, arising from the dependence of the refractive index on wavelength, which implies that long and short wavelength light will be imaged on the retina at different depths and blue colours are more blurred. Despite the fact that chromatic aberration is quite significant, around 2 diopters for the visual spectrum, its visual impact is very small for several reasons that include the strong filtering of short-wavelength (blue) light in the lens and macular pigment in the retina, and the higher sensitivity to longer wavelength light in retina photoreceptors, thus reducing the contribution of defocused colours in the blue end of the visible spectrum [2].

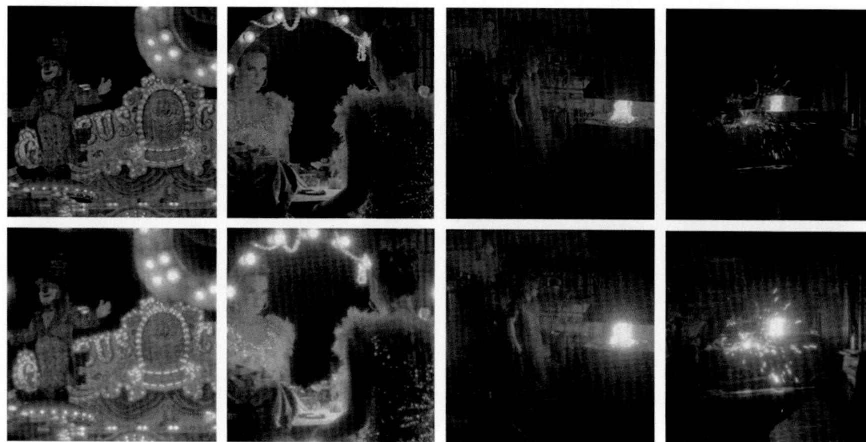

FIGURE 2.5 Blur and glare in the eye

Blur and glare in the eye. Top: original images, from [7]. Bottom: after convolution with PSF of human eye.

Media in the eye behave as a band-pass filter on the wavelengths reaching the retina, so that transmitted wavelengths match well the sensitivity of photoreceptors and blue light is absorbed more in order to mitigate chromatic aberration. Optical elements like the crystaline compensate for moderate amounts of aberration, and there is evidence of aberration compensation carried out by the visual system through neural adaptation. So even though the eye has a relatively low optical quality, as an optical system it's highly optimised [2].

2.2 Retinal structure

The retina is 0.25 mm thick and consists of five layers: there are three layers where neuron cell bodies are located, while their axons and dendrites make synaptic connections in the two in-between layers, called plexiform layers. See Fig. 2.6.

There are five main classes of cell in the retina: photoreceptors (rods and cones), bipolar cells (BCs), horizontal cells (HCs), amacrine cells (ACs), and retinal ganglion cells (RGCs); see Fig. 2.7. For each cell class there are several types, for a total of some $\sim$ 100 cell types. It's been established that there are 4 types of photoreceptors (one for rods and 3 for cones), and 3 types of HCs, but for the rest of cells we only have estimates: there are at least 12 types of BCs, 20–50 types of ACs and 20–30 types of RGCs [9–12]. The uncertainty in these numbers comes from the difficulty in identifying in a unique way the cell type's anatomical and physiological properties.

The groundbreaking work of Spanish biologist Santiago Ramón y Cajal in the late 19th century laid the foundations of modern neuroscience. By using Golgi staining techniques in cross-sections of the retina he discovered the existence of retinal layers

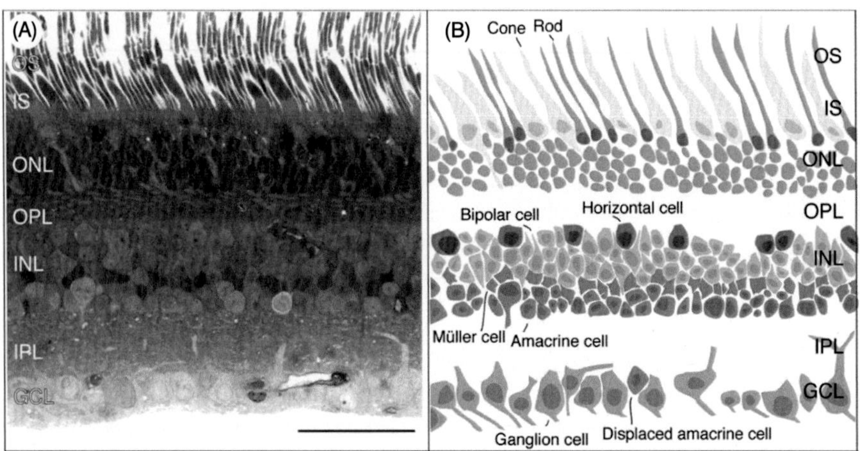

FIGURE 2.6 Retinal layers

Left: radial section through macaque monkey retina, scale bar equals 50 μm. Right: cell population in same area. Notation: OS, outer segments; IS, inner segments; ONL, outer nuclear layer; OPL, outer plexiform layer; INL, inner nuclear layer; IPL, inner plexiform layer; GCL, ganglion cell layer. Images from [8].

where the information traversed vertical pathways, from photoreceptors to BCs to RGCs, with lateral, horizontal interactions provided by HCs and ACs; he also showed how neurons were not randomly connected to one another, but rather specific cell types had synaptic connections to other specific cell types, and this happened in the plexiform layers where the corresponding dendrites and axons met at a precise depth [14]. See Fig. 2.8.

Later works used Golgi staining on sections parallel to the retinal surface, which allowed to discover new cell types and to better map the dendritic arbors. For each cell type, the dendritic fields of the cells of that type form a complete tiling of the retina, with more or less overlap. This means that every region in the retina is covered by at least one cell from each type, see Fig. 2.9. In fact, many researchers consider this to be a fundamental feature for determining a new cell type, whether or not it can tile the retina and hence ensure that visual information can be encoded by the same number of cells of that type at any point in the scene [15].

Also for each type, the dendritic field size becomes larger as the location of the cell is further away from the optical axis of the eye [12], see Fig. 2.10. The current view is that retinal neurons are arranged in semi-regular mosaics across the retinal surface [8]. All retinal cell types other than RGCs have as an ultimate purpose to process visual information to pass it along to specific types of RGCs, whose axons transmit it to the brain [15].

Going from the outer to the inner retina (i.e. from the back of the eye towards the pupil) the organisation is as follows [8], see Fig. 2.11:

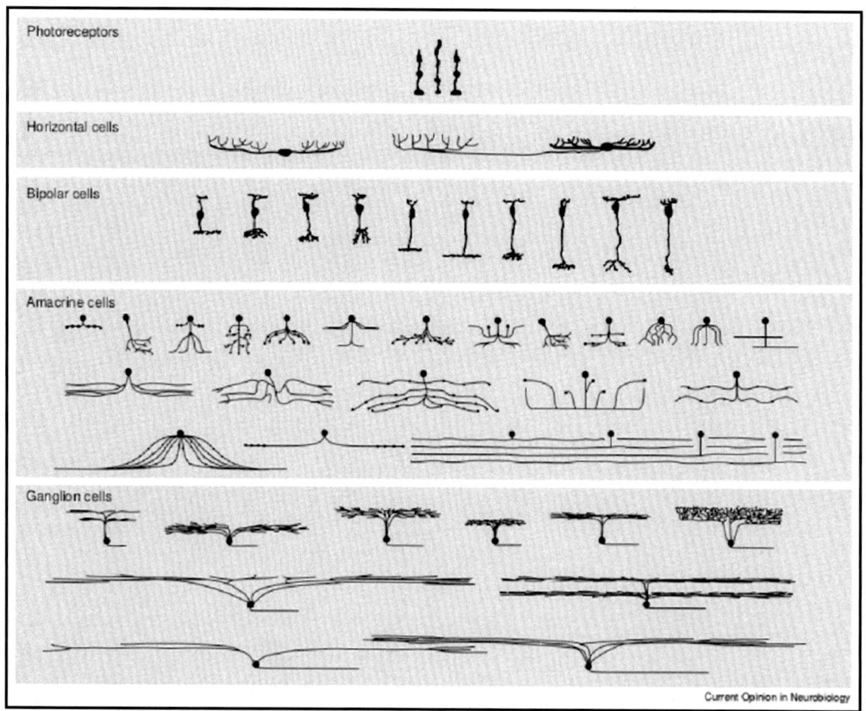

FIGURE 2.7 Retinal cell classes

Retinal cell classes: rod, cone, HC, BC, AC, RGC. Figure from [13].

- The Outer Nuclear Layer (ONL) contains the cell bodies of photoreceptors, which point *away* from the light and therefore have photosensitive segments in the deepest layer.
- The Outer Plexiform Layer (OPL) contains the synaptic terminals of photoreceptors and the dendrites of bipolar and horizontal cells.
- The Inner Nuclear Layer (INL) contains the cell bodies of bipolar, horizontal and most amacrine cells.
- The Inner Plexiform Layer (IPL) contains axon terminals of the bipolar cells, and dendrites of ACs and RGCs. The IPL has at least 10 thin sublayers, and usually the activity of any given cell type is restricted to just one sublayer.
- The Ganglion Cell Layer (GCL) contains the cell bodies of all RGCs and some amacrine cells, especially in the peripheral retina. RGC axons form the optic nerve, exiting the retina through the optic disc and connecting with the brain.

As photoreceptors are in the ONL, the light must traverse 4 retinal layers before it can stimulate rods and cones. This seems like a strange, not-too-convenient arrangement as light must travel through occluding elements like dendrites and axon terminals (other animals, like the octopus, have their photoreceptors in the inner retina and

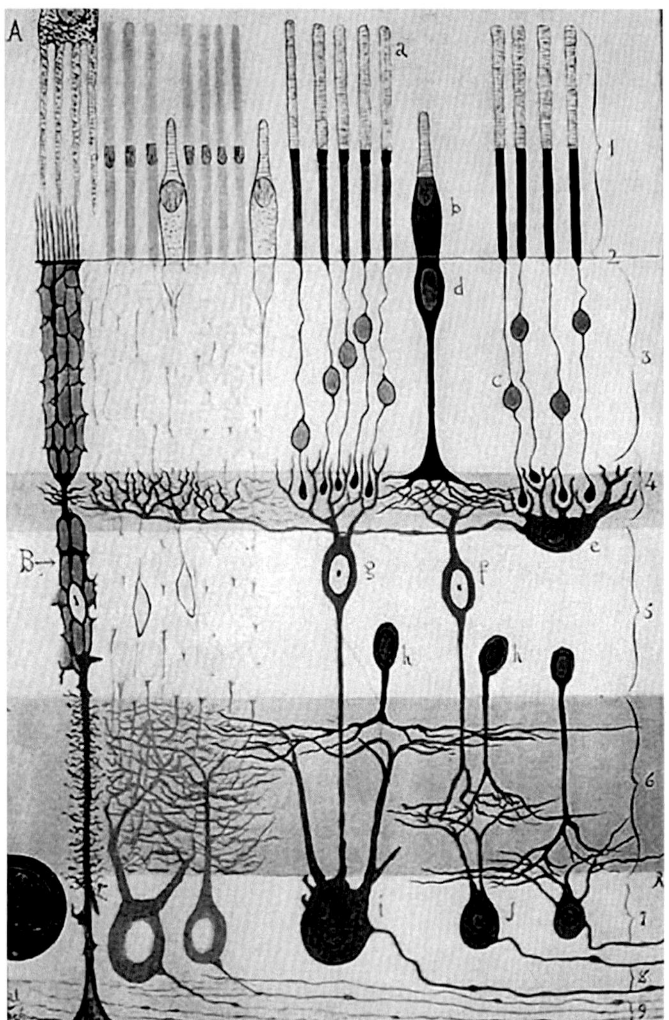

FIGURE 2.8 Drawing by Santiago Ramón y Cajal

A diagram of a vertebrate retina by Santiago Ramón y Cajal. Public domain image from "Structure of the mammalian retina", Madrid 1900.

the axons in the back), but it's been suggested [6] that a possible reason for this organisation might come from the fact that behind the photoreceptors there's a row of cells that contain the black pigment melanin, that acts as the black paint inside a camera, capturing all the remaining light that has reached the retina but not been absorbed by photoreceptors, and preventing it from bouncing back and scattering around; also, melanin helps to restore photoreceptor pigment after the chemical transduction produced by light. Furthermore, light is given more direct access to photoreceptors near

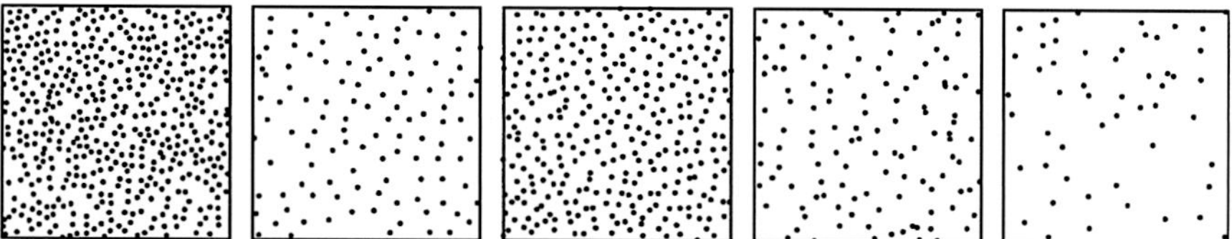

FIGURE 2.9 Each retinal cell type tiles the retina

Each retinal cell type tiles the retina: cells are distributed regularly and their dendritic trees don't leave points without coverage. From left to right, cell distributions for rods, cones, HCs and two types of ACs. Fig. from [16].

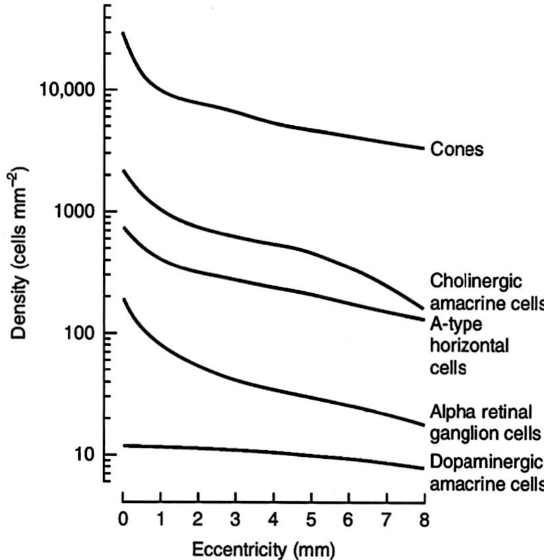

FIGURE 2.10 Dendritic field size increases with eccentricity

The dendritic field size increases with eccentricity, and as each cell type tiles the retina, cell density decreases with eccentricity. Fig. from [16].

the optical axis of the eye, where cones are most densely packed: here the primate retina has only one layer, formed just by cones, in a specialised region called the fovea that appears as a slight depression in the retinal surface (fovea means "pit" in Latin), where the neural tissue has been displaced to the sides and hence there are fewer opaque or not fully transparent elements in the path of light [3,4].

As mentioned above, the optic disc is the region of the retina where all the axons of the RGCs come together to form the optic nerve and leave the retina. Therefore, there aren't photoreceptors in the optic disc, and we have a blind spot in this area. We don't normally notice it because our visual system fills-in these gaps with information from the surround and the other eye, but there are easy experiments one can do that expose the blind spot, for instance the one shown in Fig. 2.13.

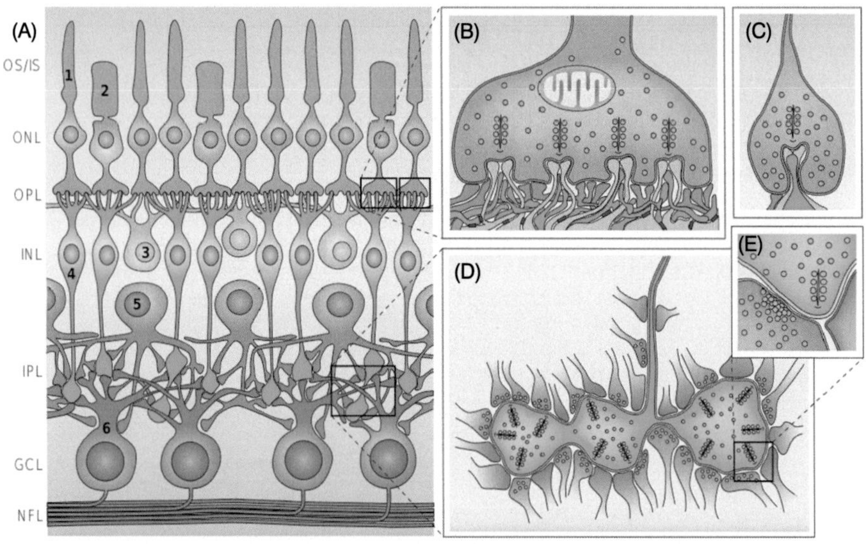

FIGURE 2.11 Schematic of the retina

(A) Retinal layers, cell classes and connectivity. The numbers denote rods (1), cones (2), horizontal cells (3), bipolar cells (4), amacrine cells (5) and retinal ganglion cells (RGCs) (6). (B) Synaptic terminal of a cone, with four ribbons apposed to the dendrites of HCs (yellow) and ON BCs (blue), and OFF BC dendrites making contact at the base (purple). (C) Synaptic terminal of a rod, with contacts to axons of HCs (yellow), dendrites of rod BCs (blue), dendrites of OFF BCs (purple). (D) The axon terminal of a cone BC (blue) contains up to 50 presynaptic ribbons, and connects to AC (orange) and RGC dendrites (purple). (E) A magnified view of a BC ribbon synapse (blue) with an AC process (orange) and an RGC dendrite (purple). Figure from [10].

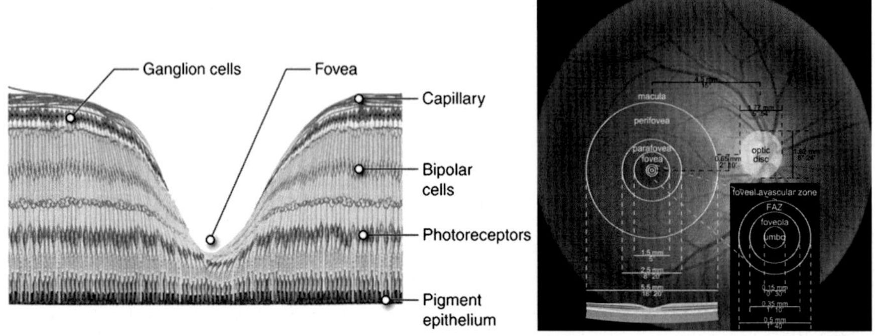

FIGURE 2.12 Fovea

Left: section through center of fovea, from [17]; notice how ganglion cells are displaced to the sides so that the light has more direct access to photoreceptors, in the bottom. Right: fovea proportions, from [18].

FIGURE 2.13 Optic disc and blind spot

If you close your right eye and look at the cross, you can adjust your distance to the page until the black circle in the left disappears: this happens when the image of the circle is formed on the blind spot of the retina.

2.3 Photoreceptors

2.3.1 How photoreceptors operate

Light enters a photoreceptor through its inner segment and then passes into the outer segment where it might be absorbed by photopigment molecules, otherwise it goes along its path and exits the photoreceptor through the other side [4]. See Fig. 2.14.

Photoreceptors are neurons and the information they need to encode is light intensity or, more precisely, the difference in light intensity with respect to an average level that depends on ambient light intensity and is determined by feedback from horizontal cells, in a crucial process called light adaptation that will be discussed in detail later. Light intensity is a quantity that varies continuously and rapidly, and as such it would not be too efficient for photoreceptors to encode this information in the way neurons usually do, by the mean number of spikes or impulses fired over a given time interval. Instead, photoreceptors use an analog form of encoding which is graded

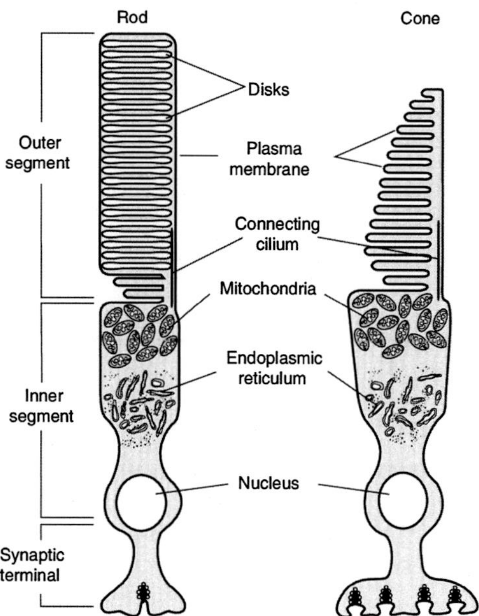

FIGURE 2.14 Schematic for photoreceptors

Schematic for photoreceptors, from [19].

changes in membrane potential. When the neuron is in its resting state, the outside of the cell has an electrical potential that is 70 mV higher than that inside the cell, the resting potential [6]. An increase of light intensity makes the photoreceptor reduce its release of the excitatory neurotransmitter glutamate, causing the membrane to reduce its potential (hyperpolarise), while darkness triggers photoreceptors to increase glutamate release, raising the membrane potential and thus producing membrane depolarisation [15].

The optic nerve is comprised of the very long axons of RGCs, and in order to travel through it the visual signal must be encoded as impulses, not graded potentials. But the signal transformation from *analog* to *digital* introduces errors and loss of information, so the retina postpones it until the very last stage, that of RGCs: all other retina cell types, i.e. photoreceptors, bipolar, horizontal and amacrine cells, do not fire impulses [20].

The ribbon is a special organelle present in photoreceptors (as well as bipolar cells). Ribbon synapses release glutamate and contribute to the first stages of vision by supporting continuous transmitter release. Unlike conventional synapses, ribbon synapses lack a protein called synapsin whose effect is to limit the mobility of synaptic vesicles. Therefore, in ribbon synapses the synaptic vesicles can be continuously released, then retrieved and recycled; ribbon synapses have two modes of transmission, one fast (after a rest period) and one slow (that continues after vesicle release). Cones have more ribbon vesicles than rods, allowing them to encode light intensity with finer resolution than rods can [20].

Cone response to a light stimulus consists of a graded polarisation and the release of glutamate at the cone pedicle (its specialised synaptic terminal), where transmitter release is high for low light levels, and is reduced by light; every cone pedicle makes up to 500 contacts with two types of horizontal cell and at least 8 types of bipolar cell [10]. Rods also transfer their signals onto horizontal and bipolar cells. Recent studies show that the signal generated by a photoreceptor may be transmitted differently at different parts of the synapse depending on the ribbon properties, e.g. smaller ribbon terminals allow higher frequencies to be transmitted [20].

2.3.2 Photoreceptor sensitivity and spatial distribution

Rods are highly sensitive to light, being able to encode the absorption of even a single photon, and are operative under low (scotopic) and intermediate (mesopic) lighting conditions, corresponding roughly to night and twilight/daybreak [12]. Cones on the other hand are much less sensitive, and while they can operate in mesopic conditions they are fully active at high (photopic) light levels, corresponding to daylight. It follows that at mesopic intensity levels both rods and cones are operative. At scotopic levels cones are inactive, and at photopic levels rods saturate so they don't provide visual information.

There are around 120 million rods and 6 million cones in the human eye. It is common in most mammalian retinas [12] for rods to outnumber cones by a ratio of 20:1 to 30:1.

The fovea is the region of the retina with higher visual acuity. It's in the center, near the optical axis of the eye, and as we saw in Fig. 2.12 it's 1.5 mm wide, which corresponds to 5 degrees of visual field. It contains some 50,000 very densely packed cones, and no rods. The absence of rods in the fovea, our center of gaze, combined with the lower sensitivity of cones, explains why we can't see very dim light sources when we look directly at them: they should be looked at "sideways" so that they can be detected by rods [4].

The width of a cone is roughly 0.5 min of visual angle, and in the fovea this is also the distance between adjacent cone centers. The density of cones in the fovea is in accordance with the amount of optical blur introduced by aberrations in the eye. Fig. 2.15 shows that photoreceptor density decays with increasing eccentricity, which makes sense given that optical aberrations make the retinal image progressively more blurred as we move towards the periphery. If the sampling density were high when the visual signal is blurry, then the image encoding would be redundant and hence inefficient, using more neurons than necessary [4]. This is one of the many instances we will be seeing of the efficient representation hypothesis, to be discussed at length in Chapter 4.

Everywhere in the retina but in the fovea, rods occupy the space between cones.

Also in Fig. 2.15 the blind spot corresponds to the optic disc, where the retina is fully traversed by the axons of the RGCs and therefore there are neither rods nor cones.

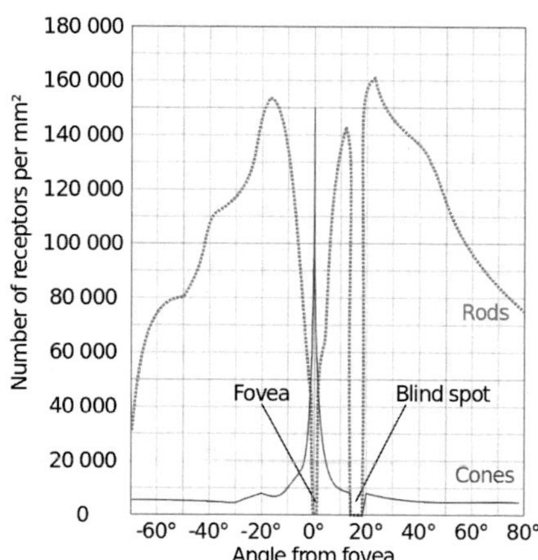

FIGURE 2.15 Photoreceptor density

Photoreceptor density. Image from [21].

In spite of the very large number of rod photoreceptors and their fine sampling of the retinal image, estimated at 250,000 rods/mm^2 [22], visual aquity in low light conditions when only rods are active is quite poor. This is due to the fact that rod signals are not processed individually but averaged instead, so as to reduce the noise inherent to the scotopic conditions and the very high sensitivity of the rod photoreceptors: the high density of rods allows to better capture little light, not to achieve high spatial resolution [4].

There is a perceptual phenomenon called the Stiles–Crawford effect, that only takes place when the visual stimulus acts on the cones, and that denotes the fact that we see the brightness of a light spot differently depending on the angle at which the light rays from the spot enter the eye [3]. A possible explanation of the Stiles–Crawford effect is that, since cones are aligned with the radial axis that goes from each retinal surface point to the center of the pupil, off-center light rays strike cones at an angle and this reduces the chance that the quanta they carry are converted by the photoreceptors.

Signals from many rods must converge onto a single RGC given that the number of rods far exceeds that of RGCs. The number of cones in the fovea, on the other hand, is approximately the same as the number of RGCs in that region.

Visual acuity analysis made under static eye assumptions provide a good match to empirical data, suggesting that the visual system does not improve resolution by integrating information across eye movements [4].

2.3.3 Visual pigments and colour perception

While it's important to bear in mind that colour is not a physical property but a perceptual quality, in many situations we can predict with good accuracy the colour a regular observer will say an isolated light source has, and this colour perception is associated to the wavelength of the light source. For instance long wavelength light will be seen as red, middle wavelength light will be seen as green, and short wavelength light will be perceived as blue.

But photoreceptors are not capable to measure nor encode wavelength, they just absorbe photons and photons don't carry wavelength information. If the response of a photoreceptor changes, it could be due to a variation of light wavelength or to a change in intensity (or both), but a single photoreceptor can't make the distinction: this is the principle of univariance [23]. In order to see colours we need to discriminate wavelengths, and this requires comparing the outputs of photoreceptors which must be of different types so that they have different spectral sensitivities.

In scotopic light conditions only rods are active, there's just one type of photoreceptor in operation, therefore night vision is achromatic. In photopic conditions cones are active, and there are three types of cones, whose spectral sensitivities are unimodal but peak at different wavelengths. S cones (the S stands for "short wavelength") peak at 420 nm, M cones (for "middle") peak at 534 nm, and L (for "long) cones peak at 564 nm. See Fig. 2.16.

There is a yellowish visual pigment covering a 5-degree-wide region centered in the fovea, called the macular pigment. Its density changes considerably among individuals, so the spectral sensitivity curves for cones are different between different people (the scotopic curves are not too affected due to the absence of rods in the fovea) [3].

Since there are three types of cones, humans are said to be trichromats and to have trichromatic vision. Some colour-defficient humans are dichromats because they have

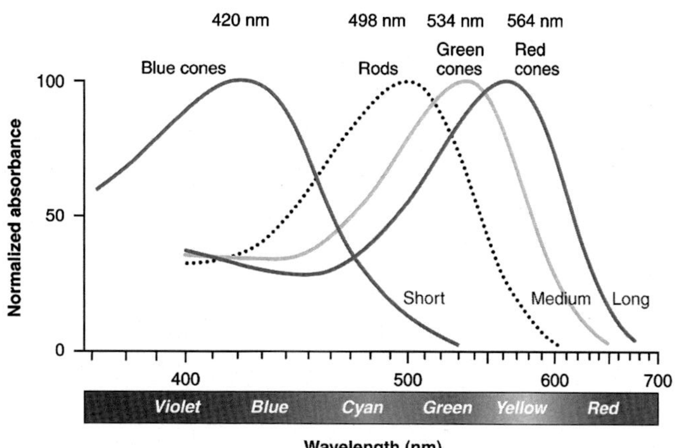

FIGURE 2.16 Spectral sensitivity of rods and cones

Spectral sensitivity of rods and cones. Image from [24].

just two types of cones, as is the case with most mammals, that only have L and S cones, while some nocturnal animals have only one type of cone and therefore can't see colours [10,23].

2.3.4 Photoreceptor evolution and the cone mosaic

The spectral sensitivities of photoreceptors are determined by the chemical properties of their visual pigments, which in turn derive from the opsin genes. For rods, the photopigment is called rhodopsin.

Vertebrate species, some 500 million years ago, already had 3 or 4 classes of cones apart from one class of rods [8]. Early mammals evolved from these vertebrates, but they were nocturnal and didn't need more than 2 cone types so these mammals were dichromats, with just S cones and L cones. About 35 million years ago, Old World primates re-evolved trichromacy via a mutation that duplicated the gene coding for the L cone opsin [25]. Rods also evolved from cones, likely from a duplication of a middle-wavelength sensitive cone class [8], so their signal is handled by retinal circuits that had already developed for cones [9].

The genes for the visual pigments for L and M cones are in the X chromosome, which explains why men are much more likely than women to be red–green colour-defficient if these genes are altered [23], since females have two X chromosomes while males have an XY pair. A mutation can produce a shift in the spectral sensitivity of the visual pigment, and in the case of primates the mutation that introduced a third cone type provided a selective advantage, for instance by allowing to better discriminate red fruits from the green leaves surrounding them, that otherwise have the same colour for a dichromat with only S and L cone types. Surprisingly, in order to exploit this foraging advantage the animal doesn't necessarily need new, evolved, retinal circuitry; the new cone type can just make use of the existing postreceptoral visual system [25], although the presence of a novel cone class can be learned from the statistics of the sensor responses to natural images. Some researchers argue that primates were able to evolve trichromacy because when the opsin mutations took place they already had a high-acuity fovea, where the spatial resolution was optimised and therefore the retinal circuits did not perform averaging of foveal cone signals: most mammals on the other hand do average cone signals, and these species could not evolve trichromacy [10]. S cones have different length and shape than L and M cones, which are morphologically identical [4].

In [25] Brainard also points out, apart from trichromacy, a number of striking features of the human cone mosaic that other species do not necessarily have:

- There are very few S cones, around 5% of the total, with an average spacing of 10 min of visual angle (6 cones per degree, vs 120 cones, L or M, per degree at the fovea) and a fairly regular organisation. They are absent from the center of the fovea, a region of 0.3 degrees in diameter.
- The ratio between the number of L cones and the number of M cones (i.e. the L:M cone ratio) varies greatly among individuals, from 0.4:1 to more than 10:1, with an average value of around 2:1. See Fig. 2.17. It was previously thought that L and M cones formed a regular lattice, but modern measurements show that they are randomly distributed [23].
- The peak spectral sensitivities of L and M cones are very close.

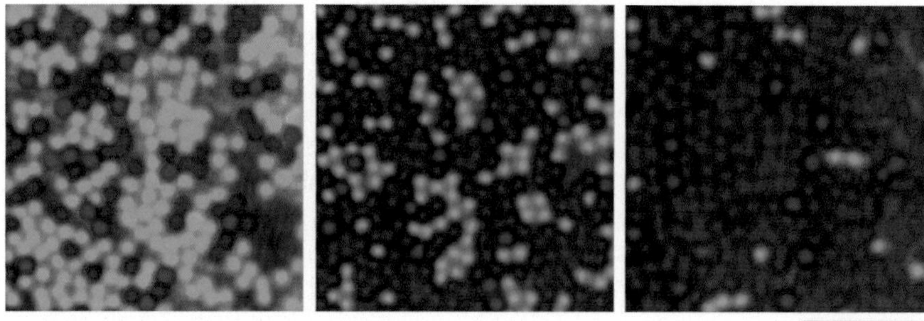

FIGURE 2.17 Cone mosaics

Cone mosaics for three different individuals. Cells have been coloured: blue for S cones, green for M cones, red for L cones. Scale bar (bottom right) corresponds to 50 μm. Figure from [8].

It turns out that these three phenomena can be explained by considering that the optical elements of the eye produce chromatic aberrations, and the cone mosaic is most efficiently adapted to these limitations. Again, this is another instance of efficient representation, the main subject of Chapter 4.

We mentioned in Section 2.1 how the spheric shape of the cornea and the difference between the refraction index of the air around us (approximately 1) and the

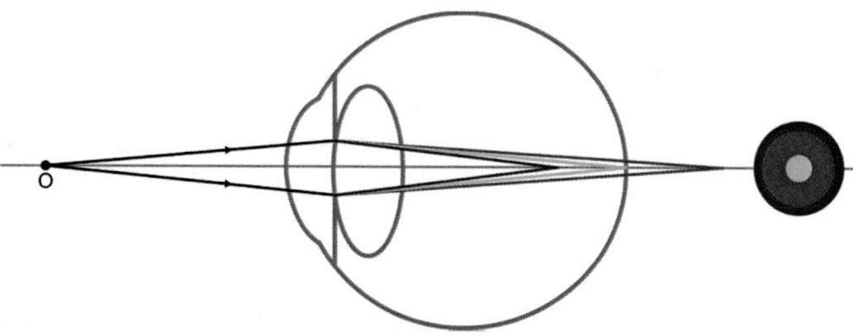

FIGURE 2.18

Chromatic aberration in the eye.

refraction index of the ocular media (around 1.3, very similar to that of water) has the effect of refracting light rays that reach our eyes and focus them on the retina. But the refraction index is a physical property (of substances and materials) that depends on wavelength, and generally the higher the wavelength the lower the refraction index is. That is why an optical system like the eye or a lens will decompose incident white light into its different wavelength components, which will be focused at different distances: the blue component will be closest to the lens, the red component furthest. And if we focus on one so that it appears sharp, the others will be blurred. See Fig. 2.18.

Thus if the spectral sensitivities of L and M cones are very similar, with a high overlap and a small difference in peak position, then the images sampled by the L and M cones can both be in fairly good focus at the same time, and in this manner to allow for high-resolution foveal vision. It also follows that if the images seen by L and M cones are in focus, then the image seen by the S cones will be blurred, which explains why there are no S cones in the center fovea (because they would reduce the visual aquity) and why there's only 5% of S cones: they are less numerous and regularly spaced because they need to sample a more blurry image than L and M cones do. The spacing between S cones matches well the maximum spatial variation that is permitted by the bandpass behaviour of the optical media of the eye, which also causes S cones to encode slower temporal variations [4]. The visual system uses L and M cone signals to fill-in information from S cones [26], and we normally don't perceive less acuity for blue objects than for red or green ones.

The closeness between the peak spectral sensitivities for L and M cones can explain as well why L and M cones are arranged randomly and the L:M ratio varies so much: the amount of visual information from natural images that these cones can carry does not depend crucially on cone ratio, so there's no advantage from an evolutionary standpoint to have more uniform mosaics where the cone ratio is more stable across individuals.

But the random distribution of L and M cones leads to clusters of cones of one type, which must limit the acuity of colour vision [23]. And the fact that colour vision is fairly similar among individuals with very different L:M ratios implies that the visual system must have circuitry that is specifically adapted to each person's cone mosaic, for instance by normalising L and M cone signals by individual variation in luminous efficiency or by calibrating colour processing by taking into account statistical regularities of the environment [25,27].

While S cones don't make many contacts with their neighbours, L and M cone pedicles are coupled to rods and neighbouring cones, allowing the network to reduce uncorrelated noise by averaging it out [10]; nonetheless, M and L cones can't have much crosstalk because they adapt independently of other cones [8].

2.4 Receptive fields and lateral inhibition

Before proceeding any further we need at this point to introduce two concepts that are key in vision science but also in imaging applications that are based on vision modelling.

The first one is that of receptive field (RF). In the case of a single visual neuron, the classical definition [6] states that the neuron's RF is the area of the retina in which a light stimulus is able to influence the response of the neuron. For instance, for a RGC the RF would be shaped as a disc of about 1 mm in diameter, corresponding to the area of the photoreceptors that feed into the RGC. A more general RF definition considers the extent of visual field that influences the neuron, instead of the retinal area, e.g. a photoreceptor is said to have a RF of a few minutes of arc.

As recounted by D. Hubel in [6], the landmark experiments of S. Kuffler in the mid 20th century permitted for the first time to characterise the RF of RGCs in a mammal. The animal was shown diffuse background light, while an electrode measured the responses of an RGC. Surprisingly, even when the background was completely dark, the RGC fired impulses at a steady rate. When a small spot of light was shown on the retina near the tip of the electrode, the firing rate increased. If the spot was shown a bit further from the electrode's tip, anywhere in a large ring-shaped region, the spontaneous firing rate of the RGC was instead reduced. And if the spot was placed further away, it had no influence on the neuron. See Fig. 2.19 (left). Therefore the RF of this RGC can be represented as a circle and subdivided into a small circular center of ON characteristics (since shining a light in this region increases the neuron's response) and a large ring-shaped OFF surround (thus called because a light stimulus in this surround reduces the firing rate). Kuffler also found that RGCs were of one of either two types, with the RGC population evenly divided among the two: ON center – OFF surround, as the one just described, or the reverse, OFF center – ON surround, which behaved in the opposite way (the spontaneous firing rate decreases when shining a spot on the center of the RF).

This, as we shall see, is the classical RF configuration of several classes of visual neurons, not just RGCs: a center-surround arrangement, one of the regions being ON and the other one OFF. See Fig. 2.19 (right). In the cortex the RFs become larger, with more complex shapes and they require more complicated stimuli to be characterised (e.g. edges with a particular orientation and direction of motion), but they still have ON and OFF regions.

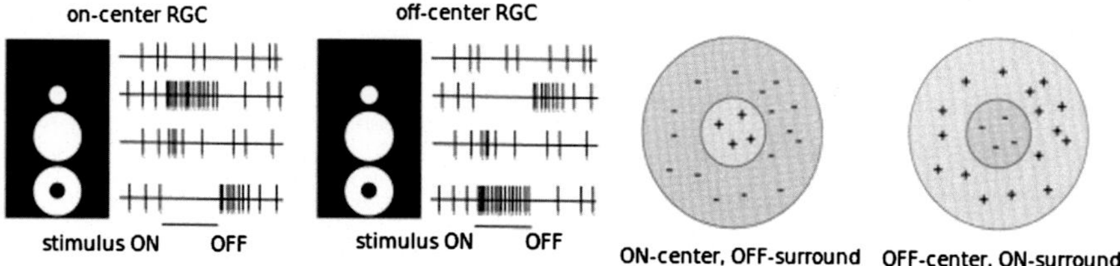

FIGURE 2.19 Center-surround RF

Center-surround RF and ON/OFF cells.

The response of a neuron that has a classical center-surround RF can be modelled as the spatial summation of the signal over the ON region minus the spatial summation of the signal over the OFF region. That is why the maximal response is obtained with a stimulus restricted to the ON area, a minimal response is produced with a stimulus limited to the OFF region, and if a uniform stimulus covers the whole RF then the response of the RGC is not affected (it remains the spontaneous firing rate) because the ON and OFF contributions cancel each other out.

This center-surround antagonism in RFs is due to lateral inhibition, a term that designates the process by which an excited neuron reduces the activity of its neighbours in the lateral direction (i.e. in the same layer in the case of the retina). This also prevents the spreading or propagation of the signal from an excited neuron to its neighbours. In terms of visual information, we can see how these two properties of lateral inhibition have the effect of preserving the sharpness of edges (preventing signal spread amounts to avoiding blur) and increasing the local contrast (defined here as the difference in response between an excited neuron and its neighbours).

There are important limitations imposed by the very concept of the RF. For instance, when the stimulus is a natural image or even a synthetic image but with a varied background, far away points can influence the response of a neuron even when they do not belong to its RF; this has led to the definition of the non-classical RF [28], essentially a classical RF with an additional non-classical surround and/or extra spatial processes for the model. Also, when we talk about a single neuron's RF we need to bear in mind that this neuron's response is in fact influenced by many other neurons that came before in the visual stream but also by many other neurons that are downstream but send their feedback to earlier stages [4]. And finally, the RF of a neuron is not a fixed, identifying property of the neuron: it's different for different stimuli, e.g. it changes with the mean level and contrast of the input [29].

2.5 Horizontal cells

In a very recent work, Chapot et al. [30] provide an overview of what is known about HC activity, mentioning that *"a comprehensive picture of HCs and their precise function [...] is still elusive"*. Nonetheless, it's made clear that HCs play a key role in

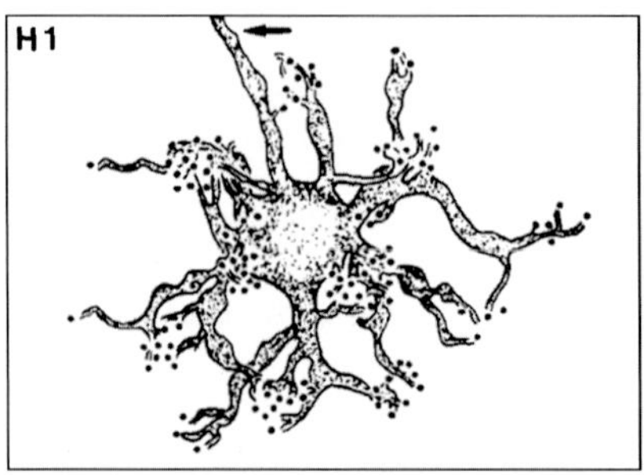

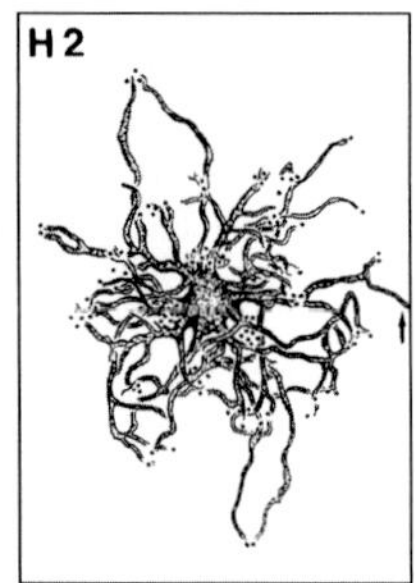

FIGURE 2.20 HCs in primate

HCs in monkey retina. Left: HCs of type HI. Right: HCs of type HII. Both cells are from 5.7 mm eccentricity, the width of each frame represents 60 μm. Axons indicated by arrows. Figure adapted from [32].

vision by providing feedback to photoreceptors and BCs, which generates center-surround RFs in them, thus contributing to perform contrast enhancement and colour opponency. They also are essential for adjusting the operational range of photoreceptors, enabling them to adapt to the ambient light level. While until fairly recently it was assumed that HCs only provide feedback to cones, it has been shown [31] that there's also feedback from HCs to rods.

Primates have HCs of two types: HI and HII. (See Fig. 2.20.)

HI cells have radiating dendrites that in the peripheral retina may contact up to 18 cones [14], almost exclusively of the L and M type. They have a long thick axon with terminals connecting to rods.

HII cells have more intricate dendritic fields [14], making stronger contacts with S cones than with L and M cones. Although it's been said that they don't have an axon [12], other researchers argue that HII cells do have a short, curled axon that makes contact with S cones only [14].

Only HI cells have axons that make connections to rods, and the HII cell is the only known retinal element that adds up L, M, and S inputs [8], producing a correlate of light intensity.

The cone synapse is very sophisticated and involves two other neuron classes: in the cleft of the cone axon terminal there are dendrites from two (possibly different) HCs and a dendrite from an ON-type BC, while at the base of the cone synapse there are dendrites from OFF-type BCs. See Fig. 2.21. Photoreceptors release the neurotransmitter glutamate and, as mentioned earlier, photoreceptor terminals have special structures called ribbons that allow for glutamate to be released in a sustained and faster way; HCs modify cone output via reciprocal feedback, altering their glutamate release. Recent studies suggest that within the same cone different ribbons may act independently, and that HCs may also provide individual local feedback to cones apart from a global signal that is common to all the cones they contact [30].

For each HC type, the dendritic fields of neighbouring cells are electrically coupled, which makes their RFs much larger than the extent of the dendritic arbour.

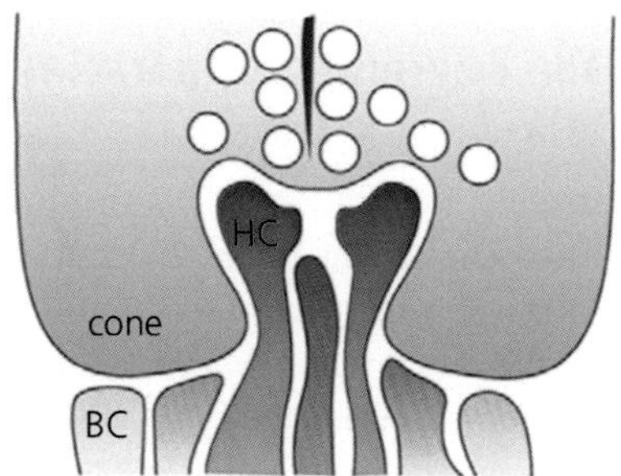

FIGURE 2.21 HC and cone synapse

HC and cone synapse. Simplified cone axon terminal (purple) with a single exemplary ribbon (black vertical line) surrounded by numerous synaptic vesicles (white), with two invaginated lateral HC dendritic processes (red) and an ON cone BC dendrite (dark grey). OFF cone BC dendrites (light grey) are located at the base of the cone axon terminal. Figure from [30].

When photoreceptors are stimulated by light they react by hyperpolarising (i.e. reducing the release of neurotransmitter glutamate). HCs hyperpolarise as well, and they sum the signals over their wide RF providing inhibitory feedback to photoreceptors and BCs, and thus creating the center-surround RF of these cell types [14].

HCs are often assumed to be key in enabling light adaptation through lateral inhibition to cones, approximating the ambient light level by adding cone signals over their extense RF and then subtracting this mean level from the cone's response.

The feedback that HCs provide to cones is very fast, with a time constant of 35–200 ms, which suggests that HCs can be involved in fast processing tasks like local adaptation to the mean and contrast in between eye movements, which we will discuss in Chapter 4.

Light adaptation to the global mean appears to rely on the amount of coupling between HCs, which in turn depends on the signals provided by a special type of RGCs called ipRGCs, to be discussed in Section 2.8.4, that are sensitive to the absolute level of light intensity instead of being sensitive to variations w.r.t. the mean light level, as all other RGC types are. When the ambient light level increases, a chain of chemical processes leads to a decrease in the electrical coupling of HCs [30]. This in turn makes lateral inhibition more local, and therefore enhances more the contrast.

Lateral inhibition and the different cone connectivities of the two types of HC show that HI cells are involved in regulating and enhancing the output of L and M cones, while HII cells shape the output of S cones instead and also generate colour opponency already at cone level by adding L and M cone signals and providing lateral inhibition to S cones. While it would appear that the HC networks for the two HC types are independent, in fact they need to have some interaction, for instance to allow for colour constancy [30].

2.6 Bipolar cells: the beginning of parallel processing in the visual system

The axons of the approximately 1.5 million RGCs form the optic nerve, and this constitutes an anatomical bottleneck that forces the retinal output to be properly condensed so as to allow for an efficient representation [11]. In mammals, this is accomplished by converting the visual input into a number of different signals, each carrying a particular, segregated aspect of the input. Each signal is transmitted by a connected series of neurons in what is referred to as a visual stream, pathway or channel. These pathways run in parallel, there are at least 12 of them, and they start already at the first synapse of the visual system, between photoreceptors and BCs. All visual signals travel through BCs.

The output of each cone is sampled by two HCs and twelve BCs of different types, through up to 500 contacts that are present in each cone pedicle, and the signal from each of the contacted BCs represents a different aspect from the cone's activity [9,10]. The axons of BCs terminate in the IPL, which is divided into ten thin sublayers. Different BC types have their axon terminals at different sublayers of the IPL, where they are contacted by RGCs whose dendrites end at the same sublayer: this is called co-stratification, and it's usually an indication that the BC type and RGC type involved belong to the same parallel pathway. Some of these pathways, like the ON/OFF channels that we'll discuss shortly, go all the way to the visual cortex, while some other channels, like the rod pathway, exist for just a few synapses and then merge with other channels after they have served their purpose [4].

There are some 12 types of BCs, one that connects with rods, the rest making synapses with cones. See Fig. 2.22.

While all cone BCs synapse directly with RGCs, rod BCs don't make contacts with RGCs, they send their signals to ACs which in turn provide input to RGCs. Rod BCs integrate the signal from many rods, up to 1500 of them, which reduces noise and allows to capture information at low light levels, at the price of reducing visual acuity [4]. The summation performed by rod BCs is very nonlinear, and in order to separate the dim light stimuli from the noise the rod signals are first temporally averaged, then thresholded and finally summed [1]. Given that rods predominate outside the fovea, rod BCs outnumber cone BCs in the human retina [14].

Cone BCs, on the other hand, behave linearly for low light levels and their response becomes progressively more nonlinear as the background light intensity increases [12].

BCs can be classified, according to different criteria, into:

- ON or OFF type, depending on their response to changes in light intensity with respect to the mean.
- Diffuse, midget or blue-cone bipolar, depending on the number and type of cones they connect to.
- Transient or sustained, depending on their temporal response.

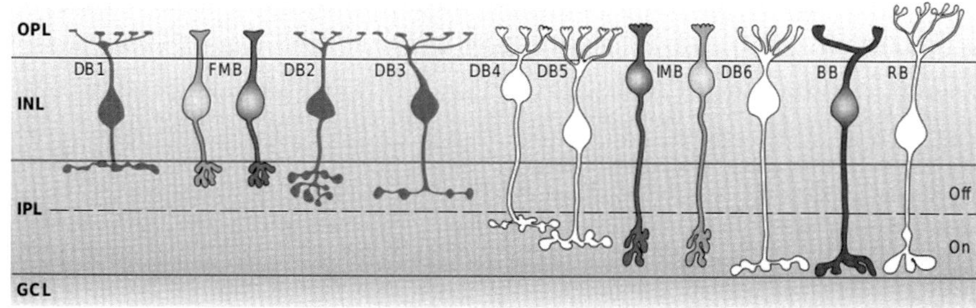

FIGURE 2.22 BC types

BC types, from [10].

These classifications overlap, e.g. there are OFF-center midget transient BCs. At any point in the retina, the dendritic arbors of 12 types of BC overlap, so that the output of each and every cone is sampled by 12 different BCs [9] and for each type of BC the dendritic fields of the cells of this type provide a complete tiling of the retina, with each BC contacting all the cone terminals it can reach with its dendrites.

Let's see some of these types of BC in a bit more detail.

2.6.1 ON/OFF

Due to feedback from horizontal cells BCs have a center-surround receptive field, and depending on the sign of the contrast they encode they can be of one of two types, ON-center when they respond to light increments above the mean, or OFF-center when they increase activity with light decrements with respect to the mean. The axons of ON BCs terminate in the inner half of the IPL, where they synapse with dendrites from ON RGcs; likewise, OFF BCs contact OFF RGCs, but at the outer half of the IPL. As a result, ON RGCs are excited by stimuli that are brighter than the background, while OFF RGCs are excited by stimuli that are darker than the background: thus, the separate sensations of black and white that were postulated by Ewald Hering in the late 19th century are originated by two types of BCs with different properties [10], the ON-center BC making contact with the photoreceptor at its interior and inverting the signal, while the OFF-center BC that makes contact at the base of the photoreceptor preserves the polarity. See Fig. 2.23.

This division between ON and OFF types continues through the thalamus to the visual cortex, forming the basis of the ON and OFF parallel pathways. The existence of these two channels has been reported in many animal species, from flies to primates, that use vision to move about in their surroundings [33].

One fundamental challenge that our visual system must tackle is imposed by the limited dynamic range of spiking neurons, of around two orders of magnitude, while in the course of a day the ambient light level may vary over 9 orders of magnitude [1]. In order to deal with this the visual system uses a number of strategies, that include encoding contrast instead of the absolute light level, and having separate pathways

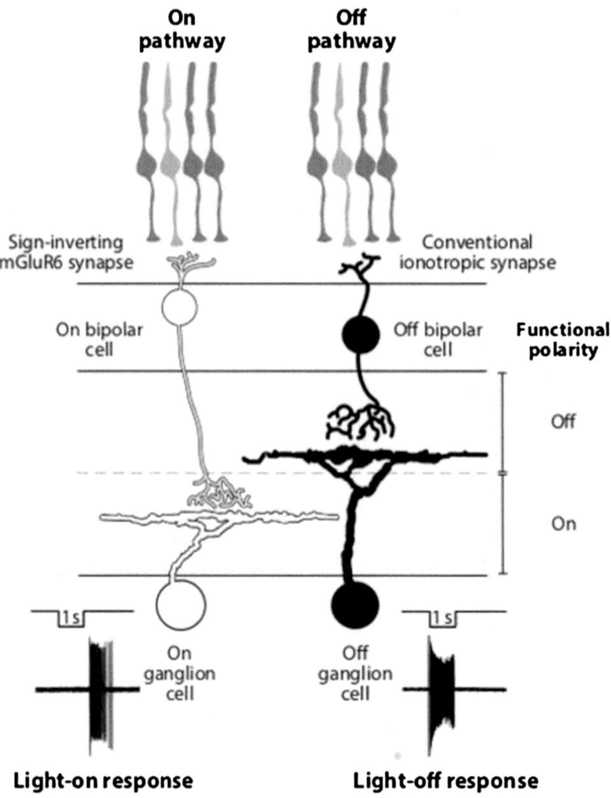

FIGURE 2.23 BCs and ON/OFF channels

BCs and ON/OFF channels. Figure from [15].

for ON and OFF information. Local contrast is defined as percent change in intensity with respect to the average or background level. In a typical scene large changes in the absolute level of illumination do not have an impact on the range of contrast values, that remain fairly constant and below two orders of magnitude [4], and therefore can be *fit* into the limited dynamic range of neurons. In [6] it's hypothesised that the reason for the existence of the ON and OFF pathways is efficiency: neuron spikes consume energy, so rather than having neurons firing at some significant level for the average light intensity so that the firing rate can decrease for low intensities and increase for high intensities, it appears more convenient to have the neurons be silent or firing at low rates for the average light intensity, with ON neurons devoting their whole dynamic range for increments over the average and OFF neurons for decrements.

2.6.2 Transient/sustained

The information encoded by transient cells corresponds to changes in light intensity around an average level, whereas sustained cells encode this average [12]. This characterisation into distinct temporal channels is carried from BCs to the RGCs they co-stratify with, and again they do it at different sublayers of the IPL: transient OFF BCs synapse with transient OFF RGCs at the middle of the IPL, sustained OFF cells co-stratify at an outer part of the IPL, transient ON BCs synapse with transient ON RGCs at the middle of the IPL, sustained ON contacts take place in the inner IPL.

Although the response of BCs is in the form of graded potentials and not of spikes, as is the case with all retinal neurons except from RGCs, when there is a step increase in light intensity there is a mechanism by which larger voltage transients in BCs can reliably be transmitted to RGCs: the time of change is encoded by a large initial transient, followed by a sustained response encoding the steady-state level [20].

2.6.3 Diffuse/midget/S-cone BC

There are at least six types of diffuse BC, half of them OFF-type and half of them ON-type [8]. Diffuse BCs combine the information from several cones, from 5 to 20, with a preference for L and M cones over S cones. This, combined with the fact that L and M cones outnumber S cones significantly, implies that diffuse BCs transmit a luminosity signal to the IPL, essentially L+M [10].

Midget BCs, up to 40 deg eccentricity, contact a single cone, and for more peripheral regions they may synapse with 2 or 3 cones. Each cone in central retina makes contact with a single ON midget BC and a single OFF midget BC, while at the fovea and up to 10 deg of eccentricity midget BCs contact a single RGC, thus providing foveal cones with a one-to-one connection with a RGC and therefore establishing what is usually called a "private-line" from the cone to the brain [34] (although this terminology is problematic, as Cornsweet points out [3]).

There are several elements that give support to the idea that midget BCs carry a colour-opponent signal:

- the private line pathway;
- the fact that there are no S cones in central retina;
- most studies indicate that the wiring between midget BCs and L and M cones is random;
- BCs have a center-surround RF given by HC feedback.

The single-cone centre has a high gain that cancels the contribution of the same cone type from the surround, producing a L versus M cone signal, either L–M or M–L [34].

As mentioned earlier, there is a theory of the evolution of trichromacy in primates that is based on this private-line pathway: only after a one-to-one connection between cones and RGCs had evolved, it was possible for a mutation of the L-cone pigment to create a different cone type that could produce a different chromatic signal, whose identity was preserved; for mammals other than primates, whose BCs average the signals of several cones, a mutation that created a new type of cone would be lost [10]. Therefore, midget BCs would be performing two tasks: signalling acuity and trichromacy.

Finally, another single-cone type of bipolar is that of the S-cone BC, also called blue cone BC. This cell is only of the ON type, no S-OFF BCs have been identified [11,34]. An S-cone BC contacts only S cones [1], but it provides a colour-opponet signal, essentially S–(M+L), due to the center-surround form of its RF, with lateral inhibition from HII cells that produce a $(M + L)$ input for the surround.

2.7 Amacrine cells

Amacrine cells are interneurons located in the IPL. Between three dozen [15] and more than 50 [10] types of ACs have been characterised according to their morphology. (See Fig. 2.24.) They are rather inaccessible and hard to identify, as most are axonless and some are polyaxonal, which makes it difficult to recognise the sites of their inputs and outputs [9]. Based on the span of their dendritic arbors they can be classified as narrow or wide field. The latter kind may extend even some millimetres [12].

ACs receive input from BCs and other ACs, and provide outputs to BCs, RGCs and ACs. Almost all ACs are inhibitory. Through lateral inhibition, ACs shape the center-surround RF of RGCs. They also are essential for making RGCs sensitive to motion and edges, as we shall see. ACs perform so-called vertical integration and cross-over inhibition, meaning that they interconnect the ON and OFF layers of the IPL, carrying information from one into the other. A special type of AC is essential for colour opponency: it performs sign inversion on a blue-ON signal thus creating a blue-OFF signal for a blue-OFF RGC. This is key as there are no blue-OFF BCs.

2.8 Retinal ganglion cells

2.8.1 Impulse generation

All the visual system neurons discussed previously produce signals in the form of graded potentials. Starting with the retinal ganglion cells though, information is transmitted in the form of trains of impulses, or spikes: the signal magnitude is usually assumed to correspond to the spiking frequency or the mean firing rate, see Fig. 2.25, although sometimes the neural code represents information in other variables of the spike train, as we will mention in Chapter 4.

In [6] it is explained how the neuron is completely covered by a porous membrane that allows the passage of chemicals, and when at rest the difference of potential between the interior and the exterior is of 70 mV, positive outside: this is the resting potential. Through its dendrites, the neuron makes synaptic contact with the axons of other neurons, which release chemicals that contribute to either decrease or increase

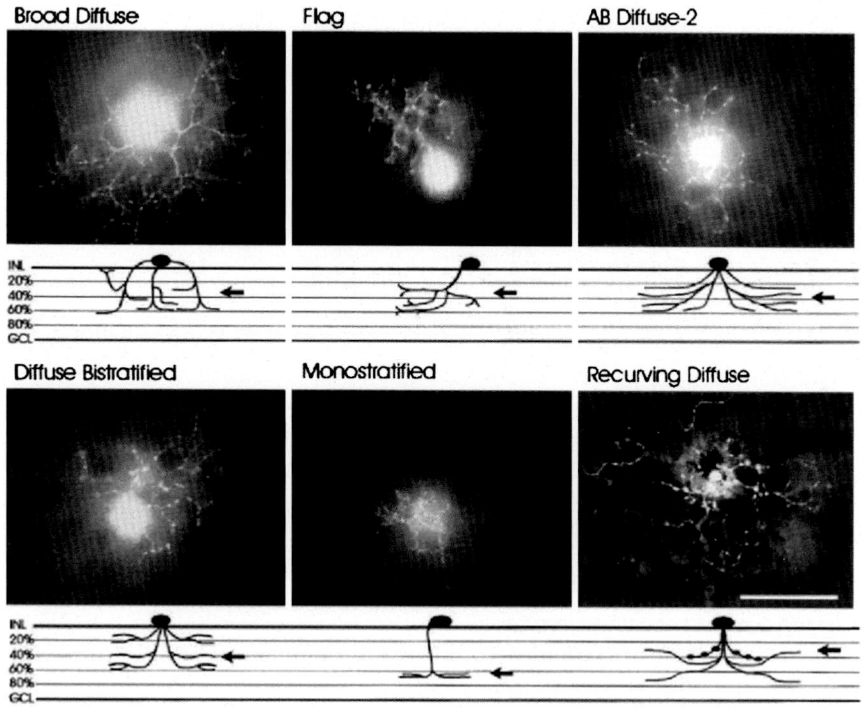

FIGURE 2.24 Amacrine cells

Amacrine cells, from [35].

the resting potential. The neuron performs a summation of all these contributions, and if the net effect is to depolarise the membrane (making the difference of potential go from the resting potential of 70 mV to any value that is under a threshold value of 40 mV), then an impulse is generated: at the base of the axon and in a short span of fiber, the difference of potential becomes negative outside. Within 1 ms the positive resting potential at this location is restored, but now it's the adjacent short span of axon which has a negative difference of potential, and thus the impulse travels along the axon at a speed of 0.1 m/s to 10 m/s. When the membrane pores are ready again, another impulse is generated. Impulse firing rates can go from one every few seconds to 1000 per second at the very maximum, with an average of 100–200 spikes per second for a very active neuron. These impulses travel along the axon to its terminals, causing some chemicals to be released at the synapse between this cell and the next neuron. In order to increase the speed of transmission of impulses, some axons (like those of RGCs) are covered by a substance called myelin that reduces the amount of charge needed to depolarise the nerve.

While the summation of the synaptic contributions at the dendrites is most often described as a linear process, essentially adding up all the excitatory (depolarising) signals and subtracting the sum of all the inhibitory (hyperpolarising) signals, recent works show that in many instances this spatial summation process is actually nonlinear [36–38].

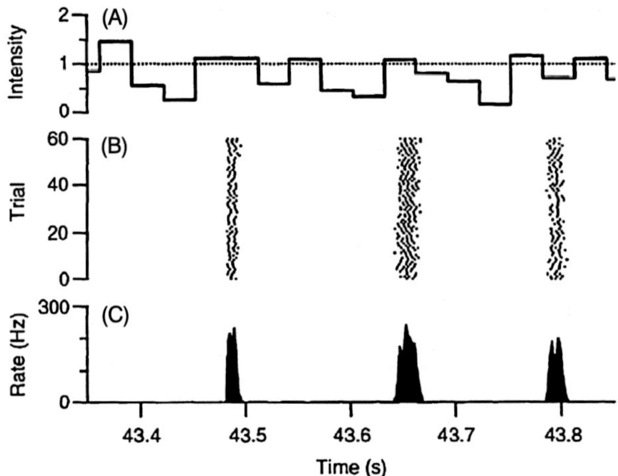

FIGURE 2.25 RGCs and spike trains

Response of a salamander ganglion cell to random flicker stimulation. (A) Stimulus intensity in a time course of 0.5 s. (B) Spike rasters from 60 trials (each row is a trial, each dot is a spike). (C) Firing rate. Figure from [39].

2.8.2 Mosaics, stratification, receptive fields

Different research works give different values for the number of distinct RGC types, with current estimates hovering around 20 [9,11]. As with BCs, the dendritic trees of the cells of each type of RGC completely tile the retina, providing a coverage factor of 1, or higher if there's partial overlap among neighbouring dendritic arbors. Therefore, a spot of light shining at any point in the retina can stimulate one RGC of each type. See Fig. 2.26.

For each cell type the tiling is quite regular, but independent from the tilings of the other types. It's possible for so many RGC types to overlap because the dendrites of different types stratify at different levels of the IPL, and in this manner they can avoid one another [10], as shown in Fig. 2.27.

RGCs can be classified according to varied criteria, like mosaicism (the more or less uniform spacing of cell bodies of a given type in the retinal plane), morphology (soma size, spread of dendritic tree, etc.) and stratification (depth at IPL where their dendritic arbors are directed). The latter is the most important in order to assess the function and possible responses of the RGC and to characterise its type, because the stratification of the RGC determines which BCs and ACs the cell gets its input from [15].

Stratification and mosaicism reinforce the idea that each RGC type serves a unique function and sends information through a specific parallel pathway. But of

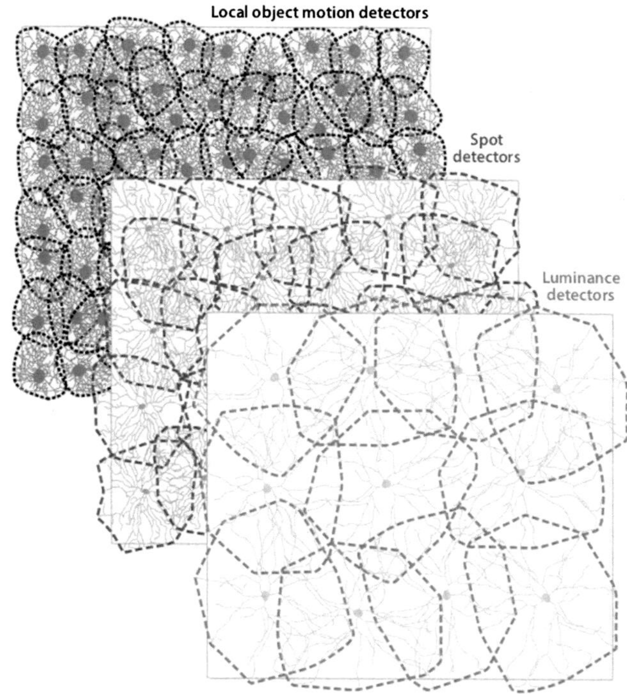

FIGURE 2.26 Retinal ganglion cell mosaics

Retinal ganglion cell mosaics, from [15].

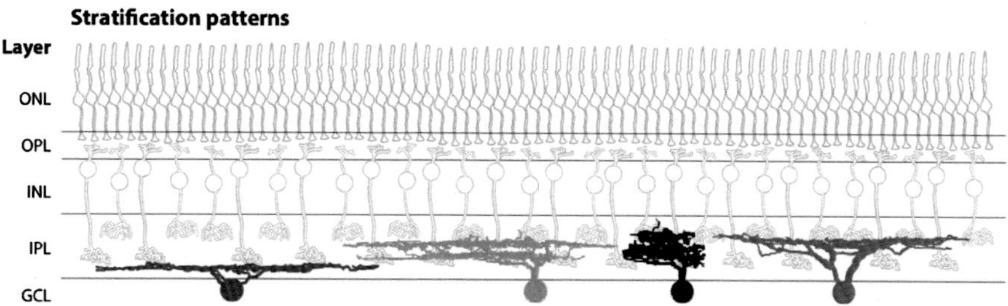

FIGURE 2.27 Retinal ganglion cell stratification

Retinal ganglion cell stratification, from [15].

the approximately 20 types of RGC that have been found, less than half of them have been characterised, the rest have unknown functions. In [9] the list of identified RGC types includes ON-transient cells, OFF-transient cells, ON-sustained cells, OFF-sustained cells, an ON-OFF cell (responsive to both increments and decrements), an ON-OFF direction selective cell of unknown function, an ON direction selective cell, blue-ON and blue-OFF cells, a local edge detector, a uniformity detector, orientation selective cells, and intrinsically photosensitive cells (ipRGCs).

Most, but not all, RGCs have a center-surround type of RF. Inhibitory surrounds are generated by wide-field ACs in a way analogous to HCs generating inhibitory surrounds for BCs. A key difference though is that ACs receive input from BCs, and are therefore sensitive to high spatial frequency stimuli [12]. The circuit connectivity for the center has an anatomical substrate and is well understood, but this is not the case for the surround [8]. The sizes and temporal dynamics of the RFs are different for ON and OFF cells [40].

2.8.3 Midget, parasol and bistratified retinal ganglion cells

According to [11] at least 13 RGC types send their signals in parallel to the lateral geniculate nucleus (LGN) of the thalamus, maintaining their anatomical segregation onto the primary visual cortex (V1). In particular, three RGC types (with their ON/OFF and transient/sustained variants) that project onto the LGN constitute around 90% of all RGC cells in the retina and have been very well characterised, to the point that several textbooks present them as the only existing RGC types: these are the midget, parasol and bistratified ganglion cells (see Fig. 2.28.) But we must stress not only that there are more RGC types, but also that the retina sends its signals

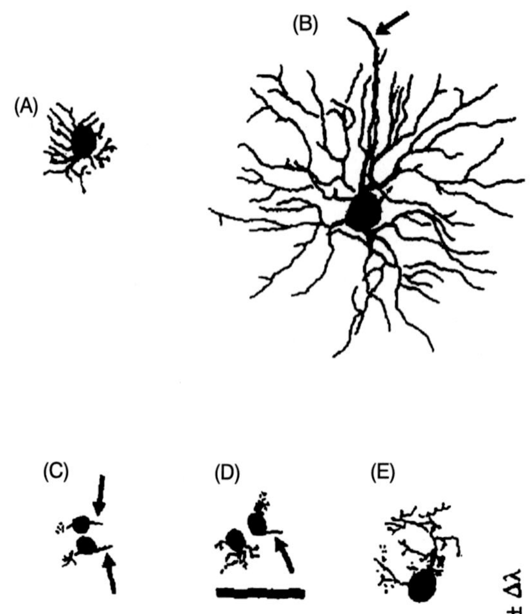

FIGURE 2.28 Parasol and midget RGCs

Parasol (A,B) and midget (C,D,E) RGCS from a macaque monkey. The arrows indicate axons. The scale bar is 50 μm. Figure from [41].

to around 45 target structures in the brain other than the LGN, most of which have roles in visual processing that are poorly understood [15].

Midget ganglion cells have small RFs, high sensitivity to low temporal frequencies and high spatial frequencies, and low sensitivity to contrast. They project onto the parvocellular (PC) layer or P-layer of the LGN, parvo meaning "small" in Latin, and therefore are the start of the PC pathway. They are monostratified and hence either ON or OFF (depending on their stratifying sublayer of the IPL). They comprise around 70% of all RGC cells, but this high number is a consequence of their small RF size and the fact that being a RGC type they have to completely tile the retina.

They can provide a spatial resolution of 60 cycles per degree, corresponding to the full sampling resolution of photoreceptors, given that from the fovea and up to an eccentricity of 10 deg they only make contact with a single BC, itself a midget bipolar that makes contact with just a single cone [4,8]. As mentioned earlier, full spatial resolution means RF centers of the size of a single photoreceptor, roughly equal to the optical blur due to diffraction [4]. With increasing eccentricity the convergence (of many cone signals onto a BC and of many BC signals onto a RGC) also becomes higher, and at an eccentricity of 30 deg–40 deg a midget RGC might get input from around 30 cones [8].

Midget RGCs carry a L/M colour-opponent signal, often called red/green although this simple relationship of retinal physiology to visual perception turned out not to be true [40]; the neural basis of the unique hues of red, green, blue and yellow isn't yet understood. As the small RF of midget RGCs can in principle encode high-resolution spatial information, textbooks describe the PC pathway as multiplexing colour information and high-acuity spatial information; the truth is, though, that the low achromatic contrast sensitivity of midget RGCs prevents them from providing high resolution spatial vision [8]. Also unknown is the wiring pattern for the surround. At the fovea the RF center of a midget RGC consists of a single cone of the L or M variety, so for colour opponency the neuron should in principle have a selective wiring, connecting only to neighbouring cells of the other cone variety, M or L. Most research though seems to support the random wiring hypothesis, in which the surround combines L and M signals, but still colour-opponency can be achieved if the center has a much higher gain that cancels the contribution from the same type of cone from the surround. The net result is that M and L have similar weight in the colour-opponent output, which might be the most efficient option for coding [8].

Parasol ganglion cells have large RFs, high sensitivity to high temporal frequencies and low spatial frequencies, and high sensitivity to contrast. They project onto the magnocellular (MC) layer or M-layer of the LGN, magno meaning "large" in Latin, and therefore are the start of the MC pathway. They also are monostratified, either ON or OFF. Parasol RGCs comprise around 10% of all RGC cells. They can provide a spatial resolution of 20 cycles per degree, with their RF center at the fovea having a diameter of 6–8 cones [8]. Parasol cells avoid sampling S-cone signals, for unknown reasons, and as they sample randomly from the L, M cone mosaic it is expected that a variability in the L/M cone weighting exists; this is consistent with the observed vari-

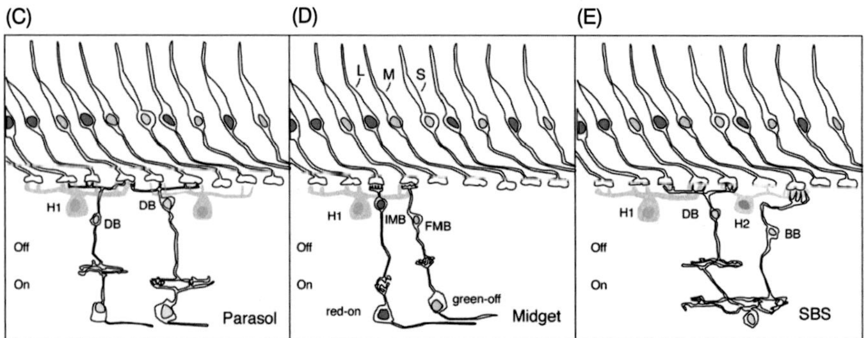

FIGURE 2.29 Midget, parasol and bistratified RGCs

(C) Parasol (MC) pathway. Excitation to parasol RGCs is through diffuse BCs. H1, horizontal cell; FDB, flat diffuse BC; IDB, invaginating diffuse BC. (D) Midget (PC) pathway. Excitation to midget RGCs is through single-cone contacting midget BCs. IMB, invaginating midget BC; FMB, flat midget BC; L, M, S: cones. (E) Small bistratified ("blue-on") pathway. On-sign excitation to blue-on cells from S cones is through blue cone BCs. Off excitation from ML cones is through diffuse BCs. Figure from [8].

ability in their spectral sensitivity [8]. Parasol RGCs carry an achromatic, luminosity signal. See Fig. 2.29.

Bistratified ganglion cells, of the small and large variety, constitute around 8% of all RGC cells. They project onto the koniocellular layers or K-layers of the LGN. Being bistratified they receive inputs from both ON and OFF BCs. They carry a blue/yellow colour-opponent signal. For blue-ON RGCs, the excitatory input comes from blue-ON BCs, and the inhibitory yellow signal from HC feedback as well as from diffuse BCs providing L and M input, see also Fig. 2.29. The case for blue-OFF RGCs is quite different: they have only been found fairly recently, the surprise being that there are no blue-OFF BCs from which they could derive their input. But, as mentioned previously, there is a special type of amacrine cell that receives blue-ON input from a BC and with this signal provides inhibitory input to a blue-OFF RGC, i.e. it performs a sign inversion on the blue-ON signal turning it into a blue-OFF input. The blue/yellow colour-opponent channel is considered to be the primordial pathway for colour vision, as most mammals have only dichromatic S/L vision. In primates, a key open question is the following: do S-cones contact midget RGC private lines?

Both for midget and parasol RGCs, with increasing eccentricity in the visual field the dendritic trees become wider, but at each eccentricity value parasol cells are larger than midget cells. On the other hand, parasol cells have more overlap with increasing eccentricity, while midget cells show no overlap even in the periphery, where they tend to become non-opponent as well. In fact the details of midget RGCs at different eccentricities have not been resolved yet [8].

2.8.4 Intrinsically photosensitive retinal ganglion cells

Intrinsically photosensitive RGCs (ipRGCs) constitute one to three percent of the total population of RGCs. ipRGCs also project to the LGN. They contain a phototransduction pigment called melanopsin that enables them to respond directly to light. They also receive input from rods and cones from BCs, but without it they still react to light. The response coming from the melanopsin transduction is slower than the one derived from the photoreceptor input. Their existence had been postulated for many years but their essential properties were formally established in the first decade of the 21st century, although many aspects of their operation remain open. Five types of ipRGC have been identified (see Fig. 2.30), four of them monostratified (ON type). Some ipRGCs can be classified as well as a 'regular' RGC, e.g. an ON-sustained-parasol cell [15]. They are most dense at the fovea.

The output of ipRGCs is proportional to ambient light intensity, an important information that provides survival advantages for the animal and which can not be derived from any other class of RGC, because they quickly adapt to the average light level [10]. A very recent work [42] shows that ipRGCs come in two variants: monotonic, in which increased radiance produces increased firing rate of the cell, and unimodal, where the firing rate peaks at a certain radiance level but decreases for higher light intensities. Different unimodal ipRGCs have different response peaks, in such a way that the whole irradiance axis, from moonlight level to direct sunlight level, is covered by different unimodal subtypes of ipRGC. This arrangement increases efficiency, as it avoids the need to have monotonic ipRGCs covering the whole irradiance axis and which would be firing at their saturated rates during most of the day. Again, we'll discuss in more detail the issue of efficiency in Chapter 4.

Very importantly for the tone mapping applications discussed in this book, ipRGCs can span and encode light intensity levels from moonlight to full daylight, directly controlling the reflex that adjusts the pupil size and also the circadian rhythm, that modulates the release of dopamine in the retina, which in turn alters the size of the receptive field of horizontal cells performing local contrast enhancement on the cone signals [30]. The key implication is that contrast enhancement will increase as the ambient level is reduced.

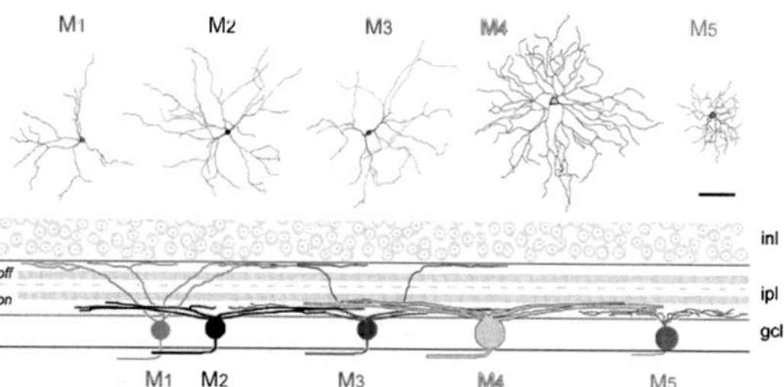

FIGURE 2.30 Main types of ipRGC

Main types of ipRGC, from [43].

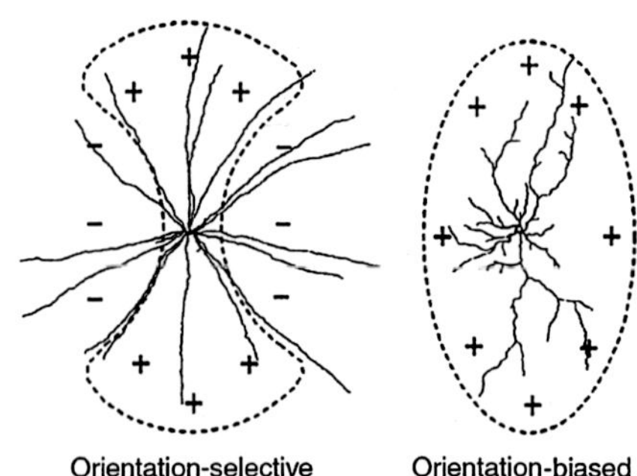

FIGURE 2.31 Orientation tuned ACs

Orientation tuned ACs, from [44].

2.8.5 Retinal ganglion cells sensitive to orientation and motion

There are a number of visual functions that have traditionally been assumed to take place mainly or exclusively in the visual cortex, like contrast adaptation, motion detection, segregation of object and background motion, sensitivity to motion direction, or sensitivity to edge orientation. In fact, it has been shown that the vertebrate retina effectively carries out all of these tasks [1,9,10,12,15,44].

For instance, RGCs with wide RFs that nonlinearly sum their BC inputs can detect global motion, without being sensitive to the direction of motion nor the location of the moving object.

Object motion on the other hand is local, not global, and can be understood as a difference in motion between a local region and the background. RGCs can be sensitive to object motion by computing motion detection in the center (adding up nonlinearly the inputs from BCs, as in the previous case) and receiving inhibitory signals from equivalent motion detectors in the surround [1].

Other RGCs are sensitive to the direction of motion because they receive inhibitory inputs from ACs with a specific orientation in their dendritic field, see Fig. 2.31. With this estimate of global motion, the retina produces signals that generate gaze-stabilising eye movements, which allow for the retinal image to be sharp by compensating for the head and eye motion. RGCs can also perform motion anticipation, that compensates for phototransduction delay and allows us to track very fast moving objects: when a moving object enters a very wide RF (like that of an AC), and relying on the fact that motion trajectories are often smooth, then RGCs that lie ahead on the motion path can already be activated [1].

Inhibition from ACs also enables some RGCs to be sensitive to edge orientation.

References

[1] Gollisch T, Meister M. Eye smarter than scientists believed: neural computations in circuits of the retina. Neuron 2010;65(2):150–64.

[2] Artal P. Image formation in the living human eye. Annual Review of Vision Science 2015;1:1–17.

[3] Cornsweet T. Visual perception. Academic Press; 1970.

[4] Wandell BA. Foundations of vision, vol. 8. Sunderland, MA: Sinauer Associates; 1995.

[5] Rhcastilhos, Jmarchn. Schematic diagram of the human eye with English annotations. https://commons.wikimedia.org/w/index.php?curid=1597930.

[6] Hubel DH. Eye, brain, and vision, vol. 22. New York: Scientific American Library; 1988.

[7] Froehlich J, Grandinetti S, Eberhardt B, Walter S, Schilling A, Brendel H. Creating cinematic wide gamut HDR-video for the evaluation of tone mapping operators and HDR-displays. In: Digital Photography X, vol. 9023. International Society for Optics and Photonics; 2014. p. 90230X.

[8] Lee BB, Martin PR, Grünert U. Retinal connectivity and primate vision. Progress in Retinal and Eye Research 2010;29(6):622–39.

[9] Masland RH. The neuronal organization of the retina. Neuron 2012;76(2):266–80.

[10] Wässle H. Parallel processing in the mammalian retina. Nature Reviews. Neuroscience 2004;5(10):747.

[11] Nassi JJ, Callaway EM. Parallel processing strategies of the primate visual system. Nature Reviews. Neuroscience 2009;10(5):360.

[12] Demb JB, Singer JH. Functional circuitry of the retina. Annual Review of Vision Science 2015;1:263–89.

[13] Masland RH. Neuronal diversity in the retina. Current Opinion in Neurobiology 2001;11(4):431–6.

[14] Kolb H, Nelson R, Fernandez E, Jones B. Webvision: the organization of the retina and visual system. https://webvision.med.utah.edu/.

[15] Dhande OS, Stafford BK, Lim JHA, Huberman AD. Contributions of retinal ganglion cells to subcortical visual processing and behaviors. Annual Review of Vision Science 2015;1:291–328.

[16] Reese B. The senses: vision, chap. mosaics, tiling, and coverage by retinal neurons. Elsevier; 2008. p. 436–56.

[17] Cenveo. https://courses.lumenlearning.com/austincc-ap1/chapter/special-senses-vision/.

[18] Hope D. Zyxwv99, https://commons.wikimedia.org/w/index.php?curid=36685094.

[19] Cote R. Photoreceptor phosphodiesterase (PDE6): a G-protein-activated PDE regulating visual excitation in rod and cone photoreceptor cells. In: Beavo JA, Francis SH, Houslay MD, editors. Cyclic nucleotide phosphodiesterases in health and disease. CRC Press; 2006. p. 165–93.

[20] Lagnado L, Schmitz F. Ribbon synapses and visual processing in the retina. Annual Review of Vision Science 2015;1:235–62.

[21] Cmglee. https://commons.wikimedia.org/w/index.php?curid=29924570.

[22] http://book.bionumbers.org/how-big-is-a-photoreceptor/.

[23] Solomon SG, Lennie P. The machinery of colour vision. Nature Reviews. Neuroscience 2007;8(4):276.

[24] College O. https://commons.wikimedia.org/w/index.php?curid=30148003.

[25] Brainard DH. Color and the cone mosaic. Annual Review of Vision Science 2015;1:519–46.

[26] Brainard DH, Williams DR. Spatial reconstruction of signals from short-wavelength cones. Vision Research 1993;33(1):105–16.

[27] Welbourne LE, Morland AB, Wade AR. Human colour perception changes between seasons. Current Biology 2015;25(15):R646–7.

[28] Graham NV. Beyond multiple pattern analyzers modelled as linear filters (as classical v1 simple cells): useful additions of the last 25 years. Vision Research 2011;51(13):1397–430.

[29] Carandini M, Demb JB, Mante V, Tolhurst DJ, Dan Y, Olshausen BA, et al. Do we know what the early visual system does? Journal of Neuroscience 2005;25(46):10577–97.

[30] Chapot CA, Euler T, Schubert T. How do horizontal cells 'talk' to cone photoreceptors? Different levels of complexity at the cone-horizontal cell synapse. The Journal of Physiology 2017;595(16):5495–506.

[31] Thoreson WB, Babai N, Bartoletti TM. Feedback from horizontal cells to rod photoreceptors in vertebrate retina. Journal of Neuroscience 2008;28(22):5691–5.

[32] Wässle H, Boycott BB, Röhrenbeck J. Horizontal cells in the monkey retina: cone connections and dendritic network. European Journal of Neuroscience 1989;1(5):421–35.

[33] Jansen M, Jin J, Li X, Lashgari R, Kremkow J, Bereshpolova Y, et al. Cortical balance between on and off visual responses is modulated by the spatial properties of the visual stimulus. Cerebral Cortex 2018;29(1):336–55.

[34] Dacey DM, Packer OS. Colour coding in the primate retina: diverse cell types and cone-specific circuitry. Current Opinion in Neurobiology 2003;13(4):421–7.

[35] MacNeil MA, Masland RH. Extreme diversity among amacrine cells: implications for function. Neuron 1998;20(5):971–82.

[36] Turner MH, Rieke F. Synaptic rectification controls nonlinear spatial integration of natural visual inputs. Neuron 2016;90(6):1257–71.

[37] Poirazi P, Brannon T, Mel BW. Pyramidal neuron as two-layer neural network. Neuron 2003;37(6):989–99.

[38] Polsky A, Mel BW, Schiller J. Computational subunits in thin dendrites of pyramidal cells. Nature Neuroscience 2004;7(6):621.

[39] Berry MJ II, Meister M. Refractoriness and neural precision. In: Advances in neural information processing systems; 1998. p. 110–6.

[40] Conway BR, Chatterjee S, Field GD, Horwitz GD, Johnson EN, Koida K, et al. Advances in color science: from retina to behavior. Journal of Neuroscience 2010;30(45):14955–63.

[41] Shapley R, Perry VH. Cat and monkey retinal ganglion cells and their visual functional roles. Trends in Neurosciences 1986;9:229–35.

[42] Milner ES, Do MTH. A population representation of absolute light intensity in the mammalian retina. Cell 2017;171(4):865–76.

[43] Cui Q, Ren C, Sollars PJ, Pickard GE, So KF. The injury resistant ability of melanopsin-expressing intrinsically photosensitive retinal ganglion cells. Neuroscience 2015;284:845–53.

[44] Antinucci P, Hindges R. Orientation-selective retinal circuits in vertebrates. Frontiers in Neural Circuits 2018;12:11. https://www.frontiersin.org/article/10.3389/fncir.2018.00011.

CHAPTER

3 The biological basis of vision: LGN, visual cortex and L+NL models

Around 90% of the optic nerve fibers terminate in the lateral geniculate nucleus (LGN), a small part of the thalamus present in both brain hemispheres. The parallel pathways starting in the retina go through the LGN towards the primary visual cortex (V1), see Fig. 3.1. There is significant feedback from the cortex to the LGN as well.

In this chapter we will describe the layered structure, cell types and neural connections in LGN and the visual cortex, with an emphasis on colour representation. We will also introduce linear+nonlinear (L+NL) models, which are arguably the most popular form of model not just for cell activity in the visual system but for visual perception as well.

An important take-away message is that despite the enormous advances in the field, the most relevant questions about colour vision and its cortical representation remain open [1,2]: which neurons encode colour, how does V1 transform the cone signals, how shape and form are perceptually bound, and how do these neural signals correspond to colour perception.

Another important message is that the parameters of L+NL models change with the image stimulus, and the effectiveness of these models decays considerably when they are tested on natural images. This has grave implications for our purposes, since in colour imaging many essential methodologies assume a L+NL form.

3.1 Lateral geniculate nucleus

3.1.1 General structure of the LGN

The LGN has some 1.5 million cells and a single synaptic stage. LGN neurons can be of two classes, relay neurons or inhibitory interneurons.

Relay neurons comprise the majority of LGN cells, and their axons join into a broad band called the optic radiations that terminates in V1. The topographic organisation from retina to LGN to cortex is preserved, meaning that if we trace a continuous path among two points in the retina, the corresponding trajectories in the LGN and the cortex are also continuous [3]. For instance, as we move from the back

Vision Models for High Dynamic Range and Wide Colour Gamut Imaging. https://doi.org/10.1016/B978-0-12-813894-6.00008-9

of the LGN towards the front, the RFs of the cells go from the fovea towards the periphery. This spatial organisation is also called retinotopic [4].

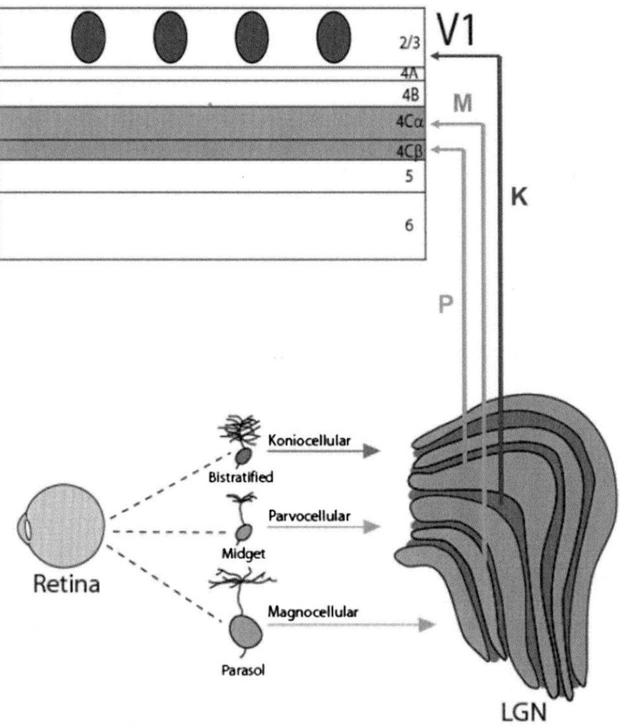

FIGURE 3.1 Retina to LGN to visual cortex

Parallel pathways from the retina through the LGN to the visual cortex, from [5].

Inhibitory interneurons constitute 20% of LGN cells. They receive monosynaptic excitation from the retina and provide local inhibition to relay neurons within the LGN, not projecting to the cortex. Relay neurons can also receive inhibitory input from neurons outside the LGN, in the thalamic reticular nucleus, whose inputs come from the LGN and the cortex, not from the retina. So despite the fact that the LGN receives most of the retinal output and more than 95% of LGN spikes are produced by retinal inputs, only 10% of LGN synapses come from retinal connections. The majority, 60%, come from the cortex, and the rest from other parts of the brain, including the LGN itself [6]. Cortical feedback and its influence on visual processing in the LGN aren't well understood yet, unlike what happens with retinal (feedforward) excitation.

The LGN is organised into a stack of layers and folded over itself along an axis. Light that comes from the right half of the visual environment forms the left half of the retinal image of both eyes (since images are inverted by refraction), and RGCs with their RFs on the left retinal image project onto the LGN in the left hemisphere.

Therefore, the right visual field projects to the LGN in the left. The converse applies to the left half of the visual environment, projecting onto the LGN in the right hemisphere. So each half of the LGN receives signals from both eyes, but just one half, the right or the left, of the visual field. See Fig. 3.2.

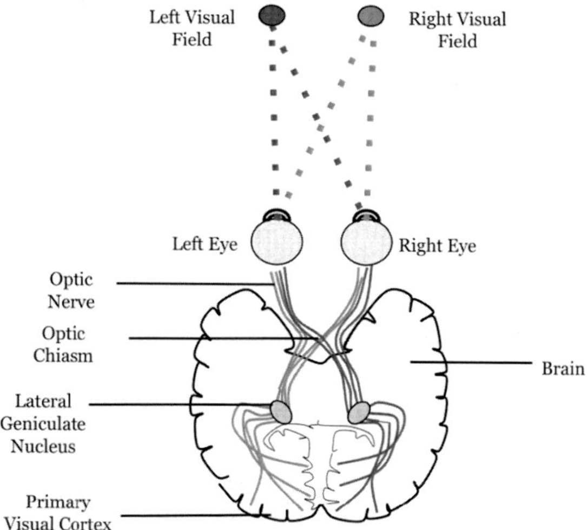

FIGURE 3.2 Left and right visual fields

Left and right visual fields, from [7].

Most LGN cells are monocular, they receive input from either the left or the right eye, and the information about the eye of origin isn't lost because the layers in LGN are segregated by eye input, alternating between same and opposite side of the head [4]. This fact, combined with the preservation of the topographic organisation, implies that if we move along a LGN layer, tangentially to it, we are changing location in the visual field, and if we move perpendicularly to the layer we stay put in the same location in the visual field but alternate from one eye to the other [3].

3.1.2 Layers of the LGN

In humans and other primates the LGN contains three types of layers, whose cells have very different physiology, project to different layers of the primary visual cortex, and represent three distinct parallel pathways for visual processing. These are the magnocellular (MC), parvocellular (PC) and koniocellular (KC) layers [6]. Each layer is 4–10 cells deep [3]. See Fig. 3.1.

There are four PC layers, located at the top of the LGN. PC neurons receive monocular input from midget RGCs and project to layer 4Cβ in V1. They respond best to chromatic stimuli of high spatial detail that doesn't move too fast. The RF centers of PC cells that get their input from near the fovea have a diameter of about 2 min of arc, while visual acuity in the fovea is 0.5 min of arc (1 mm at a distance of 8 m) which as we mentioned corresponds to the distance between two cones. Around 75% of PC cells have cone-opponent (L vs M) center-surround RFs, with all four combinations of center cone type and ON/OFF input: L+M−, L−M+, M+L−, M−L+. This type of PC neurons were called Type I by Hubel and Wiesel [3]. Another 10% of PC neurons are cone opponent but with center only, no surround. For this reason they are also called single-opponent, because they show colour opponency but not spatial opponency. Hubel and Wiesel named this kind of cell Type II. The remaining 15% of PC cells are center-surround but show no colour preference; they are called Type III [3]. See Fig. 3.3.

The MC layers are two and they are located at the bottom of the LGN. MC neurons receive monocular input from parasol RGCs and project to layer 4Cα in V1. They are larger in size than neurons in the other two types of layers. The conduction time from the retina to the MC layers is shorter than for PC layers. MC cells have a faster response to changes in stimulus contrast, and higher sensitivity to high temporal frequency and low spatial frequency stimuli. Therefore, they respond best to rapidly changing inputs without too much spatial detail, and for these reasons it's assumed that the MC pathway is best suited for motion perception [4]. MC neurons are all of Type III: they have center-surround RFs and respond to achromatic detail but have no preference for colour.

Finally, there are six KC layers, one beneath each MC and each PC layer. Neurons in the KC layers project onto the superficial layers 2 and 3 of V1. They receive input from different types of RGCs, including bistratified RGCs. Some of them have binocular responses. They can also show orientation and direction selectivity [6]. They carry S-cone colour opponent signals.

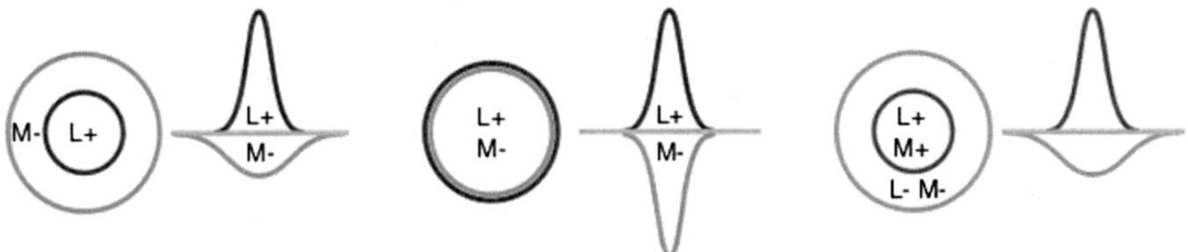

FIGURE 3.3 Types of LGN cells according to receptive field and colour properties

From left to right: Type I, Type II, Type III.

The segregation in these three channels is preserved from the retina to the cortex, and the feedback from the cortex to the LGN is segregated as well. In particular PC neurons in LGN receive feedback from a sublayer of V1 that is separate from the cortical sublayer that sends feedback to MC cells.

3.1.3 Colour representation in the LGN

Most LGN cells respond to colour, and for some time the chromatic properties of PC and KC neurons were thought to correspond closely to the perception of colour, but more recent studies have shown that this is not the case. PC cells receive opponent inputs coming from L and M cones and must therefore be involved in red/green vision, at the same time that they must support fine spatial resolution since the sampling density of PC cells is not provided by any other pathway. However, PC cells that show colour opponency are spatially low-pass, so they can signal colour but they can't provide good resolution. The reason is that each of the cone mechanisms in the RF is low-pass, and therefore a high spatial frequency red–green grating produces responses that add up in this kind of RF [8].

Also, the random distributions of L and M cones and the random sampling of midget BCs and midget RGCs lead to clusters of cone information from one class, L or M, which makes it hard to reconcile precise chromatic properties with high spatial acuity. And for KC cells, it has been shown that the chromatic properties of S-ON and S-OFF cells are different: while S-ON cells have responses that approximate well blue–yellow opponency, showing little activity for isoluminant red–green inputs, S-OFF cells have lower contrast sensitivity and a colour preference that varies in between that of S-ON cells and red–green opponent PC cells [1].

3.1.4 Visual processing in the LGN

The receptive fields of most LGN neurons have the same center-surround configuration as the RFs of RGCs, with roughly the same size and shape, but in the LGN the surround is stronger than in RGCs, which makes LGN neurons sensitive to higher spatial frequencies than retinal neurons [6]. Or, put in other terms, the LGN performs more contrast enhancement of the visual signal [9]. There is a temporal shift as well, as the surround is delayed with respect to the center.

There is convergence from the retina to the LGN, as five RGCs can send their responses to a single LGN cell. Convergence happens at every stage of the visual stream and therefore RFs tend to become larger as we proceed from the retina to the cortex [3]. In the LGN there is divergence as well, as a single RGC synapses with several LGN neurons. Convergence and divergence improve the detection of stimuli under noise, and divergence produces synchronisation in LGN responses, which allows to better transmit information to the cortex.

LGN cells also perform a high-pass temporal filtering of RGC spike trains, effectively denoising the retinal signal by discarding all impulse trains whose rate is below a threshold and thus increasing the signal to noise ratio of the signals sent to the cortex.

Therefore the LGN does quite a bit of processing that improves the quality of the visual information, contradicting the common assumption that the LGN is little more than a relay station from the retina to V1. We should also mention that alertness, sleep, and physical activity have an important influence on visual processing in the LGN.

3.2 Cortex

The cerebral cortex is a crinkled sheet of neurons that almost covers the two cerebral hemispheres. It is 2 mm thick and has an area of 1400 cm^2. The cortex is an example of gray matter and it contains neuron bodies, dendrites and axons (white matter consists of myelinated axons). It has two major types of cells, pyramidal cells and stellate cells. A ridge in the crumpled surface of the cortex is called a gyrus, and a valley is called a sulcus.

The brain is partioned into four lobes (frontal, parietal, temporal, occipital) according to the most visible sulci; see Fig. 3.4 (left). And each lobe is partitioned into areas, defined as regions with distinct anatomical connections with other regions, and where neurons function in a related way.

Most of the LGN signals arrive in the cortex at an area in the occipital lobe called primary visual cortex, or V1, or striate cortex (because in cross-section it shows layers or stripes). It has an area of 18 cm^2 and around 150 million neurons, much more than the 1 million neurons of the LGN [4].

Area V1 receives input from the LGN, and provides feedback to the LGN as well [3]. There are significant dissimilarities between the RFs of LGN neurons and those of V1 cells, though: many V1 RFs are orientation-selective, neurons in V1 receive input from many neurons in the LGN and therefore have much larger RF sizes, the majority of V1 neurons respond well to achromatic spatial detail but not to colour detail (while many LGN neurons are responsive to colour modulations) and a larger fraction of V1 neurons than LGN neurons receive S-cone inputs.

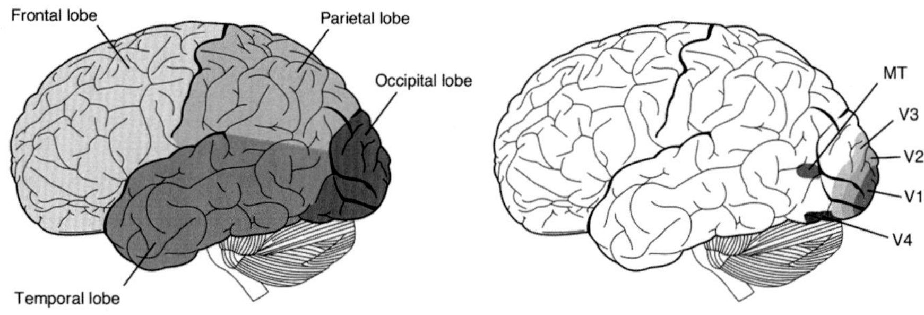

FIGURE 3.4 Brain lobes and cortical areas

Brain lobes and cortical areas.

V1 is surrounded by area V2 to which it sends most of its output. V2 projects to other areas, including V3, V4 and MT (for medial temporal), that in turn send their outputs to other areas; see Fig. 3.4 (right). In addition, each area provides feedback to the cortical area it gets its input from, and receives inputs from the thalamus.

Outside V1, the cortex has at least two parallel streams, called dorsal and ventral, which are segregated but in some cases process the same attribute (like disparity or shape) for different goals. Processes in the dorsal stream deal with motion, binocular disparity, target selection and object manipulation, while the ventral stream is concerned with colour and form [5]. But beyond V1 the knowledge becomes sketchy as it's hard to study in anesthetised animals the properties of neurons whose responses increasingly depend on signals not coming from the retina [1].

3.2.1 Layers of V1

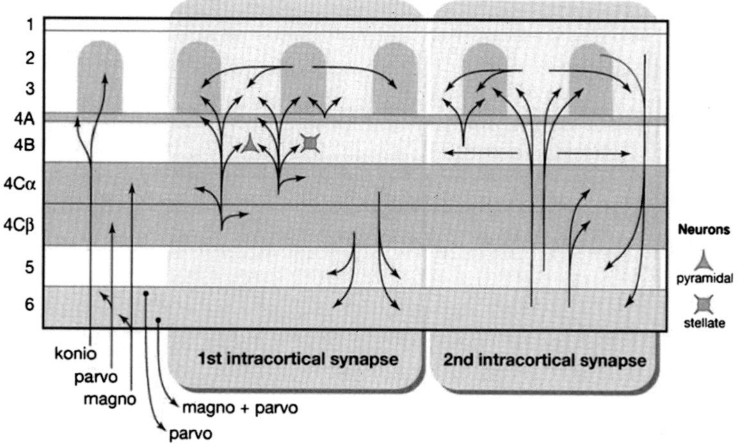

FIGURE 3.5 Layers and circuitry in V1

Layers and circuitry in V1, from [10].

V1 has six layers, numbered from 1 to 6 in increasing depth order; see Fig. 3.5. Axons from the two MC layers of the LGN terminate in V1 in the upper part of layer 4C, called 4Cα, while axons from the PC layers of the LGN terminate in V1 in the lower part of 4C, called 4Cβ. The cells in the KC layers of the LGN send their axons to different V1 layers depending on the colour signal: blue ON cells project onto V1 layers 2,3 and 4A, and blue-OFF cells onto layer 4A [2]. Therefore, the segregation of parallel pathways is preserved from the retina until arriving to the cortex, with V1 cells in layer 4Cα receiving inputs from parasol RGCs, V1 cells in layer 4Cβ receiving inputs from midget RGCs, and V1 cells in layers 2,3 and 4A receiving inputs from bistratified RGCs. But within the first few synapses in V1, these early pathways converge significantly [5]. The magnocellular pathway in layer 4Cα branches out, one part going to layer 4B and then continuing to cortical area MT: as the neurons along this path are selective for high temporal frequency as well as motion direction, it is assumed that area MT is important for motion perception. The other branch of the MC pathway goes to layer 4Cß where it merges with the PC pathway. A possible reason for the retinal segregation of magnocellular from parvocellular signals could be then to be able to send rapidly some information to MT [4]. Cells in layer 4Cβ connect to cells in layers 2 and 3, which therefore receive input from the MC, PC and KC pathways, and send their output to other cortical areas. Layer 5 projects to the midbrain (a subcortical structure) and layer 6 sends its output back to the LGN. Most pyramidal cells in layers 2 to 6 have local axon ramifications called collaterals that spread the information through V1 [3]. And many of these lateral connections remain inside a single layer without crossing into another one, supporting the view that V1 is organised into specific layers [8].

3.2.2 Simple and complex cells

The RFs of V1 cells get progressively larger and more complex along the visual path. In the input layer, layer 4, all cells in 4Cβ and some in 4Cα have a center-surround RF, while other cells in 4Cα have RFs where the ON and OFF regions are not concentric but elongated, and therefore they have a preferred orientation; see Fig. 3.6. Hubel and Wiesel [3] called these types of cortical cells "simple", arguing that for simple cells the response can be expressed as a linear combination of the inputs, with weights given by the RF values (i.e. positive in the ON regions, negative in the OFF regions). Put another way, the cell's response results from convolving the input with the RF. These RFs resemble Gabor filters, which allow for a sparse representation of signals, and therefore it's been conjectured that these cells support an efficient coding or efficient representation of the visual information [11]. We discuss efficient representation in some detail in Chapter 4.

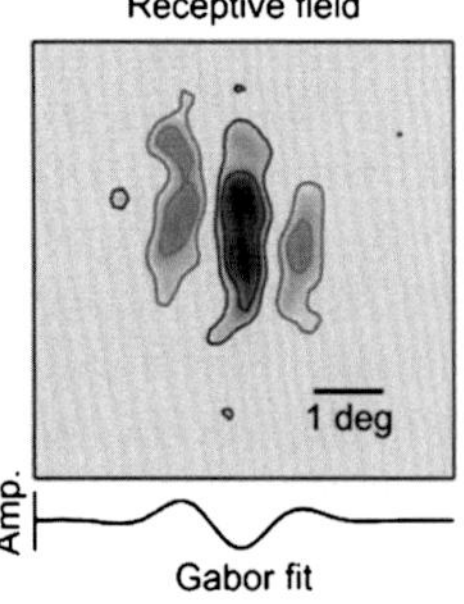

FIGURE 3.6 Oriented RF

Oriented RF, where colour blue (dark gray in print version) denotes the OFF region and colour red (gray in print version) the ON region. Image from [12].

The rest of the cortical cells were called "complex" by Hubel and Wiesel. In downstream layers, above and below 4C, most cells are complex. Complex cells, as most simple cells, have a preferred orientation, but their response is nonlinear, often modelled as a sum of nonlinear transforms applied to several linear RFs.

With these criteria, approximately half of V1 cells are simple and half complex. These definitions have had a fundamental impact in vision research, from neurophysiology to visual perception, but they are also highly controversial [13], and we'll return to this issue at the end of this chapter.

3.2.3 Ocular dominance columns and orientation columns

Unlike what happens in the LGN, where we recall that a vertical displacement across layers corresponds to an alternance between left eye and right eye input, in V1 the signals from the two eyes remain segregated (until the superficial layers, where there are binocular neurons) but the alternance is not vertical but horizontal. In a path perpendicular to the surface of V1 all cells respond to the same eye, while a path parallel to the surface shows 0.5 mm wide bands of alternating ocular dominance, where cells in a band favour one eye and cells in the adjacent band favour the other eye. This alternation is sharp in layer 4C, the input layer, but smoother in the other layers, due to the spread of information in horizontal and diagonal connections [3], and in fact many neurons in the superficial layers respond to both eyes [4]. Given that paths perpendicular to the surface preserve eye preference, they are called ocular dominance columns. (See Fig. 3.7.)

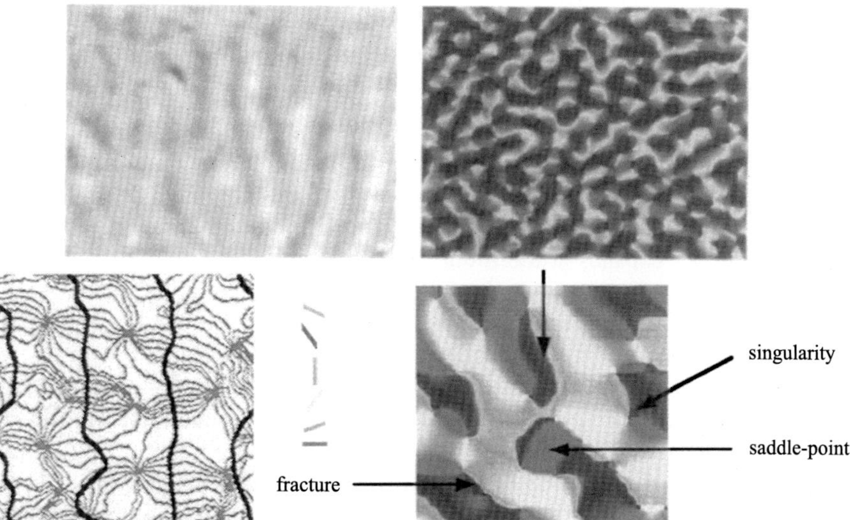

FIGURE 3.7 Ocular dominance and orientation columns.

Ocular dominance and orientation columns, from [14].

There are also orientation columns. As mentioned above, except from layer 4 the majority of cells in V1 layers show an orientation preference, be they simple or complex cells. Again a path perpendicular to the surface finds cells with the same preferred orientation, while a path parallel to the surface shows a change of preferred orientation at a rate of about 10 degrees every 0.05 mm, so a shift of 180 degrees is covered in 1 mm, same as the width of one ocular dominance column. In fact, near the edge of the ocular dominance column, iso-orientation lines are roughly parallel to each other and perpendicular to the ocular dominance edge, while these iso-orientation lines converge towards single points in the middle of the ocular dominance column called singularities [4].

3.2.4 Maps of V1

As it was the case in the LGN, signals in V1 are also organised retinotopically, so a tangential path along layer 4C finds at the beginning neurons with RFs corresponding to the fovea, and gradually the RFs become more peripheral [4]. (See Fig. 3.8.)

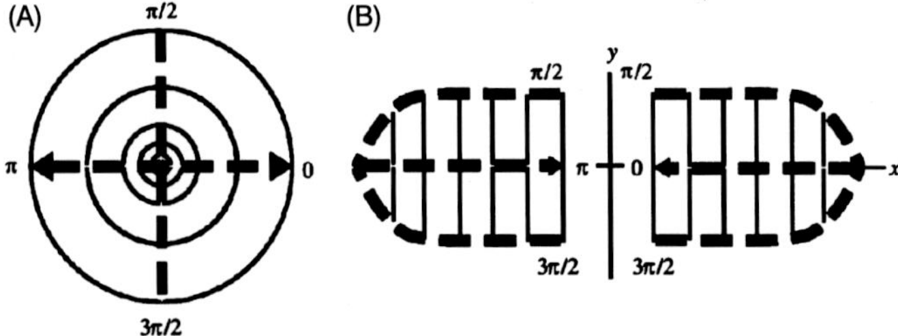

FIGURE 3.8 Retinotopic organisation

Retinotopic organisation, from [15].

The topography of ocular dominance columns and orientation columns is very uniform, but V1 is very much non-uniform in the allocation of cortical area, since magnification of the fovea is more than 30 times larger than in the periphery: one degree of visual field in the fovea corresponds to 6 mm in the cortex, while one degree of visual field in the periphery (90° eccentricity) gets just 0.15 mm in the cortex [3]. Therefore V1 expands at the fovea, while everything else (cortical thickness, the width of ocular dominance and orientation columns) remains constant. It has been suggested that this expansion of foveal representation corresponds to each RGC being allocated the same amount of cortical area. Moving tangentially to layer 4C, the RF center positions change smoothly within an ocular dominance column but shift abruptly when passing the ocular dominance edge, by an amount of half the RF size [4]. (See Fig. 3.9.)

Due to this organisation of V1, the pattern of activity on the cortex does not resemble at all the visual stimulus, and for example a line in the visual field produces in V1 responses that appear as an array of bars [3].

3.2.5 Motion and area MT

Around 25% of V1 neurons, mainly in layers 4A, 4B, 4Cα and 6, are direction selective, responding to stimulus moving in one direction but not in the opposite. Direction selective neurons are also tuned to a particular speed of motion, and a neuron can be direction selective and orientation selective at the same time, responding best to lines of a specific orientation moving in a given direction.

Area MT receives input from most of the direction selective neurons in V1, which in turn have inputs originating in parasol cells in the retina, that we recall are sensitive to high temporal frequencies. MT is the first cortical area with a majority of direction selective cells, arranged in an organised way, where neighbouring neurons respond best to similar directions of motion. Some 25% of MT cells are not only direction

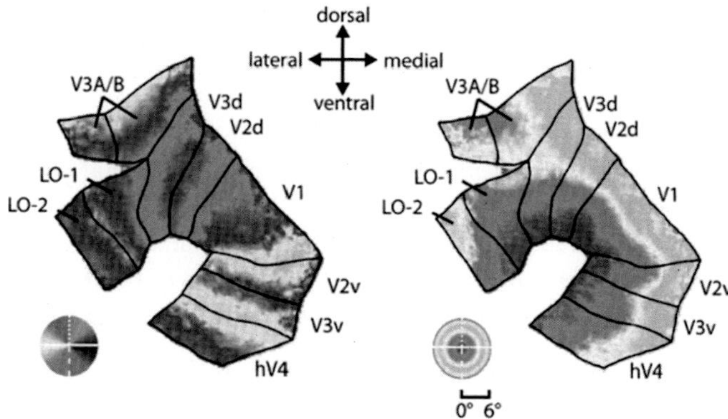

FIGURE 3.9 Visual field maps

Visual field maps, from [16].

selective but pattern selective as well, responding best when a pattern as a whole is moving in a certain direction. For these reasons the visual stream from retina to MT is called the motion pathway, and is specialised for motion perception. In fact lesions in MT don't produce losses in visual acuity or colour perception, but do impair the performance of motion-related tasks.

However, we must note that the perceived speed of an object is affected by its colour, contrast and spatial detail, an unwanted side-effect that may be due to the fact that the encoding of the visual information in the retina must simultaneously serve several different needs [4].

3.2.6 Colour representation in the cortex

There are at least two colour perception phenomena, colour contrast and colour constancy, that can't be explained with the sort of center-surround RFs that we have been discussing.

Colour contrast makes a red object appear to be a more vivid red when it's surrounded by green (and the same goes for a green object in a red surround), while colour constancy allows us to perceive the colour of an object as more or less the same despite changes in the illuminant. At first glance it would seem that a RF with an L-ON center and an M-OFF surround, as shown in Fig. 3.10 (bottom), would indeed produce a colour contrast effect, but in fact it's the opposite. Let's say we have a red spot inside a green circle. The red spot makes the L-ON center produce a positive contribution, but the green region around the red spot makes the M-OFF surround produce a negative contribution, that detracts from that of the red spot. The net result is that the red colour of the spot surrounded by green is less intense than if the red spot were inside a red circle, in which case both center and surround would provide positive contributions.

The conclusion is that a center-surround colour opponent RF can't produce colour contrast, a different type of RF would be needed for this. For instance, a double-opponent RF, where the center is colour opponent and the surround is also colour-opponent but with the opposite polarity, e.g. a M–L surround for a L–M center [3]; see Fig. 3.10 (top, with red presented as dark gray and green as mid gray in the print version). This type of cell would respond well to a stimulus with chromatic contrast (e.g. a colour edge) or to a small stimulus of uniform colour, but it wouldn't respond much to a large stimulus of uniform colour and it would not respond to an achromatic stimulus of any kind [1].

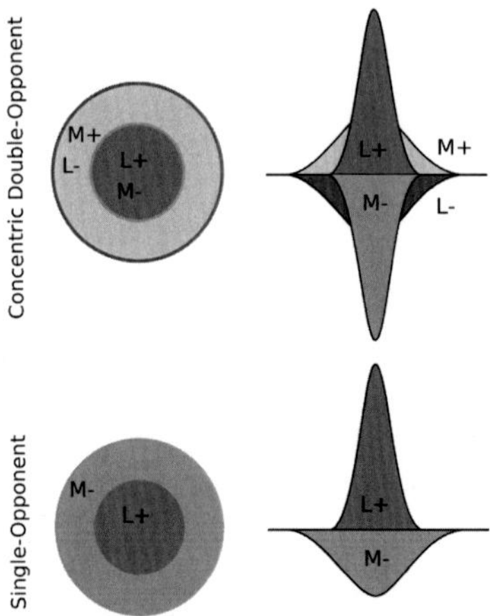

FIGURE 3.10 Single and double opponency

Single and double opponency: two-dimensional maps and spatial profiles. Adapted from [17].

Double-opponent cells can also explain colour constancy, as we will presently show. Say we have a small object that for long wavelengths reflects $l\%$ of the light, $m\%$ for medium wavelengths and $s\%$ for short wavelengths. Under an illuminant I_1 of intensity (a_1, b_1, c_1) respectively for the long, medium and short wavelengths, the light reflected by the object will be $(a_1 l, b_1 m, c_1 s)$.

Let (L_1, M_1, S_1) be the cone responses to the light coming from the object. In Chapter 5 we'll see how it's reasonable to approximate these responses with a logarithmic function, so we get

$L_1 = log(a_1 l) = log(a_1) + log(l)$; $M_1 = log(b_1) + log(m)$; $S_1 = log(c_1) + log(s)$.

If the object is in a green surround of reflectances (l', m', s'), under the same illuminant I_1 the surround will produce cone responses

$L_1' = log(a_1) + log(l')$; $M_1' = log(b_1) + log(m')$; $S_1' = log(c_1) + log(s')$.

When these signals are the input to a double-opponent cell with a (L–M) center and a (M–L) surround, the output of the cell will be:

$(L_1 - M_1) + (M'_1 - L'_1) = log(a_1/b_1) + log(l/m) + log(b_1/a_1) + log(m'/l') = log(l/l') - log(m/m')$,
which is independent of the illuminant I_1 and therefore would yield the same result for another illuminant I_2 of intensity (a_2, b_2, c_2).

To summarise, double-opponent cells would be able to explain some colour perception phenomena, like colour contrast and colour constancy, that it was assumed could not be produced with regular center-surround cells as those found in the retina, so there has been a lot of scientific effort trying to find this type of cell in the cortex. Some early studies found V1 neurons that seemed to have center-surround RFs with space and colour opponency. Later works that carried out more exhaustive experiments reported no evidence of this type of double-opponent neuron in primate V1 (many studies on cortical physiology are done on monkeys, which have been shown to have very similar colour vision to humans [3]). And the more recent neurophysiological evidence shows that there's indeed a significant proportion of double-opponent cells in V1, but they're different from what was hypothesised in the 1980s: they don't necessarily have center-surround RFs, they respond both to chromatic and achromatic contrast, and many are orientation-selective [2].

According to their colour properties there appear to be three types of cells in V1 [8]:

1. Around 60% of V1 cells do not respond to colour; this is in sharp contrast with what happens in the LGN, where some 90% of cells respond to colour. These V1 cells are achromatic because they receive L and M input with the same sign, but this doesn't necessarily mean that their input comes from MC cells, they could also be indiscriminately sampling PC cells. Their spatial frequency response is bandpass, and this corresponds to a center-surround RF (because the spatial frequency response is the Fourier transform of the RF weights).
2. Around 10% of V1 cells show strong colour opponency, and don't respond to achromatic stimuli. Their spatial frequency response is lowpass, corresponding to a single-opponent (center-only) RF. See Fig. 3.11 (top).
3. Around 30% of V1 cells respond both to isoluminant colour gratings and to achromatic gratings, which implies that their spatial frequency response is bandpass both for chromatic and achromatic stimuli and therefore they are described as double-opponent. This can be achieved in at least two ways: a double-opponent RF consisting of two single-opponent RFs, side by side but with opposite signs, or a center-surround colour-opponent RF with an extra surround that is wider and achromatic [1]. See Fig. 3.11 (bottom). But in the retina, the lateral inhibition provided by horizontal and amacrine cells also produces a surround that has both a narrow and a wide component, and in [18] we have shown how this dual-surround can explain already at retinal level some perceptual phenomena that were assumed to be taking place in the cortex. This suggests the possibility that colour contrast and colour constancy could be happening at the retina as well.

It's been argued that both single and double opponent neurons are necessary for colour representation, with single-opponent RFs better suited for uniform coloured

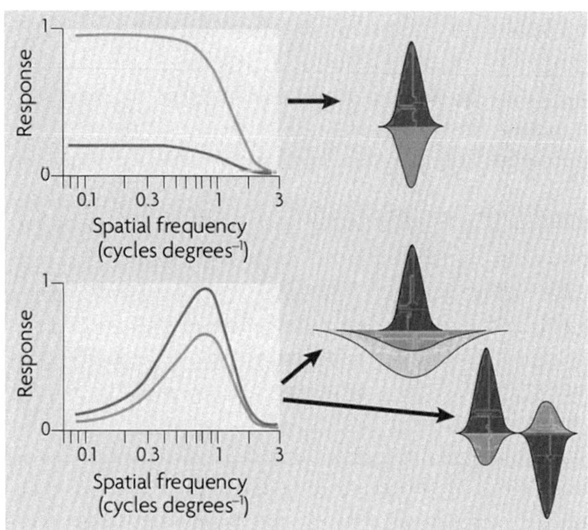

FIGURE 3.11 V1 RFs and colour response

V1 RFs and colour response, from [1].

regions and double-opponent RFs for colour boundaries, colour contrast and colour constancy [2].

As for colour processing in the cortex beyond V1, quite little is known in comparison with the understanding there is for the retina, the LGN and V1. In the inferior temporal cortex (IT) there are many neurons narrowly tuned to colour and saturation, and which could be involved in colour cognition and colour categorisation [2]. And for a while area V4 attracted attention as a region specially involved with colour vision, but later research put this into question [1].

3.3 L+NL models

In order to analyze spike trains of cells in the retina, the LGN or the cortex, most modelling techniques consist of a cascade of linear and nonlinear (L+NL) stages.

For the linear filtering part, the underlying assumption is that the visual system may behave linearly around an operating point depending on the mean level of illumination, "linearly" meaning that the visual system response to the sum $S_1 + S_2$ of two stimuli equals the sum of the responses to S_1 and S_2 independently; the advantage of this formulation is that linear systems can be probed with sinusoidal stimuli, which pass through the system simply changed in amplitude and phase [19].

We recall that the receptive field (RF) of a neuron is defined as the region of the visual field where light stimuli can raise or lower the level of neural activity [20]. In 1953, work by Kuffler in the cat retina determined that the receptive fields of retinal ganglion cells took the form of concentric circular center and surround regions of opposite polarities (ON and OFF), while in 1965 Rodieck made a key advance by modelling this RF as the sum of two overlapping Gaussians of different radius and opposite polarity, in what's now known as the "difference of Gaussians" (DoG) model [19]. The model by Rodieck begins with a linear stage that is separable in two consecutive space and time linear filtering operations. First the light stimulus at time instant t is convolved in space with a DoG filter (this is just a weighted average of the input signal, with the weights given by the RF), and then the resulting signal is convolved in time with another linear filter, in this case a delta function minus an exponential decay [21]. Next, the nonlinear stage performs a rectification, truncating to zero all negative values of the result of the linear filtering operation: the reason is that the model is intended to predict average firing rates, which can only be positive.

With this very simple model, very good predictions can be obtained, as Fig. 3.12 shows. It was later found out that these good fits of the model require some restrictions on the input, whose variance must be small compared to the mean; otherwise, the nonlinearity must be more complex than just a rectification, and better matches of the firing rate can be obtained with more involved nonlinearities that can be static (the same for all locations) or change spatially depending on the neighbours' output, as in the divisive normalisation formulation [22].

Another way to look at it is that, in a L+NL model, the linear filter represents the stimulus to which the neuron is more responsive: for example, in the case of an ON-center OFF-surround RF, the maximal output is obtained when the input has the same shape as the RF, with high above-the-mean values in the center and low below-the-mean values in the surround. And the nonlinear part includes thresholds and distortions that determine how the output of the linear filter is turned into the cell's response [23].

In practice, a simple L+NL model with a static nonlinearity is usually determined in the following way. First the linear filter F is estimated by computing the average stimulus preceding a spike, while the nonlinearity is estimated by comparing the observed firing rate with the result of convolving F with the stimulus [24].

L+NL models are used to classify neurons, according to how well their response can be predicted by the simplest form of the model or instead they require more stages, e.g. "simple" vs. "complex" cortical cells [13] or X vs. Y retinal ganglion cells in the cat [19].

The use of L+NL models, which is by far the most common paradigm in vision science from neurophysiology to perception, assumes a certain form of system function and even determines the vocabulary used to describe system responses: for instance, "sensitivity", "threshold" and "saturation" derive from an L+NL model and correspond respectively to the slope, the foot and the knee of the nonlinearity [23].

Very importantly, a L+NL model is valid at a given mean luminance and contrast, under which it may provide a good match to the firing rate. As we shall see in Chapter 4, adaptation alters the visual system response, affecting among other things the spatial receptive field and temporal integration properties of neurons, requiring changes in the linear and/or the nonlinear stages of a L+NL model in order to explain neural responses [21]. This means that, if for a light stimulus of some mean

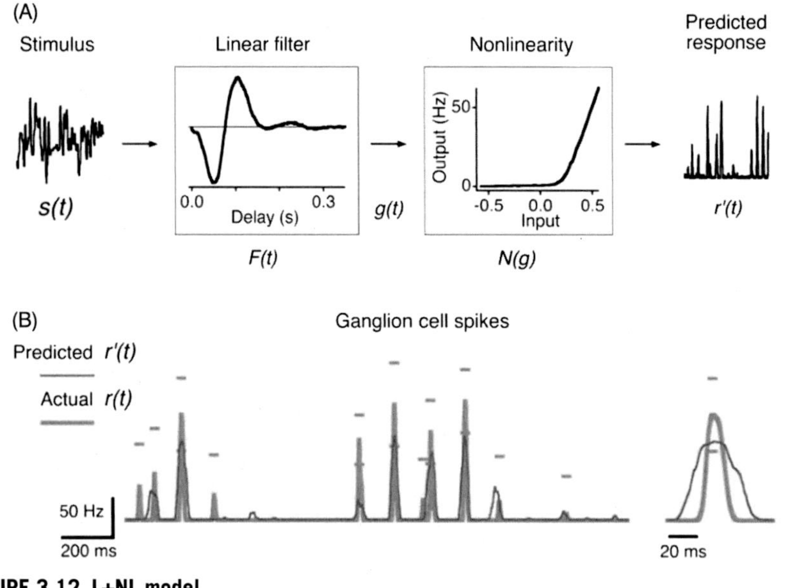

FIGURE 3.12 L+NL model

L+NL model, from [23].

level and contrast the response of a neuron can be well approximated with a linear filter F followed by a nonlinearity N, then after we change the mean and/or the contrast of the stimulus and the cell adapts to the new conditions, then the previous model with F and N is no longer valid: we need to find a new linear filter F' and/or a new nonlinearity N' for the current, post-adaptation response of the neuron. Adaptation then can simply be seen as a change in the parameters of the L+NL model [23].

The, perhaps very surprising, conclusion, is that the receptive field is not an intrinsic property of the neuron, it's not determined by its biological wiring: **the receptive field of a neuron depends on, and changes with, the input light stimulus**.

This, combined with the fact that L+NL models are much less accurate in predicting neural responses to natural images (as opposed to sinusoidal gratings, oriented bars, white noise and other types of artificial stimuli that are commonly used in vision research), has made some researchers skeptic of the actual usefulness of L+NL models and pessimistic about the possibility of upgrading these models up to the point where they become really effective for natural images [25], a topic to which we shall return in Chapter 11.

References

[1] Solomon SG, Lennie P. The machinery of colour vision. Nature Reviews. Neuroscience 2007;8(4):276.

[2] Conway BR, Chatterjee S, Field GD, Horwitz GD, Johnson EN, Koida K, et al. Advances in color science: from retina to behavior. Journal of Neuroscience 2010;30(45):14955–63.

[3] Hubel DH. Eye, brain, and vision, vol. 22. New York: Scientific American Library; 1988.

[4] Wandell BA. Foundations of vision, vol. 8. Sunderland, MA: Sinauer Associates; 1995.

[5] Nassi JJ, Callaway EM. Parallel processing strategies of the primate visual system. Nature Reviews. Neuroscience 2009;10(5):360.

[6] Usrey WM, Alitto HJ. Visual functions of the thalamus. Annual Review of Vision Science 2015;1:351–71.

[7] Mads00. https://commons.wikimedia.org/w/index.php?curid=49282011.

[8] Johnson EN, Hawken MJ, Shapley R. The spatial transformation of colour in the primary visual cortex of the macaque monkey. Nature Neuroscience 2001;4(4):409.

[9] Martinez LM, Molano-Mazón M, Wang X, Sommer FT, Hirsch JA. Statistical wiring of thalamic receptive fields optimizes spatial sampling of the retinal image. Neuron 2014;81(4):943–56.

[10] Sincich LC, Horton JC. The circuitry of v1 and v2: integration of color, form, and motion. Annual Review of Neuroscience 2005;28:303–26.

[11] Olshausen BA, Field DJ. Natural image statistics and efficient coding. Network Computation in Neural Systems 1996;7(2):333–9.

[12] Jansen M, Jin J, Li X, Lashgari R, Kremkow J, Bereshpolova Y, et al. Cortical balance between on and off visual responses is modulated by the spatial properties of the visual stimulus. Cerebral Cortex 2018;29(1):336–55.

[13] Carandini M, Demb JB, Mante V, Tolhurst DJ, Dan Y, Olshausen BA, et al. Do we know what the early visual system does? Journal of Neuroscience 2005;25(46):10577–97.

[14] Obermayer K, Blasdel GG. Geometry of orientation and ocular dominance columns in monkey striate cortex. Journal of Neuroscience 1993;13(10):4114–29.

[15] Bressloff PC, Cowan JD, Golubitsky M, Thomas PJ, Wiener MC. Geometric visual hallucinations, Euclidean symmetry and the functional architecture of striate cortex. Philosophical Transactions of the Royal Society of London Series B: Biological Sciences 2001;356(1407):299–330.

[16] Wandell BA, Dumoulin SO, Brewer AA. Visual field maps in human cortex. Neuron 2007;56(2):366–83.

[17] Johnson EN, Hawken MJ, Shapley R. The orientation selectivity of color-responsive neurons in macaque v1. Journal of Neuroscience 2008;28(32):8096–106. https://doi.org/10.1523/jneurosci.1404-08.2008.

[18] Yeonan-Kim J, Bertalmío M. Retinal lateral inhibition provides the biological basis of long-range spatial induction. PLoS ONE 2016;11(12):e0168963.

[19] Shapley R, Enroth-Cugell C. Visual adaptation and retinal gain controls. Progress in Retinal Research 1984;3:263–346.

[20] Olshausen BA, Field DJ. Vision and the coding of natural images: the human brain may hold the secrets to the best image-compression algorithms. American Scientist 2000;88(3):238–45.

[21] Meister M, Berry MJ. The neural code of the retina. Neuron 1999;22(3):435–50.

[22] Carandini M, Heeger DJ. Normalization as a canonical neural computation. Nature Reviews. Neuroscience 2012;13(1):51.

[23] Baccus SA, Meister M. Fast and slow contrast adaptation in retinal circuitry. Neuron 2002;36(5):909–19.

[24] Bonin V, Mante V, Carandini M. The statistical computation underlying contrast gain control. Journal of Neuroscience 2006;26(23):6346–53.

[25] Olshausen BA, Field DJ. How close are we to understanding v1? Neural Computation 2005;17(8):1665–99.

CHAPTER

Adaptation and efficient coding

4

In the world that surrounds us lighting conditions vary enormously, and from dark night to bright noon light intensity covers a range of 10 orders of magnitude; for instance, the same piece of white paper is one million times brighter under sunlight than under moonlight [1]. This supposes a fundamental challenge for our visual system, since all neurons have a limited range of responses, of around two orders of magnitude, so there is an intrinsic trade-off between encoding the full range of the input signal and encoding its gradations [2].

In order to deal with this issue the visual system uses a number of strategies, and a key one is that of adaptation. Adaptation is an essential feature of the neural systems of all species, a change in the input–output relation of the system that is driven by the stimuli [3]. Through adaptation the sensitivity of the visual system is constantly adjusted taking into account multiple aspects of the input stimulus, matching the gain to the local image statistics through processes that aren't fully understood and contribute to make human vision so hard to emulate with devices [4].

Adaptation happens at all stages of the visual system, from the retina to the cortex, with its effects cascading downstream [5].

4.1 The purpose of adaptation

Adaptation increases the sensitivity when the light intensity is low so that the signal-to-noise ratio improves, and decreases it when the light intensity is high so as to avoid saturation [6]. For instance, if the ambient light level decreases the visual system adapts by, among other things, dilating the pupils, increasing the gain of photoreceptors, and reducing the size of the surround of the receptive fields of RGCs [7]. Without adaptation, we would become blind in bright daylight as all reflecting objects would appear white. Decreasing the sensitivity when there is ample signal also allows for a faster visual response, as high sensitivity relies in part on long integration times; therefore, adaptation also allows to sharpen responses in time [8].

Adaptation provides us with the key survival advantage of discriminating slight differences in colour or contrast in our environment, detecting changes in these properties under varying lighting conditions, on longer timescales as the sun moves in the sky and on very short timescales as our eyes scan our surroundings [9]. Thus an important goal of adaptation is to make visual objects produce a retinal response

Vision Models for High Dynamic Range and Wide Colour Gamut Imaging. https://doi.org/10.1016/B978-0-12-813894-6.00009-0

that remains fairly constant under changes in the illumination, so that when processing visual information the brain doesn't need to take into account whether or not the average light level is changing [1]. This is termed light adaptation and is fully accomplished by the retina. It starts already in the photoreceptors, whose sensitivity declines inversely with the ambient light level over a wide range of intensities [1], with the net result that retinal output can become relatively independent from illumination. Light adaptation continues with bipolar cells and retinal ganglion cells, that make the magnitude of the retinal response depend on local contrast instead of absolute light level. Local contrast is defined as percent change in intensity with respect to the (local) average or background level and to a good approximation it depends only on surface reflectance, a physical property of objects. In a typical scene large changes in the absolute level of illumination do not have an impact on the range of contrast values, that remain fairly constant and below two orders of magnitude [10], and therefore can be *fit* into the limited dynamic range of neurons. In this way, adaptation allows the visual system to approximately encode the reflectance of objects, which provides a key survival ability [11].

In the early visual system, at the retina and LGN, adaptation is concerned mainly with changes in two statistical properties of the light intensity: its mean and variance [6]. Adaptation to the mean is what we discussed above, termed light adaptation. Contrast adaptation tailors the performance of the visual system to the range of fluctuations around the mean, i.e. the contrast, and for instance when the contrast increases (under constant mean) the RGCs become less sensitive [7]. By adapting to the statistical distribution of the stimulus, the visual system can encode signals that are less redundant and this in turn produces metabolic savings by having weaker responsiveness after adaptation, since action potentials are metabolically expensive [12]. In this view adaptation is one of the ways the visual system follows the efficient representation principle, to be discussed in the following section.

Some potential advantages of adaptation have been questioned. For instance, in the same way that light adaptation improves the discrimination of light levels around the mean, for which there is ample psychophysical evidence, contrast adaptation should improve contrast discrimination as well, but there is little experimental support for this. It has also been argued that adaptation can improve the detectability of novel stimuli [5], where common inputs are ignored to highlight new ones, although this seems to be contradicting the whole point of adaptation making the visual system more sensitive to common rather than rare stimuli [12].

Finally, adaptation comes with a number of shortcomings. On one hand, adaptation can produce ambiguities in decoding the neural signal [3]. Also, by making the visual system rather insensitive to the illumination, our ability to estimate the mean light level is truly poor, and since adaptation relies on estimates of mean and variance that are bounded in space and time, noise becomes an issue and the sensitivity might change (due to a poor estimate) even when the input stimulus does not [12].

4.2 Efficient representation

The efficient representation principle, introduced in 1961 by Horace Barlow [13], is based on viewing neural systems through the lens of information theory. The retina, the LGN, the cortex, can be seen as systems that process and transmit information, and the efficient representation principle states that the organisation of the visual system in general and neural responses in particular are tailored to the statistics of the images that the individual typically encounters, so that visual information can be encoded in the most efficient way, optimising the limited biological resources.

Natural images (and this goes for images in urban environments as well) are not random arrays of values, they present a lot of statistical structure and nearby points tend to have similar values. As a result, there is significant correlation among pixels, with a redundancy of 90% or more [14], and it would be highly inefficient and detrimental for the visual system to simply encode each pixel independently. For instance the number of axons in the optic nerve is bounded and each of them consumes energy [15], to name just two biological constraints, so having several optic nerve fibers sending the same signal would be a clear waste of limited resources: the retina must compress the visual input so that it can be encoded in low dynamic range signals and transmitted through a small number of axons [16]. Another very important reason to remove statistical regularities from the representation is that the statistical rules impose constraints on the image values that are produced, preventing the encoded signal from utilising the full capacity of the visual channel, which is another inefficient or wasteful use of biological resources. By removing what is redundant or predictable from the statistics of the visual stimulus, the visual system can concentrate on what's actually informative [17], and this underlies Horace Barlow's proposal that neural processes must encode information in a way that minimises redundancy, thus maximising the statistical independence of the neural outputs.

The efficient representation principle is a general strategy observed accross mammalian, amphibian and insect species, where visual processing considers the statistics of the visual stimulus and adapts to its changes [2]. In fact, efficient representation requires that the statistics of the image input are matched by the coding strategy, and while a global part of this coding strategy must have evolved on long timescales (development, evolution), in order to be truly efficient the coding must also adapt to the local spatio-temporal changes of natural images occurring at timescales of hours (e.g. from daybreak to dawn), seconds (e.g. we move from one environment into another), or fractions of a second (e.g. our eyes move around). Put another way, efficient representation involves matching the neural code to global and local statistics at different timescales.

Atick [14] presents three potential benefits of efficient representation:

- In the presence of an information bottleneck (limited dynamic range, limited bandwidth, limited number of fibers etc.) efficient representation implies data compression so as to fit the huge rate of incoming signals into the limited capacity of the visual pathways.

- In order to learn a new association between events, the brain must store prior probabilities of conjunctions and the only practical representation would be an efficient one in which all elements are statistically independent.
- An efficient representation of visual stimuli decomposes images into features that are independent, and whose combination allows to assemble natural images in the most economical way. These elements might be similar to the patterns that the individual needs to recognise from the environment in order to acquire knowledge of its location and surrounding objects.

There is abundant experimental evidence that supports that neural processing is in fact embodying the efficient representation principle. To name just a few: single neurons adapt their action potentials in real time to efficiently encode complex dynamic inputs, maximising information transmission [18]; the processing of contrast closely matches the statistical properties of natural scenes [19]; cortical activity has been shown to perform decorrelation, with adaptation producing a sort of signal equalisation that neutralises changes in the statistics of the environment [20].

But even more importantly, we can use the efficient representation principle as an "ecological approach" to predict neural processing from the statistical properties of the environment [14]. Given some statistic of natural images (e.g. that the power spectrum has a $1/f^2$ form), a biological constraint (e.g. that retinal activity might to some extent be approximated as a convolution with a linear filter following light adaptation), and making use of concepts and techniques from information theory (like entropy, redundancy, information maximisation), one can find the *theoretical* neural coding that is optimal in removing redundancies and then compare it with the *actual* neural coding that can be observed through neurophysiology experiments. Remarkably, this ecological approach has correctly predicted a number of neural processing aspects and phenomena like the photoreceptor response performing histogram equalisation, the dominant features of the receptive fields of retinal ganglion cells (lateral inhibition, the switch from bandpass to lowpass filtering when the illumination decreases, and, remarkably, colour opponency, with L, M and S signals being highly correlated but $L+M$, $L-M$ and $S-(L+M)$ having quite low correlation), or the receptive fields of cortical cells having a Gabor function form [14,15]. See Fig. 4.1 for an example.

Efficient representation is the only framework able to predict the functional properties of neurons from a simple principle, and given how simple the assumptions are it's really surprising that this approach works so well [16].

In information theory, redundancy is quantified in the following manner [14]. Given an ensemble M of messages w, the information $I(w)$ of a message w is a measure of how unexpected or improbable the message is, and the entropy $H(M)$ of the ensemble is the average information per message. The upper bound on $H(M)$ gives a value C that is achieved when all code words are used with the same frequency. The redundancy R is defined as $R = 1 - H(M)/C$, and the most efficient representation is that which brings R closer to zero. This optimal compression might be impractical, because reducing the representation to its minimal size might be computationally very costly, but also because in real scenarios noise is always present and

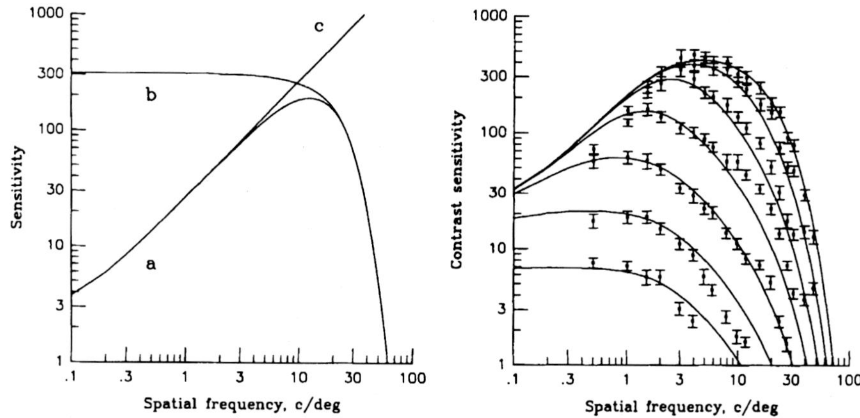

FIGURE 4.1 Efficient representation predicts neural processing

Efficient representation predicts neural processing. Left: the optimal linear filter, in the sense of information maximisation, for a visual signal with noise and for a given mean illumination level; being bandpass, the filter corresponds to a center-surround receptive field. Right: the optimal filter found theoretically provides an excellent fit to perceptual data (the contrast sensitivity function; each curve corresponds to a different illumination level and therefore a different filter). Images from [21].

redundancy is what allows us to tell apart the signal from the noise, so we wouldn't want to remove it completely.

Very importantly for relating efficient coding principles to adaptation in neural processing, and in particular for designing tone mapping algorithms as we will see in Chapter 9, Atick [14] makes the point that there are two different types of redundancy or inefficiency in an information system like the visual system:

1. **If some neural response levels are used more frequently than others.** For this type of redundancy, the optimal code is the one that performs histogram equalisation. There is evidence that the retina is carrying out this tye of operation [15]: Laughlin showed in 1981 how the photoreceptors of a fly had a response curve that closely matched the cumulative histogram of the average luminance distribution of the fly's environment.
2. **If neural responses at different locations are not independent from one another.** For this type of redundancy the optimal code is the one that performs decorrelation. There is evidence in the retina, the LGN and the visual cortex that receptive fields act as optimal "whitening filters", decorrelating the signal. It should also be pointed out that a more recent work [17] contends that decorrelation is already performed by the rapid eye movements that happen during fixations, and therefore the signal arrives already decorrelated at the retina: the subsequent spatial filtering performed at the retina and downstream must have other purposes, like enhancing image boundaries.

4.3 The neural code

In order to analyze spike trains, and not only for retinal responses, most modelling techniques consist of a cascade of linear and nonlinear (L+NL) stages, as mentioned in Chapter 3.

For a number of reasons there has been a considerable amount of research in trying to decipher the neural code used by the retina in the RGC spike trains, where the application of efficient coding principles is expected to be very important:

- What we see is fully represented in the signals travelling through the optic nerve.
- In [16], Meister and Barry remark that the retina performs the bulk of the signal compression in the visual system, going from 10^8 photoreceptor axons to 10^6 RGC axons in the optic nerve (this number then goes up again, by about 40 times, at the first synapse in the visual cortex).
- Retinal structure (number of layers, types of cells, basic circuits, etc.) is common across many species, so that a successful model for RGC firing rate for a fish might be applied to primates with slight modifications and different parameter values but with the same basic elements in place.
- RGC and optic nerve signals can be recorded with relative simplicity via electrodes.

In previous discussions we have been assuming that the aspect of the spike train that corresponds to the neural response is the firing rate, i.e. the average number of spikes over a given time interval. But adaptation happens starting at very short timescales of a few tens of milliseconds, e.g. RGCs may be active for no more than 20 ms at a time, the lapse between miniature eye movements, and over this very slim period of time just a handful of spikes are generated by the cell [16]. This very low number of spikes suggests that the firing rate is not what's carrying the signal information in the spike train. But then, what is the neural code?

First of all, this is not an easy question to answer, as the neural code must be inherently ambiguous given that it's adaptive: the code changes with time in a way that depends on how the context changes with time, so the system must encode information about the context, and this information consists on estimates that take some time to be made while simultaneously the context is changing [18]; as a result there is a trade-off between optimality of the code and how context-specific it can get.

It turns out that there are several other aspects of the spike train that also carry information. Spike trains are often modelled as Poisson processes, i.e. random occurrences of discrete events (the spikes) where the mean equals the variance, but in fact the spike variance is much lower than the mean, and therefore the timing of the impulses can be very precise, with a variability as low as 1 ms [22]. The timing of individual spikes for very short timescales has been shown to carry information about stimulus features that are normalised to the stimulus ensemble, while for slightly longer time periods it's the time interval between consecutive spikes what's encoding information about the stimulus ensemble [18]. Yet another element of the coding scheme carrying information is the synchronised firing among cells. There are several mechanisms by which cells become synchronised, including direct excitation among adjacent cells by electrical gap junctions, and shared input from an excitatory neuron. This synchronisation allows for a multiplexing scheme of the signals that might help better compress the visual information for transmission over a reduced number of optic nerve fibers, at the same time that the resolution is increased because the RF of a synchronous pair of RGCs is located at the intersection of the RF of the individual RGCs and it's smaller than any of them, so the brain could in principle achieve a resolution that is finer than that given by the RGC population [16]. This is precisely what a psychophysical study by Masaoka et al. reports [23]: observers judged images to be closest to reality when the resolution approached 150 cpd (cycles per degree), whereas the theoretical limit imposed by photoreceptor spacing is 60 cpd.

4.4 Mean and contrast adaptation

To recap, the purpose of adaptation is to optimise visual coding and improve stimulus discriminability, allocating the reduced range of neural signals to the range of input intensities expected from the near past by dynamically adjusting the sensitivity [24]. These changes in sensitivity happen at different timescales, going up to months [25], but at least some amount of very fast adaptation does need to occur, for the following reasons.

On any given natural scene, the local statistics of light intensity might change considerably. If we consider small regions of size corresponding to the size of receptive fields of RGCs, and compute on these regions the mean and contrast of the image signal, the values might change from one part of the image to another by a factor of 1000 [9]. Our eyes are constantly moving, and apart from the voluntary shifts in gaze there are involuntary microscopic movements whose purpose is to refresh the neural signal preventing it from fading, and also to turn spatial contrast into temporal contrast performing decorrelation so as to improve the efficiency of the subsequent retinal processing [17]. When the eyes go from fixating on one point of the visual field to another point in the scene, the local mean and contrast statistics may change considerably, and the time a fixation lasts is around 200 ms. As a consequence, efficient coding in this scenario would require that adaptation to mean and/or contrast changes takes place faster than changes in fixation [4], and in fact this is the case, with fast mean and contrast adaptation happening in the retina in less than 100 ms.

The efficient representation principle specifies that in order for a cell's response to be optimal the stimulus-response relation should match the input signal distribution, and while natural images have statistical structure in different statistical moments, the first two moments, the mean and the contrast, are particularly important, especially for retinal processing [9].

In natural images *of moderate dynamic range*, it has been shown [19] that mean and contrast of the stimulus luminance are statistically independent, where contrast is defined in the usual way as standard deviation divided by the mean. In other words, luminance and contrast are largely uncorrelated and the joint distributions of luminance and contrast are approximately separable, see Fig. 4.2. If we consider as well natural images of high dynamic range, mean and contrast are not so clearly independent but the retinal processing acts to compensate for this [26], as we shall discuss in Section 4.5.

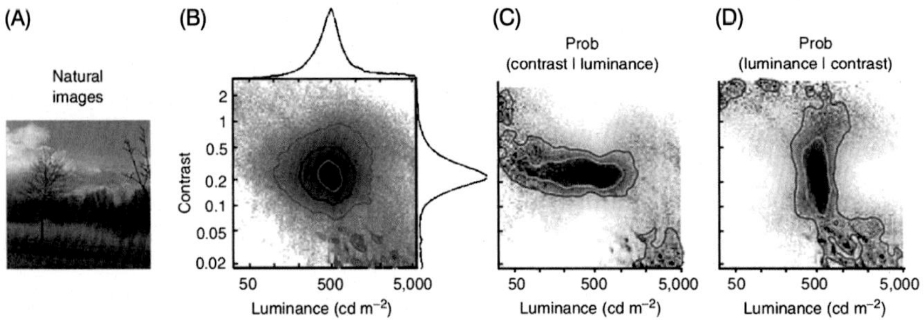

FIGURE 4.2 Luminance and contrast are independent

Luminance and contrast are independent in natural images. Image from [19].

Fast adaptation between eye fixations would require rapid adaptation mechanisms for mean and contrast, and given that mean and contrast are largely independent in natural images, the adaptation mechanisms for luminance and contrast should be functionally independent, and in fact it turns out that this is the case: for the luminance and contrast ranges common in natural scenes, light adaptation (adaptation to the mean) should have the same behaviour at all contrasts, and contrast adaptation should behave in the same way at all mean luminances [19]. In practice mean and contrast adaptation can't be separated perfectly, independent adaptation to both statistical moments requires moderate light levels and spatial averaging at least [4].

Light adaptation is fully accomplished in the retina and has the effect of dividing luminance by the mean luminance, so that the retina has to encode a signal that has a much smaller range; contrast adaptation begins at the retina, is strengthened at the LGN and the visual cortex, and has the effect of dividing the response by the local contrast [19].

The stimulus-response relationship becomes shifted and broadened with an increase in mean intensity, while an increase in contrast matches the dynamic range of the neuron to the dynamic range of the input stimulus by broadening the stimulus-response curve [4], see Fig. 4.3. Remarkably, the same behaviour is also observed on the perceptual response curve estimated from brightness perception experiments, as we will see in Chapter 5.

4.4.1 Adaptation to the mean (light adaptation)

One of the main purposes of light adaptation is to allow the retina to encode contrast, a quantity that remains constant as the mean level of the illumination changes.

We should point out, though, that there are several definitions of contrast. Up to now we've used RMS contrast, which is the standard deviation divided by the mean.

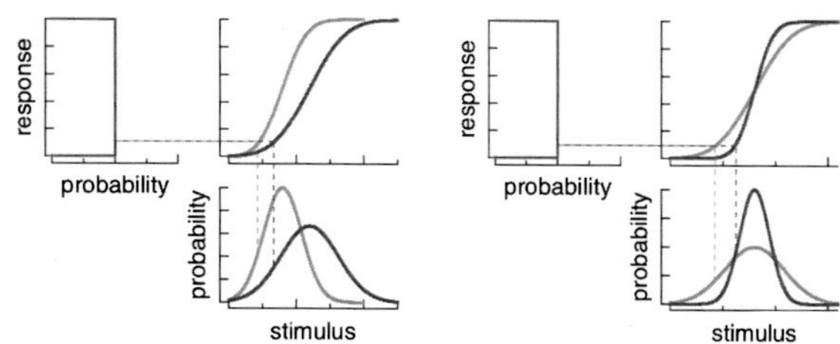

FIGURE 4.3 Neural adaptation to mean and variance

Neural adaptation to mean and variance. Left: neural response to higher (in blue (dark gray in print version)) and lower (in red (mid gray in print version)) mean luminance. Right: neural response to higher (in red (mid gray in print version)) and lower (in blue (dark gray in print version)) contrast. Image from [9].

There is also Weber contrast, used for describing the contrast of a single object with respect to its background: if L_o is the luminance of the object and L_b the luminance of the background, the Weber contrast is defined as the ratio $(L_o - L_b)/L_b$. We will discuss in detail Weber contrast and Weber's law in Chapter 5.

For periodic patterns like sinusoidal gratings, with maximum luminance L_M and minimum luminance L_m, a common definition of contrast is the Michelson contrast, that equals the ratio $(L_M - L_m)/(L_M + L_m)$ [1].

Now we follow the derivation made by Shapley and Enroth-Cugell [1] in order to show that light adaptation does in fact allow the visual system to encode contrast, a magnitude that is related to the reflectance, a physical property of objects that does not change with illumination. When a diffuse surface A of reflectance R_a is under a uniform illumination I, the light reflected by A is approximately IR_a; likewise, for the background B the light intensity is IR_b. If the receptive field of a neuron has been for a while on a region of the visual field corresponding to the background, then the neuron is adapted to IR_b. If there is an eye movement and the RF of the neuron suddenly lies over surface A, the stimulus for the neuron is the change $IR_a - IR_b$, and light adaptation adjusts the response to this stimulus by changing the sensitivity, making it inversely proportional to the mean level IR_b at which the cell is adapted. As a result, adaptation makes the visual system encode object A with a signal proportional to $(IR_a - IR_b)/IR_b$, which is the Weber contrast, and that is clearly independent of the illumination I, it just reduces to the ratio of reflectances: $(R_a - R_b)/R_b$.

This Weber regime of adaptation implies that two visual displays where one is simply a proportional scaling of the other will approximately produce the same retinal response [16]. Therefore, adaptation to the mean allows for brightness constancy, i.e. the neural response to an object remains roughly the same despite changes in the illumination. So for instance a paper will appear to us equally bright at daytime whether we are outdoors or indoors, despite the fact that the intensity of the light coming from the paper outdoors may be several orders of magnitude higher than the intensity of the light reflected by the paper indoors; in fact, the black print of the paper may have a higher luminance outdoors than the white background of the paper has indoors, and nonetheless we perceive the print as being black and the background as being white.

Light adaptation is cone-type specific, meaning that adjustments in the sensitivity of one cone type do not affect other types [4], so brightness constancy and the Weber regime imply that light adaptation is allowing the retina to encode reflectance ratios for the long, medium and short colour wavebands. This is consistent with the Retinex theory of colour vision [27] and therefore light adaptation substantially contributes to colour constancy [28], which as we discussed in Chapter 3 is our ability to perceive the colours of objects in a way that approximately matches their reflectances, independently of the colour of the illuminant.

Brightness constancy (also termed contrast constancy [29]) requires that the object and its surround are under the same illumination, and that the neural gain is the inverse of the mean illumination. Ideally, then, when the ambient illumination increases the sensitivity of the neurons is reduced in the exact same proportion: this is known as Weber's law behaviour, to be discussed in detail in Chapter 5.

Sensitivity can be defined as the inverse of the minimum amount of stimulus increase that is needed to produce a change in response. When, at a given mean intensity, the stimulus changes above a certain threshold, there is a change in neural response: the sensitivity is the inverse of this threshold. Weber's law behaviour requires then that the threshold be proportional to the mean intensity.

Through experimental procedures in which animal retinas are stimulated by short flashes on a constant and uniform background, the threshold values that are needed to elicit a response are estimated; this is done for a range of mean background luminances, creating threshold-versus-intensity or TVI curves. For TVI curves, the Weber regime requires that the plots are linear of slope 1.

As sensitivity is the inverse of these thresholds, Weber's law would imply that, in log-log coordinates, the plot of sensitivity versus intensity is a line of slope -1. This is in fact the case for photoreceptor responses for all the photopic range, as Fig. 4.4 shows: in this figure we can see that contrast constancy holds for around 7 log units of background intensity [29].

These flash-over-background experiments allow to plot the curve for the photoreceptor response versus flash intensity, at a given background luminance level. Fig. 4.5 shows a family of these curves for primate cones, obtained at different background levels by Valeton and Van Norren [30]. Each curve in the figure corresponds to a different mean level, with a small horizontal line in each curve denoting the response level for the background. Increment flashes (flashes brighter than the mean level) produce responses that are above the small horizontal line, while the opposite happens for decrement flashes.

As we can see, when the mean level increases the response curves maintain their shape and simply shift to the right; at the same time, there is less range for increments, and more for decrements. The operating range of the cones, defined as the intensity range where the response isn't flat (either at low or high saturation levels), is around 4 log units for all background levels, and is therefore independent of the state of adaptation.

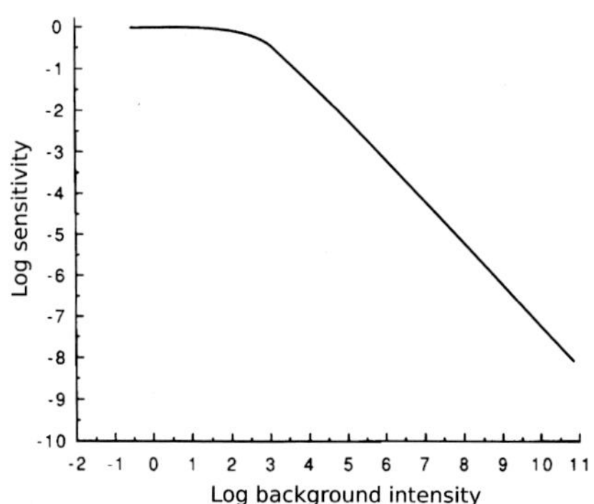

FIGURE 4.4 Weber regime in photoreceptor response

Weber regime in photoreceptor response, adapted from [29].

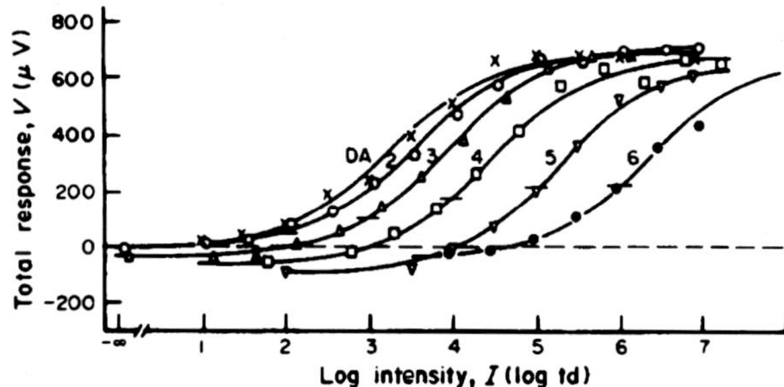

FIGURE 4.5 Photoreceptor response curves

Photoreceptor response curves for primate cones, for different background levels. Image from [30].

Photoreceptor response curves can be approximated very well with the Naka-Rushton equation:

$$R(I) = R_{max} \frac{I^n}{I^n + I_s^n}, \tag{4.1}$$

where R is the response, R_{max} is the maximum or saturation response, I is the intensity, n is a fixed exponent of around 0.75, and I_s is the so-called semi-saturation value, the intensity at which the response is one-half of its maximum value. If we increase I_s and plot R in linear-log coordinates, as in Fig. 4.5, then the curve moves to the right, the same curve-shifting phenomena observed when the background level increases. Therefore, light adaptation can be seen as changing the semi-saturation constant in the Naka-Rushton equation [1]. Furthermore, from Eq. (4.1) and if $n = 1$, we can obtain Weber's law. For this and other factors, it appears that the perceptual effects of light adaptation can be mostly accounted for by retinal processing [16].

The total voltage span in Fig. 4.5 was normalised. In other study of the same kind performed on turtle cones [29], the responses were not normalised and in this case we can see in Fig. 4.6 that while the curves had the same form and behaviour as in Valeton and Van Norren's study, the total range of the stimulus response varies with background illumination, and therefore the effect of light adaptation can't just be modelled as a curve-shift as it's usually done.

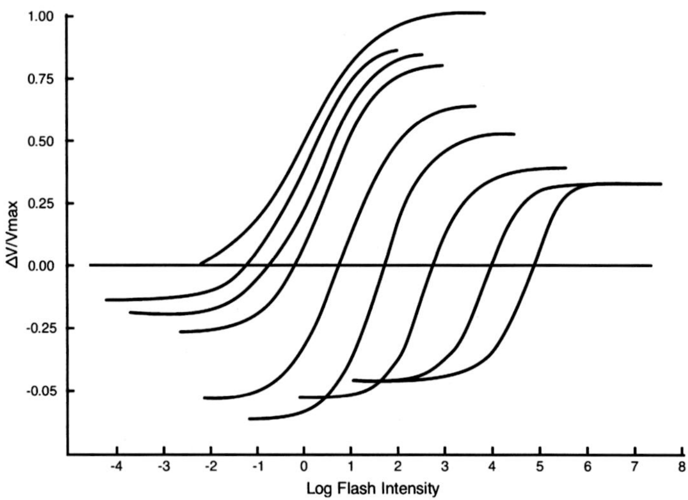

FIGURE 4.6 Photoreceptor response curves

Photoreceptor response curves for turtle cones, for different background levels: as the background level increases, the curves shift right and downwards. Adapted from [29].

We must remember though that all the above results, as is the case in most vision science studies, were obtained with synthetic stimuli in laboratory conditions; it's expected that with natural images and in a free viewing scenario the behaviour of the visual system might be different, possibly in a substantial way.

From what we've shown, light adaptation starts with photoreceptors but all other retinal cell types adapt to the mean level as well. Adaptation is dynamic, and the visual system needs some integration time in order to estimate the statistical parameters of the visual signal at the same time that it's varying, so the adaptation mechanisms introduce some amount of noise in the retinal signals due to the statistical estimates not being perfectly accurate. In low light conditions, the photoreceptor signal is itself quite noisy, so light adaptation at the photoreceptors could be so inaccurate as to be detrimental rather than a beneficial process.

For this reason the main site of light adaptation changes with the mean illumination level: when it's medium to high (indirect sunlight levels) light adaptation takes place at photoreceptors and the response is very fast, but when mean illumination is lower (interior lighting levels) it's better to first average the photoreceptor signals (so as to remove noise) and then adapt. As a result, in low ambient conditions light adaptation is done by RGCs, that are more reliable than cones under noise since they pool signals over many photoreceptors [31]. This allows RGCs to adapt at levels 10 times dimmer than cones [4]. Light adaptation then becomes more accurate, although slower. Also, in dark situations the receptive fields of RGCs lose their surround, so the spatial resolution suffers as well [16].

Finally, we would like to point out the following. As noted above, the operational range of cones changes with the mean illumination, and this can be modelled as a modification of the semi-saturation value I_s that causes the response curve to shift. The semi-saturation value depends then on the mean intensity, and here comes a crucial aspect for biology-based tone mapping algorithms (as we shall see): what is the extent of the visual field over which this mean is estimated? In their seminal work [1], Shapley and Enroth-Cugell state that the size of the adaptation pool should be roughly the same as the receptive field size of the corresponding retinal neuron. But photoreceptors have really small RFs, so according to this view the estimated mean would be a very local value. If one were to use the Naka-Rushton equation to perform tone mapping, and the semi-saturation value relied on a mean estimate computed in this way, the image result would present clear halo artifacts [32]. Later works (e.g. see [33] and references therein) propose instead that it's horizontal cells, with their very wide RFs, who set the operational range of cones, through feedback. Not only are the dendritic fields of HCs very large, they are electrically coupled with those of neighbouring HCs which makes the RFs of HCs be 25 times larger than their dendritic arbours, with the net result that HCs provide cones with an estimate of average ambient luminance [34]. A tone mapping method based on a Naka-Rushton curve with a global estimate of the mean can provide excellent results, and in fact this is a basic technique in vision-based tone mapping as we shall discuss in Chapter 9.

4.4.2 Contrast adaptation

For a long time it was assumed that contrast adaptation was mainly a cortical phenomenon, but it's made clear now that contrast adaptation starts at the retina and happens at every stage of the visual pathway, where it gets progressively strengthened.

When scene contrast is low contrast adaptation magnifies sensitivity, and when scene contrast is high the sensitivity is decreased so as to allow the visual system to encode the broader intensity range, at the same time that the response becomes faster [1,2]. The effect of contrast adaptation is to divide responses by their standard deviations, an operation also called "normalisation by RMS contrast", whose effect it is to maximise information transmission, in accordance with the efficient representation principle [35]. It turns out that local contrast adaptation removes redundancy in a more effective way than center-surround filtering [9]. While some studies have shown that contrast adaptation just considers the variance of the signal and ignores higher order statistical moments [36], others contend that there might be adaptation to higher order statistics downstream from the LGN or encoded in the temporal form of the signal [9].

As it was the case with light adaptation, there is a wide range of timescales for contrast adaptation. Fig. 4.7 shows an example from [12] where contrast adaptation takes seconds. If you look at the white bar at the center of the left image for some 30 seconds, slowly moving your eyes over it, and then suddenly look at the image on the right, then for a brief period of time a low contrast part of the right image should appear invisible, because you've become adapted to the higher contrast image of the left.

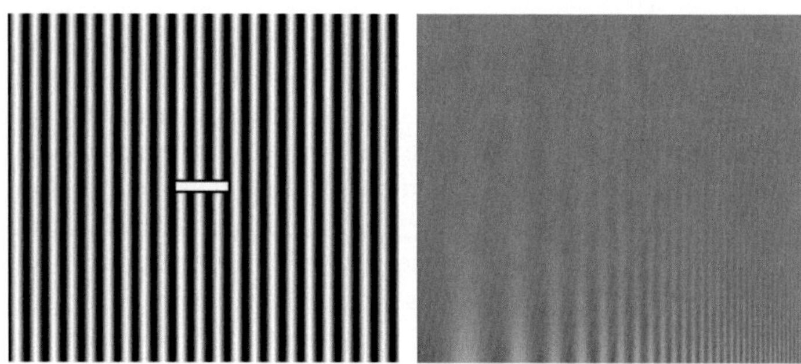

FIGURE 4.7 Contrast adaptation example

Contrast adaptation example, see text. From [12].

But rapid eye movements, saccades and microsaccades, change the contrast (as we mentioned, they turn spatial contrast into temporal variations), and adaptation must match the local statistics during a single fixation [4]. The duration of the adaptation process does not modify the nature of the perceptual effect, it just determines its strength; after a change in contrast, most of the adaptation takes place in less than one second [12]. After a switch in contrast (while the illumination mean remains constant), adaptation takes $\sim$ 100 ms, or less if there has been an increase in variance, and this is the same amount of time that it takes for the information transmission rate to recover.

As for the neural code, the standard deviation or overall dynamic range of the signal (the normalisation factor of the contrast adaptation process) is encoded by slower variations of the spike rate, and after just a few spikes following a contrast change the inter-spike intervals become typical of the new variance [18]. For slow contrast adaptation, the time courses of retina and cortex are similar to those in perception.

Contrast adaptation considers a sampling region (to estimate the contrast) that's usually much larger than the RF of the cell, from hundreds of μm to 2 mm, or 2 degrees, in the retina [16,1]. The contrast sampling window can be inherited from cells that are upstream, as it happens to LGN neurons that inherit their adaptation from RGCs [37].

Most of the vision literature on dynamic range focused on how objects appear invariant over changing lighting conditions, thus omitting irradiance information [38]. But tracking changes in irradiance is advantageous in terms of survival, and we recall that the outputs of ipRGCs are proportional to scene irradiance; individual ipRGCs activate within a range of irradiance values so that collectively they can span and encode light intensity levels from moonlight to full daylight. ipRGCs directly control the pupillary reflex and the circadian rhythm, that modulates the release of dopamine in the retina, which in turn alters the size of the receptive field of horizontal cells performing local contrast enhancement on the cone signals [33]. The electrical coupling among horizontal cells is strong under intermediate luminance levels, and poor otherwise; when the coupling is reduced, the size of the RF of the HCs decreases, and as they provide lateral inhibition the result is that there is more local contrast enhancement, and a higher acuity [34].

Neurons that receive stronger stimulation, for example because they pool over many signals, need more intense adaptation so as to avoid saturation and maintain sensitivity to small inputs [39]. In fact, contrast adaptation becomes stronger as the pooling of signals gets larger, i.e. the size of the RF increases, which is what happens when we go down the visual pathway and there's progressively more input convergence: photoreceptors do not adapt to contrast, BCs adapt a little, RGCs adapt some more, LGN cells even more, etc. For this same reason, in the retina and the LGN, M cells (with large RFs) adapt much more than P cells (with small RFs) [8,37]. In fact, M cells are capable of adapting to high contrast stimuli whose temporal frequencies are above the limits of perception [37].

Performing a L+NL analysis of different retinal cell types before and after a change in contrast shows that, except for photoreceptors, all other retinal cell types adapt to contrast and this means that both the linear filter and the nonlinearity must change in order to fit the data after a change of contrast, see Fig. 4.8. When the contrast increases, the peak of the linear filter decreases, and it becomes faster as well; the nonlinearity may get shifted and change in slope.

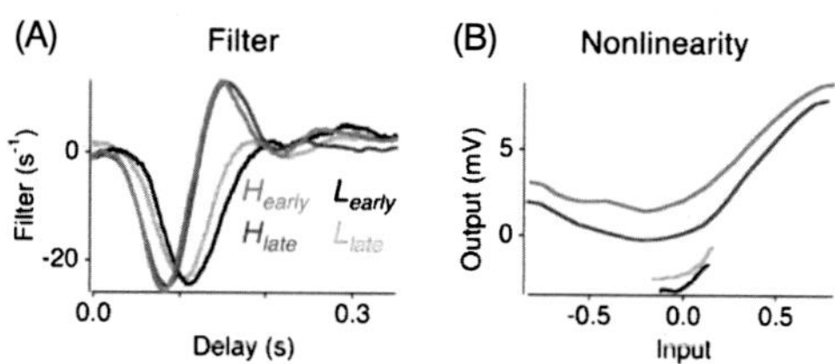

FIGURE 4.8 L+NL model changes with contrast adaptation

Linear filter and nonlinearity change with contrast adaptation, figure from [7].

An L+NL analysis also shows that the ON and OFF channels are asymmetric for contrast adaptation, the sensitivity of the ON cells is reduced more than in OFF cells [24]; see Fig. 4.9. We will see in Chapter 9 how this asymmetry proves fundamental for the design of a very effective tone mapping algorithm. And in Chapter 5 we discuss how brightness response curves may also present the same form of asymmetry, in a phenomenon called crispening.

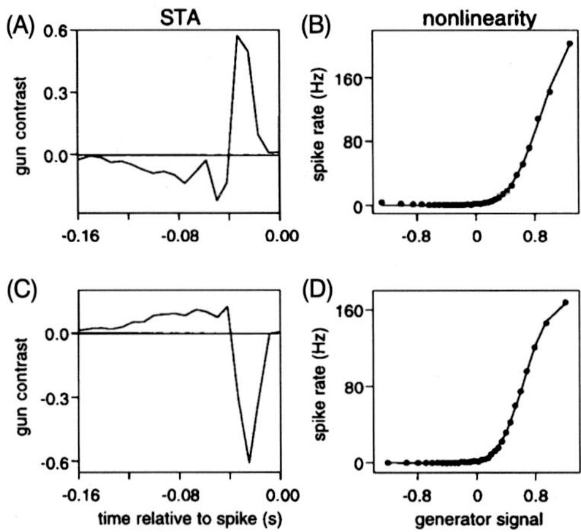

FIGURE 4.9 ON/OFF asymmetry for contrast adaptation

ON and OFF channels are asymmetric for contrast adaptation. The figure shows the linear filter and the nonlinearity for an ON cell (top) and an OFF cell (bottom) of macaque monkey. Image from [24].

4.5 Statistics of natural images as a function of dynamic range

Following the efficient representation principle, the identification of statistical regularities in natural images and the concept of redundancy reduction have proven most relevant not just to vision science, but to applied disciplines like image processing as well [40]. While a large number of studies investigating the properties of natural images can be found in the literature, nearly all of them use regular, single-exposure photography, that only allows to capture low dynamic range images, while common natural scenes are high dynamic range.

In consequence the statistics derived from the natural image databases most used in vision research, like those of [41] and [42], can be biased due to the exclusion of HDR scenes, and the conclusions gathered from these statistics with respect to vision processes might not be valid in general.

In this last section we review the recent results by Grimaldi, Kane and Bertalmío that appeared in [26]. The motivation of that work was to assess whether or not, by previously ignoring HDR scenarios, vision research has misrepresented or misunderstood the statistics of natural images. The contributions of the work are, firstly, to show that the statistics of natural images vary appreciably with the dynamic range (DR) of the underlying scene, using for this purpose the recent Southampton-York Natural Scenes (SYNS) HDR image database created by [43], and secondly, to highlight that early visual processes greatly reduce the impact of these variations. These results are consistent with efficient coding theory, in that by removing the dependence of intensity distributions on DR already at retinal level, the visual system gains in efficiency at not having to perform its processes for signals that widely vary in their statistical properties.

4.5.1 Methods

The SYNS dataset includes scenes captured from both rural and urban locations in and around the city of Southampton in the United Kingdom. Panoramic images in HDR format were recorded with a SpheroCam HDR using a Nikkor 16 mm fish-eye lens covering a range of 26 f-stops.

The database was divided into five different DR categories. The range was chosen such that each DR bracket contains either 73 or 74 images. The median DR and range of each DR bracket are shown in Table 4.1.

Table 4.1 DR values for the different categories.

DR category	median DR	min DR	max DR
1	63	19	130
2	194	130	294
3	418	294	585
4	1123	585	2685
5	13295	2685	3248876

Throughout [26] we investigate how the image statistics vary in each DR bracket, and also how these statistics change as they undergo three transformations corresponding to some very well-established processes taking place in the eye (this is clearly not intended to be a complete model of retinal processing).

Each input image from the database is considered as the real-world illumination reaching the eye. This signal is convolved with a point spread function (PSF) filter, simulating the optics of the human eye, as seen in Chapter 2. Next, this new image is passed through the Naka-Rushton (NR) equation, putatively modelling the response function of photoreceptors. Finally, we pass the image resulting from the previous step through a center-surround filter (CSF), in the form of a difference of Gaussians, to emulate the impact of lateral inhibition on contrast sensitivity.

For the statistical analysis, we chose to perform studies on single-pixel and derivative statistics with the exact same methodology used in [44], which is a very well-established work on natural image statistics based on the LDR database of [41]. This allows for a better comparison of the new results we obtain here for HDR images, and also to check the validity of our study since the statistics for the lower DR category of the database we are using should be consistent with those reported in [44].

4.5.2 Results

4.5.2.1 Statistical moments

We begin by investigating how the statistical moments of median, standard deviation, skewness and kurtosis vary as a function of DR. The results are presented in

Fig. 4.10; each subplot reports a separate moment, each dot represents an individual image and each colour denotes a different transform, with blue (dark gray in print version) for the original images, red (mid gray in print version) for the results of the Naka-Rushton transform and yellow (light gray in print version) for the result of the center-surround modulation. For clarity we have omitted the plots for optical scatter as the PSF transform was deemed to have a very small impact upon the statistical moments. Each image is normalised (divided by its maximum computed with 99.99 percentile) prior to computing the moments.

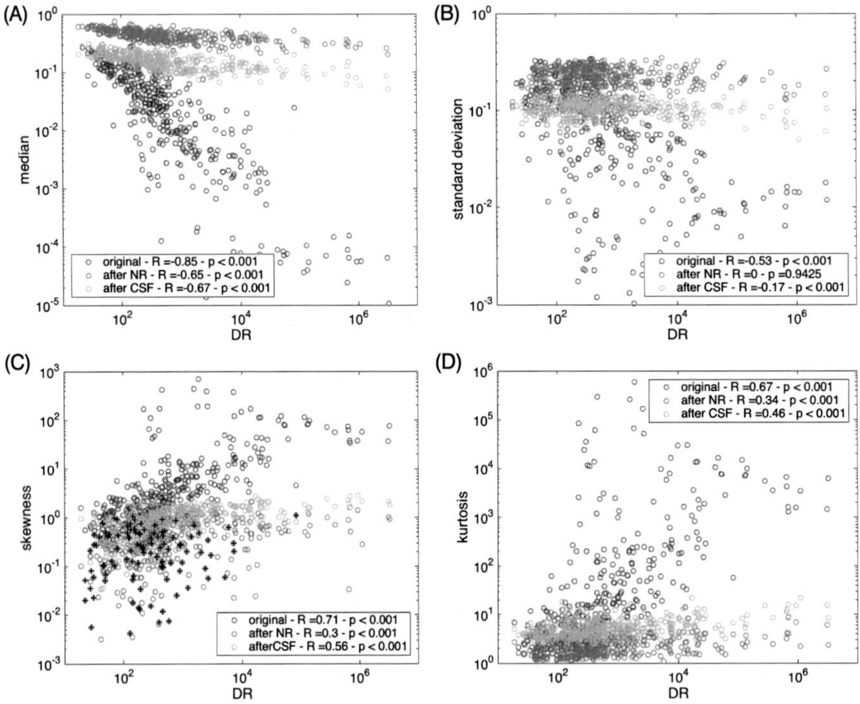

FIGURE 4.10 Statistical moments as a function of DR

Statistical moments of normalised intensity distributions as a function of DR: median (A), standard deviation (B), skewness (C), kurtosis (D). Figure from [26].

The main observation is that the linear images, represented by the blue (dark gray in print version) dots, have statistical moments that vary strongly with DR for each of the statistical moments evaluated, but that this effect size is greatly reduced by the application of the Naka-Rushton equation, as depicted by the red (mid gray in print version) dots. Overall, we can conclude that the application of the Naka-Rushton equation greatly reduces the impact of DR on the statistical moments of the illuminance distributions. The bandpass filtering of the CSF maintains this relative independence from DR while producing other effects that can be better appreciated when we compute the single pixel and derivative statistics, in the following sections.

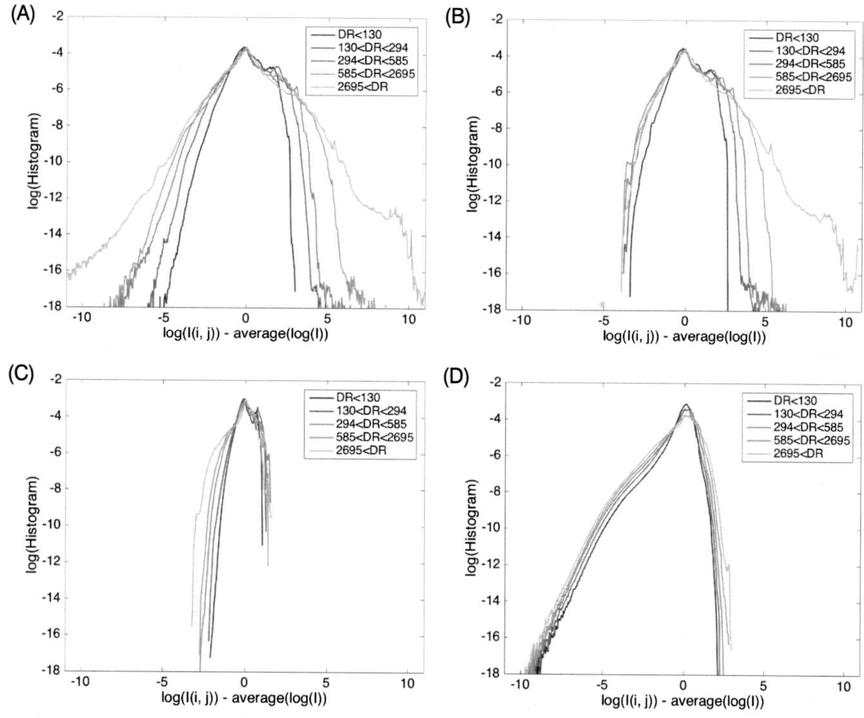

FIGURE 4.11 Histograms by DR categories

Median-subtracted histograms in log-log axes, by DR categories. (A) For the original values. (B) After light scatter. (C) After photoreceptor response. (D) After center-surround modulation. Figure from [26].

4.5.2.2 Single pixel statistics

In the previous section we evaluated the image statistics as a function of their DR on an image-by-image basis. In this section we shall examine the average shape of the histogram for each of the five DR brackets described in the methods section (Table 4.1). To do so, we use the methodology of [44] in which each image is passed through a log transform before the median is subtracted, as follows, $H = log(I(i, j)) - median(log(I))$. Average histograms are then computed for each DR bracket and the results are plotted on log-log axes. In Fig. 4.11 we plot the average histograms of the original linear image in subplot A and the remaining transforms in subplots B, C and D. Each colour represents a different DR category.

The width of the histograms for the original linear images unsurprisingly increases with DR. Interestingly, we only observe straight edges in the histogram for the highest DR bracket we evaluate. This contrasts with the work of [44], in which straight edges are found for the average histogram of the low DR database of [41]. As the tails of the histograms are better represented by the value of the kurtosis, the similarity of the kurtosis found by [44] ($k = 4.56$) and the kurtosis found for the histogram of the highest DR bracket ($k = 4.89$) confirms this observation. We speculate upon why this is the case in the discussion section where we replicate the original findings.

Regarding the impact of the eye's PSF on illuminance distributions, represented in Fig. 4.11(B), one can observe that it is limited to raising the low illuminance values while having very little impact upon values above the median. This can also be seen as an increase of the skewness values of the histograms after applying the transform. The PSF models imperfections in the eye's medium that cause light scatter, and whose net effect is to increase the minimum intensity level of the light reaching the retina photoreceptors [45].

Fig. 4.11(C) shows that after passing through the non-linear transform of the photoreceptors, the average histograms for all DR categories become much more similar and also their DR is greatly reduced (the width of the histograms is much smaller). Standard deviation values are decreased and the effect of the DR category on the moments appears to be reduced for the standard deviation, the skewness and the kurtosis. Standard deviation values decrease after the Naka-Rushton transform, and these results can be linked to the fact that the photoreceptor transform is a compressive nonlinearity that increases low intensity values much more than high intensity values, therefore reducing the DR (by increasing the minimum).

Finally, in Fig. 4.11(D) we can see that application of the center-surround filter produces images where the similarity of histograms across DR brackets is preserved, even enhanced, but now the DR has been notably increased. The skewness is now significantly negative for all the categories and the standard deviation has increased substantially, meaning that the distributions now have more weight in the low intensity values and they have drifted further away from the high intensity values, i.e. the lower values have been reduced further and therefore the DR increases.

4.5.2.3 Derivative statistics

Derivative statistics correspond to the difference between two adjacent pixels. We find similar histograms for both horizontal and vertical derivative statistics, and as such we only plot results for the horizontal derivative distributions (computed with the formula $D = log(I(i, j)) - log(I(i, j+1)))$, represented in Fig. 4.12.

For the original images, Fig. 4.12(A), we see that as the DR increases the adjacent pixels typically become more different and the width of the difference histograms widens. This result corresponds to the standard deviation globally increasing with DR. In the same figure, very high kurtosis values for all DR categories are consistent with the fact that in natural images neighbouring pixels are highly correlated and therefore difference histograms present a high peak at zero [44]. The kurtosis value for the lower DR category, $k = 16.81$, is very similar to the one reported by [44], which was $k = 17.43$.

Light scattering has the effect of making these distributions more similar for all categories and also closer to zero values, see Fig. 4.12(B). This result was expected since the PSF filtering affects neighbouring pixels and reduces differences between them.

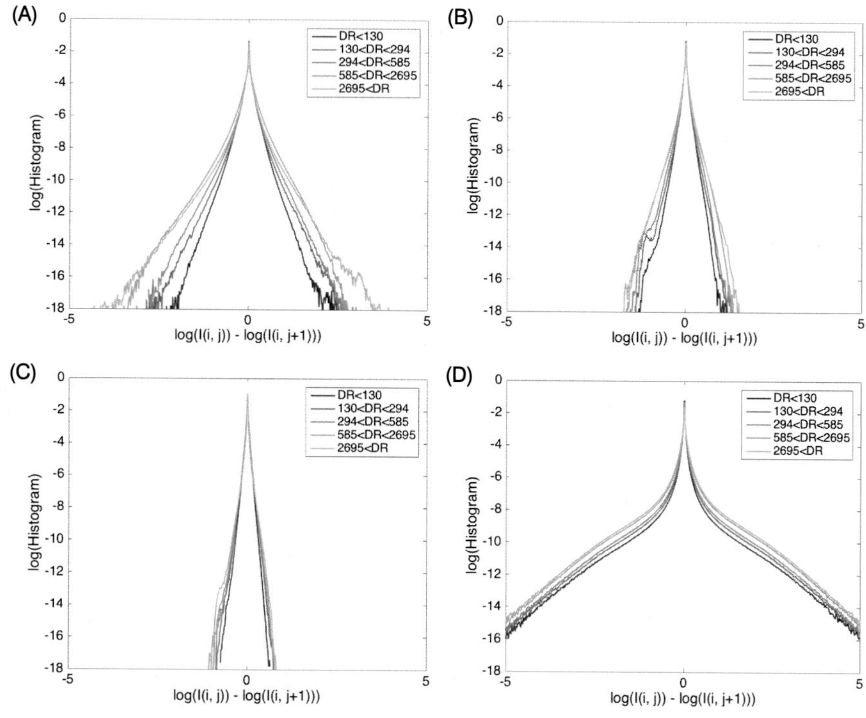

FIGURE 4.12 Difference histograms

Histograms of the difference between adjacent pixels in linear-log axis. (A) For the original values. (B) After light scatter. (C) After photoreceptor response. (D) After center-surround modulation. Figure from [26].

The distributions for the five DR brackets become even more similar and their spread is reduced even further after applying the Naka-Rushton transform, see Fig. 4.12(C). This is consistent with what we observed in Fig. 4.11(C), that showed that the DR was reduced as well as the variability among the distributions in the different categories.

Finally, the contrast enhancement effect of the center-surround modulation can be clearly seen in Fig. 4.12(D), where the distributions remain very similar for all DR categories but the spread is now greatly increased.

4.5.2.4 Power spectra

The literature about natural image statistics has consistently demonstrated that the power spectra of natural scenes is linear when plotted on log-log axes. This corresponds to a $1/f^{2+\eta}$ power law relationship, with η called the "anomalous exponent", usually small. An implication of this feature is that natural images are scale-invariant [46–49,41].

In the previous works of [50] and [51], it was found that this rule was not valid for some HDR images. They associated this finding with the presence of direct light sources and/or specularities in the image. In [26] we investigated the effect of the DR on the averaged power spectrum of each image after collapsing across orientation.

Fig. 4.13(A) shows the mean of power spectra for the five DR brackets. Substantial curvature is observed for the highest DR category of the dataset, corresponding to a flattening of the distribution at low spatial frequencies. Some flattening may also be seen in the second-highest DR category. The effect of the eye's PSF, Fig. 4.13(B), results in a small reduction in the amplitude of the high spatial frequencies. After application of the Naka-Rushton equation we can see in Fig. 4.13(C) that the $1/f^2$ relationship is recovered. Finally, Fig. 4.13(D) depicts the result of the center-surround transform, which decreases the energy in the lower frequencies and also in the frequencies over 100 cycles/image. The CSF acts like a band pass filter resulting in the effect observed in Fig. 4.13(D).

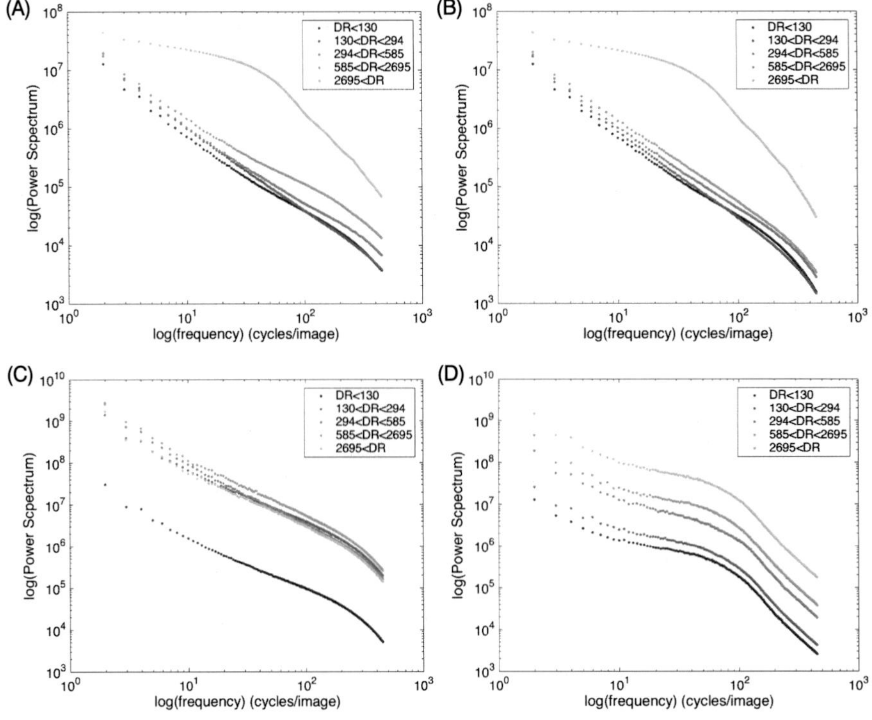

FIGURE 4.13 Power spectrum vs. spatial frequency

Power spectrum as a function of frequency on log-log axes. (A) For the original values. (B) After light scatter. (C) After photoreceptor response. (D) After center-surround modulation. Figure from [26].

We wanted to focus on the recovery of the scale-invariance property of natural scenes. As mentioned in [50], the curvature present in the power spectrum of HDR images can be due to strong localised light sources producing very high illuminance values over small regions. To study the image-per-image evolution of this statistic, a second order polynomial was fit to the averaged power spectrum of each natural image, $P(x) = ax^2 + bx + c$, where $a = 0$ would correspond to the scale-invariant case where the power spectrum can be approximated by a linear function, and $a \neq 0$ implies some curvature in the fit. These fits were performed up to 200 cycles/image to avoid an overestimation of the fitting error that would be produced by the drop on the high frequencies due to the PSF.

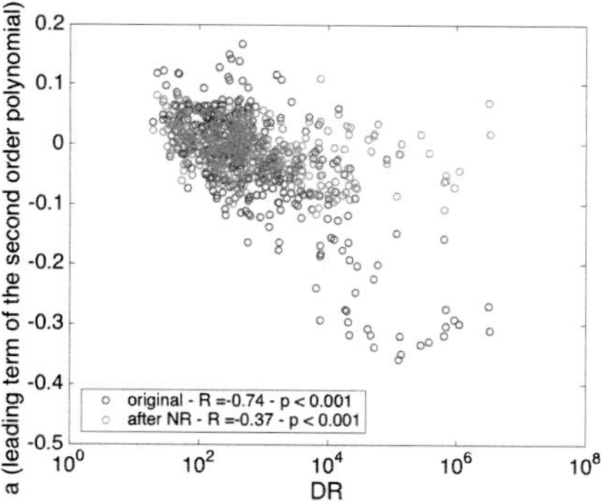

FIGURE 4.14 Leading term coefficient of 2nd order polyfit

Leading term coefficient of the second order polyfit as a function of DR for the original values (blue (dark gray in print version)) and at the photoreceptors' output (red (mid gray in print version)). Figure from [26].

Fig. 4.14 plots the term a as a function of DR, where negative values of a correspond to an inverted U-shape. The blue (dark gray in print version) dots show the fits to the original images, and it can be seen that the coefficient a becomes more negative as the DR increases. The correlation coefficient plotted in Fig. 4.14 is significant and the effect is clear on the figure. The red (mid gray in print version) dots show the fits to the data after the image has been passed through the eye's PSF and the Naka-Rushton equation. As the figure shows, the fits do not exhibit the same degree of negative curvature at high DR values. The correlation coefficient is greatly reduced and one can observe that the a values remain around zero. In Fig. 4.15(A), we plot the fitting error for a first and second order polynomial fit respectively in blue (dark gray in print version) and red (mid gray in print version) for the camera sensor output, and in Fig. 4.15(B) we plot the error for the images after modelling the effect of the light scatter and the photoreceptor response. The main observation that can be made is that the curvature of the power spectra is reduced after the application of the Naka-Rushton equation. Indeed, in Fig. 4.14 the absolute value of the a coefficients are below 0.1. And, although Fig. 4.15(A) shows that even a second-order fit may have significant error, the application of the Naka-Rushton equation allows for the power spectra to be well approximated by a first-order polynomial, as the small errors of Fig. 4.15(B) suggest.

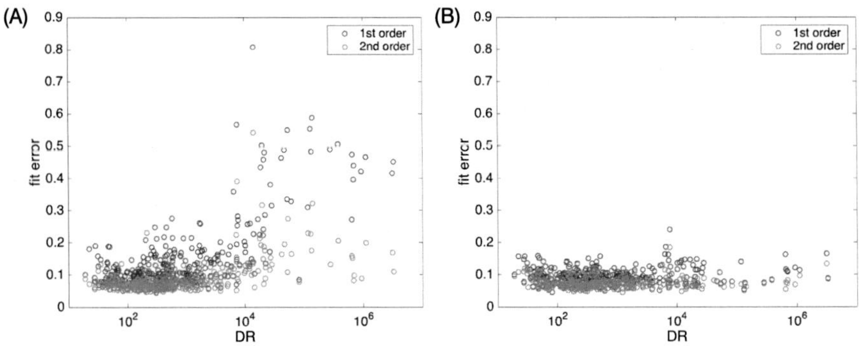

FIGURE 4.15 Fitting error for the second and first order polyfit

Fitting error for the second and first order polyfit as a function of DR for the original images (A) and the photoreceptors' output (B). Figure from [26].

In Fig. 4.14, one can see that the curvature is evolving continuously as a function of the DR for the original values. HDR scenes tend to have a curved power spectrum, and while this curvature can be due to high specularities as reported before ([50]), that isn't necessarily the case. For instance, in Fig. 4.14 some images with a high DR have a value of a close to zero or a small error of the first order polynomial fit in Fig. 4.15(A), showing that some HDR images comply with the $1/f$ rule. An example will be shown in the Discussion section.

4.5.3 Discussion

The results in [26] demonstrate that the statistics of the linear image, that is, the image received at the camera sensor, change dramatically with DR. The strongest effects are noted in the four statistical moments evaluated here and upon the one over frequency relationship for the power spectrum which breaks down for images with a very high DR. Effects are also noted in the derivative statistics, the single pixel histograms and the Haar wavelet analysis.

The second conclusion from [26] is that the early transformations of the human visual system greatly reduce the impact of DR on the statistics, turning the four statistical moments virtually independent from DR and recovering the $1/f^2$ behaviour for the power spectrum.

These results are in agreement with the efficient coding hypothesis. If image distributions have statistics that vary widely, the visual system should perform quite elaborated processes in order to directly adapt to these inputs. If on the other hand the inputs have their dependence on DR removed early on, already at retinal level, the visual system gains in efficiency at having to perform simpler processes, tailored to a more limited range of statistical variation.

The process that appears to be contributing more to make the signal distributions independent of DR is the nonlinear photoreceptor transform, which we have modelled with the Naka-Rushton equation. This result was to be expected, as one of the main purposes of adaptation at photoreceptor level is to increase efficiency by handling *"the very extensive range of light levels presented to the eye by nature"* [1], i.e. to reduce the DR of the signal. In fact, in the image processing and computer graphics fields, very successful methods for reducing the DR of images are based on the Naka-Rushton equation, as we shall see in Chapter 9.

Center-surround modulation, on the other hand, *increases* the DR, as observed in Fig. 4.11(D), which is consistent with contrast enhancement properties of the visual system [52] and psychophysical studies of the DR of the visual system [53]. Lateral inhibition also widens the derivative histograms (Fig. 4.12(D)), flattens the power spectra for frequencies below 100 cycles/image (Fig. 4.13(D)) and reduces the similarity among neighbours. These properties then support the efficient representation theory by showing that the transform performs signal decorrelation [14,46].

Assumptions regarding the nature of HDR scene illumination distributions can also be made with these results. When observing the evolution of the moments in Fig. 4.10, one can see that illumination maps go to highly positively skewed and leptokurtic distributions when the DR increases, that could correspond to images with highlights. The curvature of the power spectra of most of the HDR images can also support the hypothesis that HDR scenes are characterised by specularities and bright light sources, that flatten the low frequency part of the power spectrum as pointed by [50]. Although it is important to notice that some HDR images do not present a curved spectrum. In Fig. 4.14, they correspond to blue dots in the top-right corner, with a relatively high a and a high DR. An example of this image with its power spectrum is presented in Fig. 4.16. In order to be displayed properly the image is tone mapped so one can't see the specularities in the car and the truck on the background. This scene may abide by the $1/f$ law because of the multiple specularities and their distribution in space, opposed to a local light source that would indeed flatten the power spectrum.

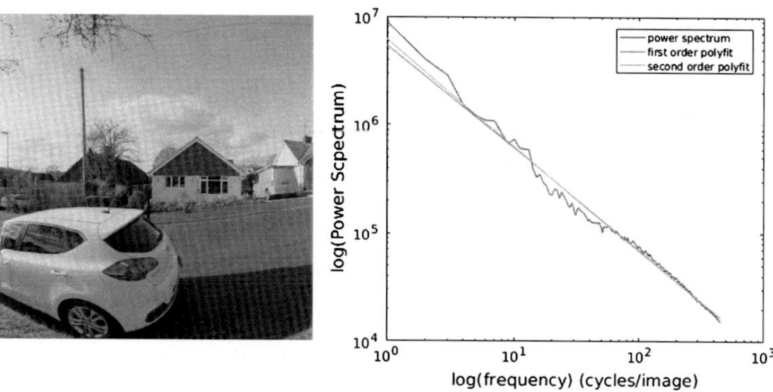

FIGURE 4.16 HDR image with $1/f$ power spectrum

Example of an image (tone mapped version on the left) with a high DR ($DR = 92683$) and a power spectrum (on the right) following the $1/f$ rule. Figure from [26].

Some HDR images can have also a more complex shape than this flattenning in the low frequencies. Most images have a power spectrum well represented by a second order polynomial but some exceptions with a relatively high fitting error are noticed in Fig. 4.15(A). Fig. 4.17 shows one example of this complex power spectrum. Here, the sun is present in the top-left corner of the image forming a local light source. The geometry of the scene can affect the power spectrum and make it take a shape that can only be approximated with a third order polynomial.

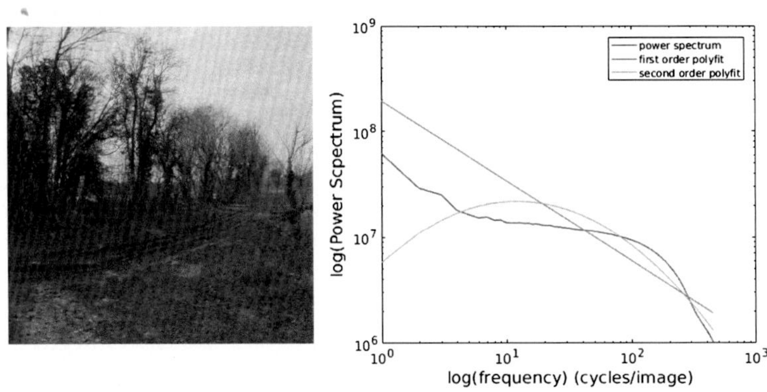

FIGURE 4.17 HDR image with power spectrum not well approximated by 2nd order polynomial

Example of an image (tone mapped version on the left) with a high DR ($DR = 453400$) and a power spectrum (on the right) that can't be well approximated by a second order polynomial. Figure from [26].

The histogram analysis conducted in Fig. 4.11 followed the methodology used in [44] to compute and visualise the average illuminance histogram of the van Hateren image database [41], showing that it was well approximated in piece-wise linear formulation when plotted on log-log axes. There was however some ambiguity as to how the original histogram was computed, as [44] stated only that images underwent the following transformation: $log(I) - average(log(I))$. Thus it was not clear whether average referred to the mean, the median, or some other computation. Additionally, the van Hateren database contains two images sets denoted *.iml* and *.imc*, the former being images that are linearly related to the sensor values and the latter having a correction for the optics of the camera applied. In Fig. 4.18, we tested the four possible combinations using the van Hateren image database. The results demonstrate that we only obtain the characteristic linear slopes when we use the *.imc*, optically corrected image dataset and subtract each image using the image median. As such, we used the median to compute the histograms in Fig. 4.11. The gray dashed box highlights the region illustrated in the original study by [44] which is substantially smaller than the region we plot here, even if one can argue that these parts of the histograms contain a very small amount of pixels. When plotted over this greater range we do not obtain a straight line over the full range of positive values. In the SYNS dataset, images are not optically corrected and it may be a reason why we do not observe linear parts on the histograms of small DR categories. It is to be noticed that the number of images

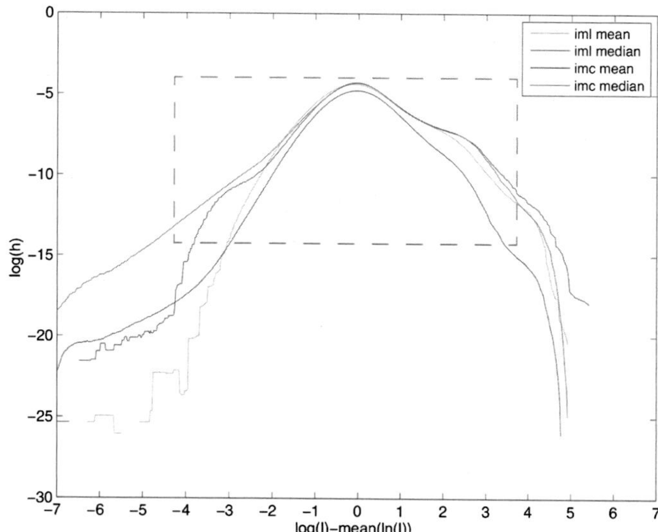

FIGURE 4.18 Histograms from LDR image database

Histograms computed from the van Hateren image database. Figure from [26].

per category is way smaller in this study than in the previous one, we have around 70 images for each category while in [44] the dataset contains more than 4000 natural scenes. The type of scenes can then affect the histograms, for instance in Fig. 4.11(A) one can observe a peak in the positive part of the histograms. This peak may correspond to the pixels forming the sky as the SYNS dataset contains a lot of scenes with the horizon creating bimodal intensity distributions.

However, a piece-wise linear histogram is observed for the HDR category and one explanation would be that HDR images are less affected by optical scatter regarding the distribution they have. It is hard to make a strong conclusion about the real world illuminations given the limitations imposed by the capture devices.

References

[1] Shapley R, Enroth-Cugell C. Visual adaptation and retinal gain controls. Progress in Retinal Research 1984;3:263–346.

[2] Smirnakis SM, Berry MJ, Warland DK, Bialek W, Meister M. Adaptation of retinal processing to image contrast and spatial scale. Nature 1997;386(6620):69.

[3] Wark B, Fairhall A, Rieke F. Timescales of inference in visual adaptation. Neuron 2009;61(5):750–61. https://doi.org/10.1016/j.neuron.2009.01.019.

[4] Dunn FA, Rieke F. The impact of photoreceptor noise on retinal gain controls. Current Opinion in Neurobiology 2006;16(4):363–70.

[5] Solomon SG, Kohn A. Moving sensory adaptation beyond suppressive effects in single neurons. Current Biology 2014;24(20):R1012–22.

[6] Demb JB. Functional circuitry of visual adaptation in the retina. The Journal of Physiology 2008;586(18):4377–84.

[7] Baccus SA, Meister M. Fast and slow contrast adaptation in retinal circuitry. Neuron 2002;36(5):909–19.

[8] Shapley R. Retinal physiology: adapting to the changing scene. Current Biology 1997;7(7):R421–3.

[9] Rieke F, Rudd ME. The challenges natural images pose for visual adaptation. Neuron 2009;64(5):605–16.

[10] Wandell BA. Foundations of vision, vol. 8. Sunderland, MA: Sinauer Associates; 1995.

[11] Gollisch T, Meister M. Eye smarter than scientists believed: neural computations in circuits of the retina. Neuron 2010;65(2):150–64.

[12] Kohn A. Visual adaptation: physiology, mechanisms, and functional benefits. Journal of Neurophysiology 2007;97(5):3155–64.

[13] Barlow HB, et al. Possible principles underlying the transformation of sensory messages. Sensory Communication 1961;1:217–34.

[14] Atick JJ. Could information theory provide an ecological theory of sensory processing? Network Computation in Neural Systems 1992;3(2):213–51.

[15] Olshausen BA, Field DJ. Vision and the coding of natural images: the human brain may hold the secrets to the best image-compression algorithms. American Scientist 2000;88(3):238–45.

[16] Meister M, Berry MJ. The neural code of the retina. Neuron 1999;22(3):435–50.

[17] Rucci M, Victor JD. The unsteady eye: an information-processing stage, not a bug. Trends in Neurosciences 2015;38(4):195–206.

[18] Fairhall AL, Lewen GD, Bialek W, van Steveninck RRdR. Efficiency and ambiguity in an adaptive neural code. Nature 2001;412(6849):787.

[19] Mante V, Frazor RA, Bonin V, Geisler WS, Carandini M. Independence of luminance and contrast in natural scenes and in the early visual system. Nature Neuroscience 2005;8(12):1690.

[20] Benucci A, Saleem AB, Carandini M. Adaptation maintains population homeostasis in primary visual cortex. Nature Neuroscience 2013;16(6):724.

[21] Atick JJ, Redlich AN. What does the retina know about natural scenes? Neural Computation 1992;4(2):196–210.

[22] Field GD, Chichilnisky E. Information processing in the primate retina: circuitry and coding. Annual Review of Neuroscience 2007;30:1–30.

[23] Masaoka K, Nishida Y, Sugawara M, Nakasu E, Nojiri Y. Sensation of realness from high-resolution images of real objects. IEEE Transactions on Broadcasting 2013;59(1):72–83.

[24] Chander D, Chichilnisky E. Adaptation to temporal contrast in primate and salamander retina. Journal of Neuroscience 2001;21(24):9904–16.

[25] Welbourne LE, Morland AB, Wade AR. Human colour perception changes between seasons. Current Biology 2015;25(15). https://doi.org/10.1016/j.cub.2015.06.030.

[26] Grimaldi A, Kane D, Bertalmío M. Statistics of natural images as a function of dynamic range. Journal of Vision 2019;19(2).

[27] Land E, McCann J. Lightness and Retinex theory. Journal of the Optical Society of America 1971;61(1):1–11.

[28] Bertalmío M. Image processing for cinema. CRC Press; 2014.

[29] Burkhardt DA. Light adaptation and photopigment bleaching in cone photoreceptors in situ in the retina of the turtle. Journal of Neuroscience 1994;14(3):1091–105.

[30] Valeton J, Norren DV. Light adaptation of primate cones: an analysis based on extracellular data. Vision Research 1983;23(12):1539–47. https://doi.org/10.1016/0042-6989(83)90167-0.

[31] Dunn FA, Lankheet MJ, Rieke F. Light adaptation in cone vision involves switching between receptor and post-receptor sites. Nature 2007;449(7162):603.

[32] Reinhard E, Heidrich W, Debevec P, Pattanaik S, Ward G, Myszkowski K. High dynamic range imaging: acquisition, display, and image-based lighting. Morgan Kaufmann; 2010.

[33] Chapot CA, Euler T, Schubert T. How do horizontal cells 'talk' to cone photoreceptors? Different levels of complexity at the cone–horizontal cell synapse. The Journal of Physiology 2017;595(16):5495–506.

[34] Bloomfield SA, Völgyi B. The diverse functional roles and regulation of neuronal gap junctions in the retina. Nature Reviews. Neuroscience 2009;10(7):495.

[35] Brenner N, Bialek W, Van Steveninck RdR. Adaptive rescaling maximises information transmission. Neuron 2000;26(3):695–702.

[36] Bonin V, Mante V, Carandini M. The statistical computation underlying contrast gain control. Journal of Neuroscience 2006;26(23):6346–53.

[37] Solomon SG, Peirce JW, Dhruv NT, Lennie P. Profound contrast adaptation early in the visual pathway. Neuron 2004;42(1):155–62. https://doi.org/10.1016/s0896-6273(04)00178-3.

[38] Milner ES, Do MTH. A population representation of absolute light intensity in the mammalian retina. Cell 2017;171(4):865–76.

[39] Baccus SA, Meister M. Retina versus cortex: contrast adaptation in parallel visual pathways. Neuron 2004;42(1):5–7.

[40] Simoncelli EP, Olshausen BA. Natural image statistics and neural representation. Annual Review of Neuroscience 2001;24(1):1193–216.

[41] Hateren JHv, Schaaf Avd. Independent component filters of natural images compared with simple cells in primary visual cortex. Proceedings: Biological Sciences 1998;265(1394):359–66.

[42] Geisler WS, Perry JS. Statistics for optimal point prediction in natural images. Journal of Vision 2011;11(12).

[43] Adams WJ, Elder JH, Graf EW, Leyland J, Lugtigheid AJ, Muryy A. The Southampton-York natural scenes (syns) dataset: statistics of surface attitude. Scientific Reports 2016;6.

[44] Huang J, Mumford D. Statistics of natural images and models. In: Computer vision and pattern recognition. IEEE computer society conference on., vol. 1. IEEE; 1999.

[45] Stiehl WA, McCann JJ, Savoy RL. Influence of intraocular scattered light on lightness-scaling experiments. JOSA 1983;73(9):1143–8.

[46] Field DJ. Relations between the statistics of natural images and the response properties of cortical cells. JOSA A 1987;4(12):2379–94.

[47] Burton G, Moorhead IR. Color and spatial structure in natural scenes. Applied Optics 1987;26(1):157–70.

[48] Tolhurst D, Tadmor Y, Chao T. Amplitude spectra of natural images. Ophthalmic and Physiological Optics 1992;12(2):229–32.

[49] Ruderman DL, Bialek W. Statistics of natural images: scaling in the woods. Physical Review Letters 1994;73(6):814.

[50] Dror RO, Leung TK, Adelson EH, Willsky AS. Statistics of real-world illumination. In: Computer vision and pattern recognition, 2001. CVPR 2001. Proceedings of the 2001 IEEE computer society conference on, vol. 2. IEEE; 2001.

[51] Pouli T, Cunningham D, Reinhard E. Statistical regularities in low and high dynamic range images. In: Proceedings of the 7th symposium on applied perception in graphics and visualisation. ACM; 2010. p. 9–16.

[52] Martinez LM, Molano-Mazón M, Wang X, Sommer FT, Hirsch JA. Statistical wiring of thalamic receptive fields optimises spatial sampling of the retinal image. Neuron 2014;81(4):943–56.

[53] Kunkel T, Reinhard E. A reassessment of the simultaneous dynamic range of the human visual system. In: Proceedings of the 7th symposium on applied perception in graphics and visualisation. ACM; 2010. p. 17–24.

CHAPTER

Brightness perception and encoding curves

5

In this chapter we discuss brightness perception, the relationship between the intensity of the light (a physical magnitude) and how bright it appears to us (a psychological magnitude).

It has been known for a long time that this relationship is not linear, that brightness isn't simply proportional to light intensity. But we'll see that determining the brightness perception function is a challenging and controversial problem: results depend on how the experiment is conducted, what type of image stimulus are used and what tasks are the observers asked to perform.

Furthermore, brightness perception depends on the viewing conditions, including image background, surround, peak luminance and dynamic range of the display, and, to make things even harder, it also depends on the distribution of values of the image itself.

This is a very important topic for imaging technologies, which require a good brightness perception model in order to encode image information efficiently and without introducing visible artifacts.

5.1 The Weber–Fechner law: brightness perception is logarithmic

The quantitative study of perception is a discipline called psychophysics, that began with the seminal work of German physicist Gustav Fechner in the mid 19th century. Fechner considered that there are inner processes that correspond to the outer, physical magnitudes of the external world, and he named this mapping between psychological magnitudes and physical magnitudes *psychophysics*.

Ernst Weber, a colleague at Fechner's university, is another founding figure of experimental psychology. Weber studied touch and found that, when holding a weight of some amount I, the minimum amount ΔI that must be added to I in order to tell that there has been a change of weight is proportional to the base stimulus I, with some constant of proportionality k

$$\frac{\Delta I}{I} = k, \tag{5.1}$$

Vision Models for High Dynamic Range and Wide Colour Gamut Imaging. https://doi.org/10.1016/B978-0-12-813894-6.00010-7

which is known as Weber's law. The threshold difference ΔI is called *just noticeable difference* or JND, and the constant k is known as Weber fraction. For weights the Weber fraction is around 5%, which means that if we're holding a 100 g weight we need an increment of at least 5 g to detect a change, but if we're holding a 1 kg weight then the increment must be of 50 g or more.

Webner also found that the value of the Weber fraction is different for each type of sensory stimulus, i.e. k has one value for weights, another for loudness, another for brightness, another for electric shock pain, etc.

And very importantly, from Eq. (5.1) it follows that the lower the value of k is, the more we are *sensitive* to that particular stimulus modality.

Fechner assumed that all JNDs are equal and therefore that each JND produces the same increase ΔS in the amount of sensation S. From Eq. (5.1) this gives

$$\frac{\Delta I}{kI} = \Delta S. \tag{5.2}$$

Fechner also assumed that the total magnitude S of the sensation can be computed by adding up the contributions of all the ΔS increments, from the threshold stimulus I_T upwards. This corresponds to treating Eq. (5.2) as a differential equation and computing the integral, called in this context *Fechnerian integration*, giving

$$S = k' log\left(\frac{I}{I_T}\right), \tag{5.3}$$

where k' is a constant. As we can see this equation, known as Weber–Fechner's law, implies a logarithmic relationship between physical stimulus and perceived magnitude.

In the case of achromatic brightness perception, where I is light intensity and S the perceptual magnitude of brightness, some experiments show that the Weber fraction is $k \sim 1.4\%$. It is commonly assumed that this is only valid for an intermediate range of luminance values. Outside this range, for very dark as well as for very bright stimuli, the Weber fraction departs considerably from 1.4% so it is not constant, hence Weber's law is not valid, and therefore the Weber–Fechner law is not valid either.

Fig. 5.1 plots JND vs light intensity in log-log coordinates (this is called a "threshold vs intensity" or TVI curve, as we saw in Chapter 4). In this setting, Weber's law would imply a linear plot of slope 1, because $log(\Delta I) = log(I) + log(k)$. We can see that there are four different regimes for brightness perception.

For the lowest luminance levels the threshold is constant: this is called the "dark light" region, where it is assumed that light intensity is below the level of the neural noise that is always present in neural processes. Consequently, Fechnerian integration gives for this regime a brightness sensation S that is proportional to light intensity I: $S = k'I$.

Next, marked as "fluctuations" in the figure, comes the DeVries-Rose region, where it is proposed that JNDs are proportional not to intensity I but to the square

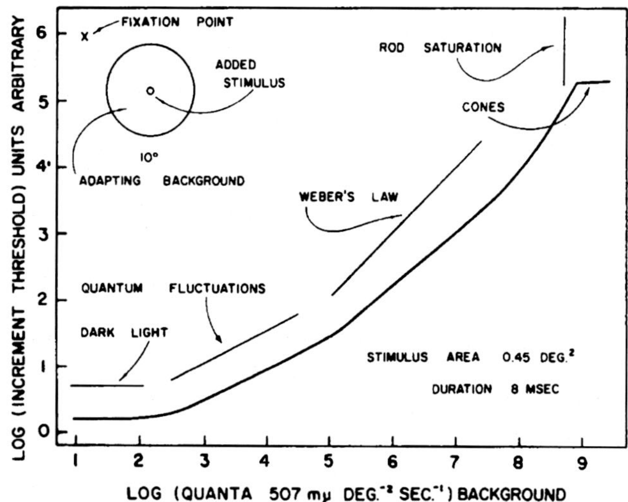

FIGURE 5.1 JND as a function of background luminance

JND as a function of background luminance. Figure from [1].

root of it:

$$\frac{\Delta I}{I^{0.5}} = k, \tag{5.4}$$

which by integration yields a brightness sensation that is a power-law, $S = k' I^{0.5}$.

In the following regime, for intermediate light levels, JNDs are proportional to light intensity, the Weber fraction is constant so Weber's law holds, and therefore Weber–Fechner is valid giving the expected logarithmic relationship between brightness and luminance: $S = k' log(I/I_T)$.

In the final regime there is saturation, and the JNDs become independent from I again.

But is important to remark that this conclusion, about k not being constant in general, is derived from experiments in which observers were looking at a uniform field of luminance I and had to detect the presence of a short (e.g. 4 msec) flash of intensity ΔI superimposed on the uniform background: with this experimental setup, as the luminance decreases k becomes increasingly larger. Let's call this Experiment 1. However, if the experiment is changed slightly and the observer is adapted to the base luminance I but the adapting field is turned off 5 sec before the flash (this would be Experiment 2), then the Weber fraction is shown to be constant for a 100,000:1 range of luminances [2] starting at very dark levels. Therefore in this case Weber–Fechner's law is actually quite accurate over a surprisingly wide range of light stimuli.

Cornsweet and Pinsker [2] argue the following:

- Some experiments yielding an accurate estimate of the Weber fraction for brightness consist in comparing a disc of luminance I with another disk of luminance $I + \Delta I$, and the observer has to detect which is the disk with the superimposed flash of magnitude ΔI. Therefore, when detecting a flash the visual system acts as if it's computing the ratio $\frac{I+\Delta I}{I} = 1 + \frac{\Delta I}{I}$ and comparing that ratio with some threshold value in order to make a judgement. A plausible physiological explanation for the computation of ratios in the visual system is the computation of differences, via lateral inhibition, of logarithmic transforms of the stimuli, as approximated by the photoreceptor responses. This is simply because a difference of logarithms is the logarithm of the ratio, and hence comparing photoreceptor responses gives an estimate of the ratio of luminances.
- The very different results between Experiments 1 and 2 can be explained by the fact that the output of the retina is not instantaneous but relies on some integration time. We know that, as the overall luminance increases, temporal summation is reduced. In Experiment 1 the retina has been continuously active for some time so the integration time is shorter than in the case of Experiment 2, and this is very relevant because the duration of the flahes is very short. Cornsweet and Pinsker speculate that if the flashes were 8 times longer, then the estimated Weber fraction would be constant in Experiment 1 as well.

5.2 Stevens' law: brightness perception is a power law

S. Stevens [3] argued strongly against using threshold studies for making judgements on perception magnitude (so-called suprathreshold appearance), so he conducted a very different kind of psychophysical experiments in which subjects had to provide, verbally, an estimate of the magnitude of the sensation they were having under some physical stimulus. He found that the perceptual sensation S was related to the physical stimulus I through a power law:

$$S = kI^a, \tag{5.5}$$

where k is a proportionality constant and the exponent a is a constant that depends on the stimulus modality, e.g. $a = 0.67$ for loudness, $a = 1.4$ for the taste of salt, etc.

Stevens' law can be obtained by considering that the JNDs are not all equal (as Fechner assumed), but rather that equal stimulus ratios produce equal perceptual ratios [4]:

$$\frac{\Delta I}{I} = k' \frac{\Delta S}{S}, \tag{5.6}$$

which by integration gives Eq. (5.5).

For brightness perception, the power-law exponent a has a value of 0.33 for a small target in the dark and of 0.5 for a brief flash.

With $a = 0.33$ Stevens' law for brightness is a cube root function, and this corresponds to the standard scale for lightness proposed by the CIE: the lightness value L^*, defined as brightness relative to the brightness a white diffuse surface, is essentially $L^* = (I/I_W)^{1/3}$, where I_W is the luminance of the white reference. The CIE L^* scale was proposed as it modelled well the appearance of a sequence of Munsell papers, going from black to white.

With $a = 0.5$ Stevens' law for brightness is a square root function, as in the DeVries-Rose law, and this is essentially the nonlinearity used in the TV industry and the most common display standard sRGB: by being a nonlinearity that approximates well the perception of brightness, it allows for a better and more efficient representation of the luminance signal, at least for moderate dynamic ranges, as we will discuss in Section 5.9.

5.3 Connecting the laws of Weber–Fechner and Stevens

As pointed out in [5], a connection between the laws of Weber–Fechner and Stevens can be made so that they become consistent.

A concept that we've mentioned already in the book, and that we shall reiterate a number of times because it is quite relevant for the topics we will be covering, is that neural responses are adequately modelled as sigmoidal functions, typically with some variant of the Naka-Rushton equation:

$$R = R_{max} \frac{I^n}{I^n + I_s^n}, \tag{5.7}$$

where R is the response to input signal I, R_{max} is the maximum response, I_s is the semisaturation value, i.e. the input value that produces a response that is half the maximum, and n is some fixed exponent. The exponent n was not in the original model of Naka and Rushton, it was added later in oder to better fit neurophysiological data: in the retina $n \sim 1.0$, while $n \sim 1.1$ in the LGN and $n \sim 2$ in V1. A sigmoid is approximately linear for small inputs, like a power law for moderate inputs, and like a logarithm for high inputs.

By measuring the minimum threshold of light intensity needed to change the response of a neuron that is adapted to a uniform field, it has been shown that these increment thresholds can be modelled with the Naka-Rushton equation and that they closely follow Weber's law for an input range of several orders of magnitude. If we schematically represent the visual system as the interaction of two pathways, a bottom-up one that transmits the input I and a top-down pathway that attempts to adapt to the input with a sensation magnitude S, the feedback process eventually converges to a solution where the two processes match, and at equilibrium we have that $S = k_1(I - I_0)^a + k_2$, a simple variation on Stevens' law. See [5] for details.

5.4 Crispening: brightness perception is neither logarithmic nor a power law

Heinemann [6] conducted yet another type of threshold detection experiment. The stimulus arrangement was a very small disk of luminance ΔI (the "increment field"), superimposed on a small disk of luminance I (the "test field"), surrounded by a large ring of luminance I_s (the "inducing field"). Subjects in the experiment had to detect the luminance difference ΔI.

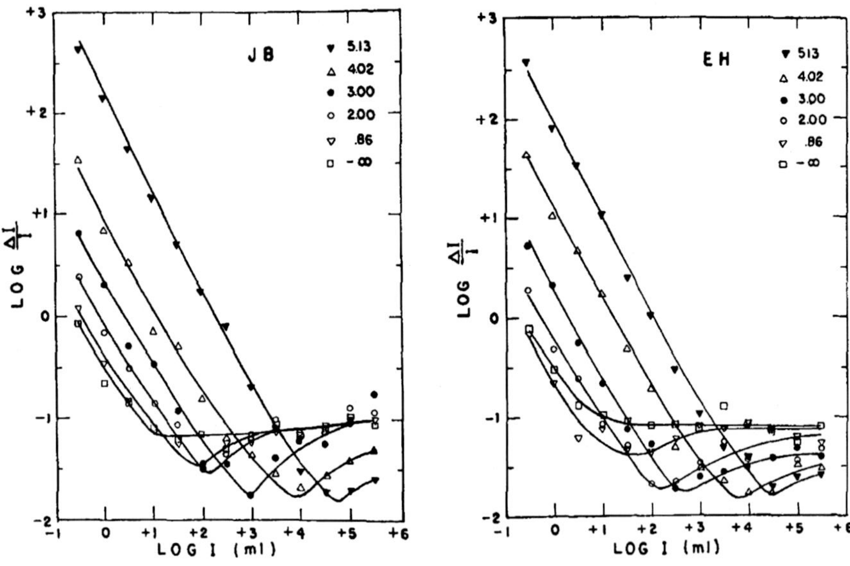

FIGURE 5.2 Heinemann's experiment

Weber fraction vs luminance, in log-log coordinates. Each image corresponds to one observer. Each curve corresponds to a different surround luminance, whose values are given in the upper corner. Figure from [6].

Fig. 5.2 shows the results for two observers, plotting the Weber fraction as a function of the luminance of the test field, in log-log axes. Each curve corresponds to a different surround luminance I_S. There are several important observations we can make from these results:

- The surround luminance clearly affects brightness perception, as each surround luminance value yields a different curve.
- Weber's law, i.e. a constant Weber fraction, would require the plot to be a horizontal line, which is only the case when the surround is totally dark and for luminance values above some threshold. Also as the luminance I becomes very large the curves do seem to converge to a horizontal asymptote, so brightness perception would approach logarithmic form as the luminance increases.

- Straight lines in these log-log plots correspond to a power law for the brightness sensation:

$$log(\frac{\Delta I}{I}) = k_1 log(I) + k_2 \rightarrow \frac{\Delta I}{I} = k_3 I^{k_1} \rightarrow \frac{\Delta I}{I^{k_4}} = k_3, \tag{5.8}$$

$$\frac{\Delta I}{k_3 I^{k_4}} = \Delta S \rightarrow S = k_5 I^{k_6}. \tag{5.9}$$

- We recall that the lower the Weber fraction gets, the higher our sensitivity. Therefore, the peak sensitivity happens at the curve minimum, which is at a value of I that is close to the luminance of the surround I_s. This phenomenon is called *crispening* [7].

5.4.1 Two studies by Whittle

Paul Whittle made a very important contribution to the study of brightness perception, by proposing a model that can account for crispening and fits very well experimental data. In what follows we reproduce our analysis from [8], by Kane and Bertalmío, where we re-model and re-evaluate the data from two papers by Whittle.

The first study [9] was a luminance discrimination experiment. The stimulus was achromatic and consisted of two small uniform patches presented horizontally either side of a central fixation point. The background was uniform and a schematic is illustrated in Fig. 5.3. The task was to discriminate between the luminance of the reference (I_r) and the test ($I_t = I_r + \Delta I$) patches. Luminance discrimination thresholds (ΔI) were investigated using a reference luminance that could be above or below the background luminance I_b.

A subset of Whittle's discrimination data is shown in Fig. 5.3 and it can be seen that substantially different results are obtained depending on whether the reference luminance I_r is larger or smaller than the background luminance I_b. If $I_r > I_b$, thresholds increase in proportion to the luminance difference with respect to the background, $I_r - I_b$. If $I_r < I_b$, a more complex pattern of thresholds is observed with the form of an inverted 'U'.

To account for the data Paul Whittle proposed a contrast term W,

$$W = \frac{|I_r - I_b|}{min(I_r, I_b)}, \tag{5.10}$$

for which a form of Weber's law holds:

$$\frac{\Delta W}{W} = k, \tag{5.11}$$

with k a constant that is nonetheless different for values above the background ($I_r > I_b$) than for values below.

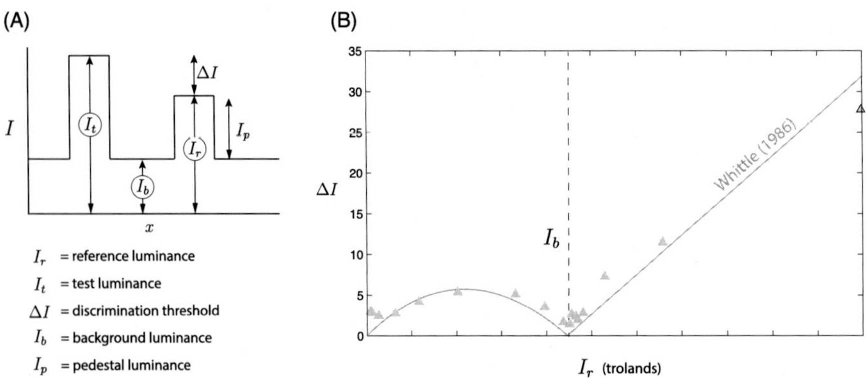

FIGURE 5.3 Luminance discrimination experiment by Whittle [9]

(A) Stimulus schematic and key for the luminance discrimination experiment. Schematic of [9]. (B) Discrimination thresholds plotted over the range of a CRT monitor. The solid line denotes the predictions of [9]. Figure from [8].

From these equations the threshold values ΔI would be:

$$\Delta I = \begin{cases} k_1|I_r - I_b|, \ I_r \geq I_b \\ k_2|I_r - I_b|I_r/I_b, \ I_r \leq I_b \end{cases} \tag{5.12}$$

In the case $I_r \geq I_b$, the formula is a simple Weber's law on the contrast dimension, while for $I_r \leq I_b$ the product of the ascending variable I_r and the descending variable $|I_r - I_b|$ produces an inverted 'U'. The predictions from these two equations are shown by the solid line in Fig. 5.3, where we can see that this model fits very well the experimental data.

A second paper by Whittle [10] investigated the perception of brightness, asking subjects to create a uniform brightness scale by adjusting the luminance values of a series of circular patches – in a uniform surround – until there appear to be equal brightness steps from black to white. In Fig. 5.4 we reproduce the data from two conditions. In the leftmost figure, the test patches were yellow (light gray in print version) and the background green (dark gray in print version) and the resulting nonlinearity is compressive with no inflexion. This form is common to most of the nonlinearities reported in the literature. However when the test patches and the background were both achromatic, the function steepens around the background luminance level: this is the crispening effect, and it mirrors the high sensitivity around the background luminance noted in the discrimination threshold dataset. The reason is that a high slope for the brightness perception curve means that a small change in luminance produces a high change in brightness, therefore our sensitivity is high, and hence the discrimination threshold must be small. This can be simply expressed by saying that sensitivity is the derivative of the brightness function, and that the threshold is the inverse of the sensitivity.

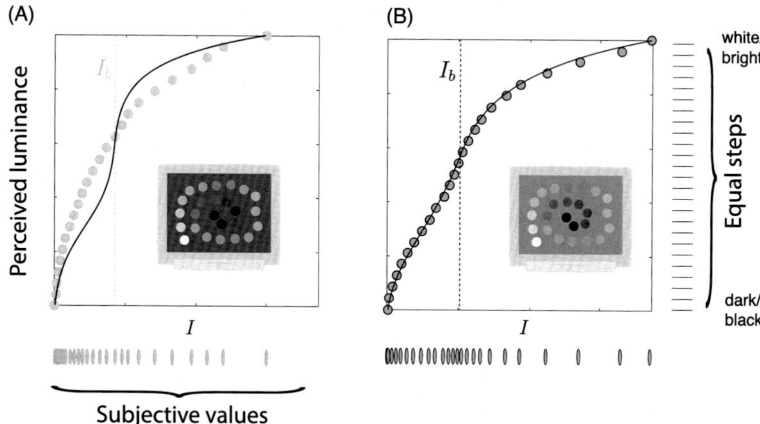

FIGURE 5.4 Brightness perception experiment by Whittle [10]

The insets illustrate the two experimental conditions from [10]. (A) A simple compressive form; (B) The 'crispening effect'. Figure from [8].

But, in what was a key development, Whittle showed that the luminance intervals set by subjects in [10] produced a brightness scale that was fully consistent with the discrimination thresholds of [9]. To make this point Whittle plotted the discrimination thresholds alongside the luminance intervals and we re-plot this data in Fig. 5.5. The luminance intervals (blue (dark gray in print version) dots) are just a scaled version of the discrimination thresholds (green (light gray in print version) triangles).

This strong correlation indicates that, contrary to the objections posed by Stevens, threshold data can be used to measure appearance (the *suprathreshold* data of the brightness scale): as mentioned above, sensitivity is the slope of the brightness curve and the inverse of the discrimination threshold, so if we have the threshold data, we can simply invert it and then integrate and we should arrive to a brightness function. Doing so from Eq. (5.12) we get [12]

$$B = \begin{cases} B_b + S_b(1+c)log(1+\frac{W}{W_0}), \ W \geq 0, \\ B_b - S_b log(1-\frac{W}{W_0}), \ W \leq 0, \end{cases} \tag{5.13}$$

where B is the brightness, B_b is the brightness of a uniform field of luminance I_b and alongside the slope S_b it's a free parameter, W_0 is a normalising contrast and c allows for different slopes for increments and decrements.

An example curve obtained in this manner is shown in Fig. 5.4(B) and provides an excellent fit. However, no parameter combination could allow the prediction of the data in the yellow–green condition as shown in Fig. 5.4(A). In short, the crispening effect is hard-coded into the model of Whittle.

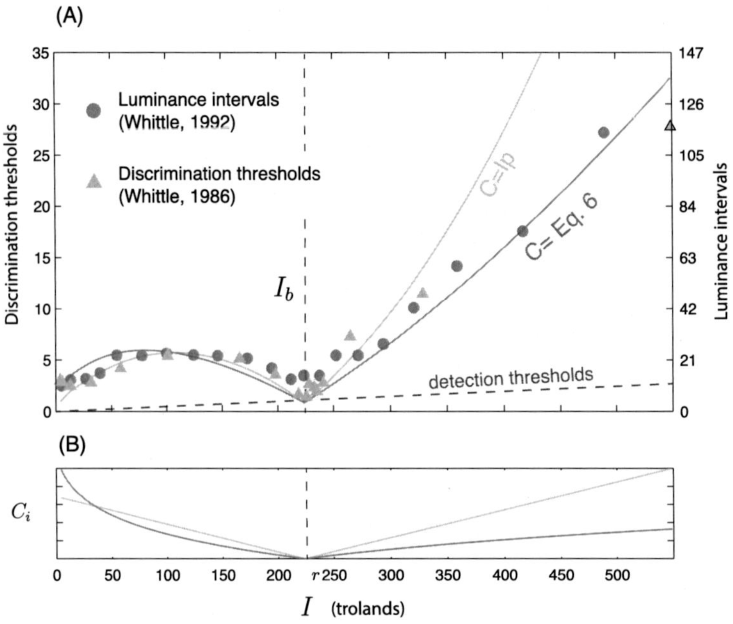

FIGURE 5.5 Threshold and suprathreshold data are consistent in [9,10]

(A) The discrimination thresholds from [9], are plotted with green (light gray in print version) triangles, whilst the luminance intervals set by subjects in [10], are denoted by blue (dark gray in print version) circles. A detection threshold function obtained from [11] is shown by the dashed line. The predictions from our proposed model are shown with solid lines, using two models of contrast gain, plotted in (B). Figure from [8].

5.4.2 Whittle's model for brightness perception: advantages and limitations

Carter points out in [4] a number of advantages and limitations of Whittle's model for brightness perception.

Among the many virtues of the model he cites that it matches both threshold and suprathreshold data very well, provides an expanded gray scale on brighter surrounds as required by the data from Heinemann [6], and matches psychophysical data both from reflective surfaces and from self-luminous displays.

The main disadvantages are, firstly, that it always produces crispening, while it's been shown that crispening might be greatly reduced depending on the context (nonetheless, Moroney [13] points out that *"while crispening can be reduced in certain spatial configurations of stimuli, it should not be ignored with respect to digital imaging"*). Secondly, that the model is based on experimental data using either threshold increments or threshold decrements over a uniform surround, but natural images have regions that are at the same time increments and decrements from neighbouring areas.

In 2018, the CIE recommended Whittle's formula as a brightness scale for self-luminous devices [14].

5.4.3 Two alternatives to Whittle's model

In [15], Kingdom and Moulden criticised the model of Whittle because it required a different formulation for incremental and decremental pedestals. As such they introduced an alternative model in which thresholds could be predicted from a single equation. This was based upon a logarithmic formulation for contrast G,

$$G = ln(I/I_b), \tag{5.14}$$

and the observation that (like Whittle's formation of W), G could also predict the pattern of discrimination thresholds. By relaxing the Weber's law constraint to include the square root-behaviour of detection thresholds at low luminance levels, this model predicts thresholds as follows:

$$\Delta I = k(I_r - h) \left| ln \left(\frac{I_r + h}{I_b + h} \right) \right|^n, \tag{5.15}$$

which can now fit data both with or without crispening.

Nonetheless, neither of the models of Whittle [9] or Kingdom and Moulden [15] can predict another phenomenon called the *dipper effect*, which is a counterintuitive increase of the thresholds for reference luminance values that are very close to the background level.

It is well established that integrating over the inverse of *detection* thresholds (e.g. detecting an increment flash over a target) produces a compressive nonlinearity with no inflexion and provides a reasonable estimate of the global perception of luminance under many circumstances [16], but as described above *discrimination* thresholds (e.g. discriminating between the target and a reference) are more appropriate for the stimulus configurations investigated by [10,9]. As such, a model that can account for both types of phenomena would be able to unify the two sets of results and this was the idea behind the work that is discussed in [8], which proposes a relationship between detection thresholds and discrimination thresholds as follows,

$$\Delta I(I_r) = \Delta I_{det}(I_r)\,(e + cC(I_r)), \tag{5.16}$$

where ΔI is the discrimination threshold, e and c are constants to be fit to the data and C some function of the contrast between I_b and I_r.

The basic idea behind the model is that when $I_r = I_b$ discrimination thresholds reduce to detection thresholds, but when $I_r \neq I_b$ contrast gain decreases sensitivity and gives rise to the crispening effect. In the original model $C = I_p$, however we can see in Fig. 5.5 that while it captures the basic effect it does not produce a particularly accurate fit. The alternative is to consider nonlinear processing of luminance and, more specifically, that there is good reason to expect contrast to be computed differently depending on the polarity of the reference patch. To estimate the initial contrast function we begin with a detection threshold function ΔI_{det} derived from the psychophysical data by [11], which is used to estimate the initial model contrast as the integral of one over thresholds, as follows,

$$Z(I_r) = \frac{1}{s} \sum_{I_{min}}^{I_r} \frac{1}{\Delta I_{det}(I_i)} \tag{5.17}$$

where s is a normalisation constant. The contrast C in the modified model is then

$$C(I_r) = |Z(I_r) - Z(I_b)|. \tag{5.18}$$

This nonlinear computation of contrast is shown in Fig. 5.5(B). Critically, contrast is now an expansive function when the reference is dark and a compressive function when the reference is bright and this contrast term allows for an accurate prediction of thresholds. A second modification is made to allow the capture of the dipper effect:

$$\Delta I(I_r) = \Delta I_{det}(I_r) \left(\frac{e_1}{e_2 + c_2 C(I_r)} + c_1 C(I_r) \right), \tag{5.19}$$

please see [8] for details.

5.5 Detecting thresholds depends on frequency: the Contrast Sensitivity Function

The seminal work of Campbell and Robson [17] showed that our ability to detect changes in contrast depends on the spatial frequency of the stimulus. This led to the idea that the visual system has a number of spatial channels, each selective to a different spatial frequency, and that can be modelled by multiple spatial filters. Each channel has its own filter, and the signal that the channel carries is computed by convolving the input stimulus with a kernel that is often thought of as the receptive field corresponding to the filter [18].

Characterising the visual system, then, would consist of finding the convolution kernels. It is equivalent, and more practical, to find the Fourier transform of a convolution kernel, and this the contrast sensitivity function (CSF).

The CSF expresses contrast sensitivity as a function of spacial frequency. It consists of several, around 5 to 10, contrast sensitivity values measured at spatial frequencies that are placed at equal intervals over a logarithmic scale measured in cycles per degree [19]. Contrast sensitivity is the inverse of the contrast threshold, which is the minimum amount of contrast that a target on a uniform background must have in order for us to see it.

We've seen that contrast is a measure of the difference between the luminance of a target and the luminance of its background, and that there are several definitions of contrast. The systems-theory approach to vision pioneered by the work of Campbell and Robson [17] assumes that, at a given background luminance, the visual system behaves linearly and therefore in order to characterise it it's easier to probe it with sinusoidal gratings, because the linear transform of a sinusoid is also a sinusoid [1]. In the case of sinusoidal gratings it is common to use the Michelson contrast, defined as

$$C_M = \frac{I_{max} - I_{min}}{I_{max} + I_{min}}, \tag{5.20}$$

where I_{max} is the maximum luminance of the target and I_{min} its minimum luminance. In experiments measuring contrast thresholds, the target has the same average luminance as the background.

A common experimental setup for characterising the CSF consists in the two-interval forced choice (2IFC) protocol. With 2IFC, the observer is presented with two images, one after the other. One of the images consists simply of the uniform background, the other image has the background with a small sinusoidal grating superimposed on it. The observer has to judge which of the images has the grating, and the contrast threshold is the value of the Michelson contrast at which the observer response is correct at least 75% of the trials [19].

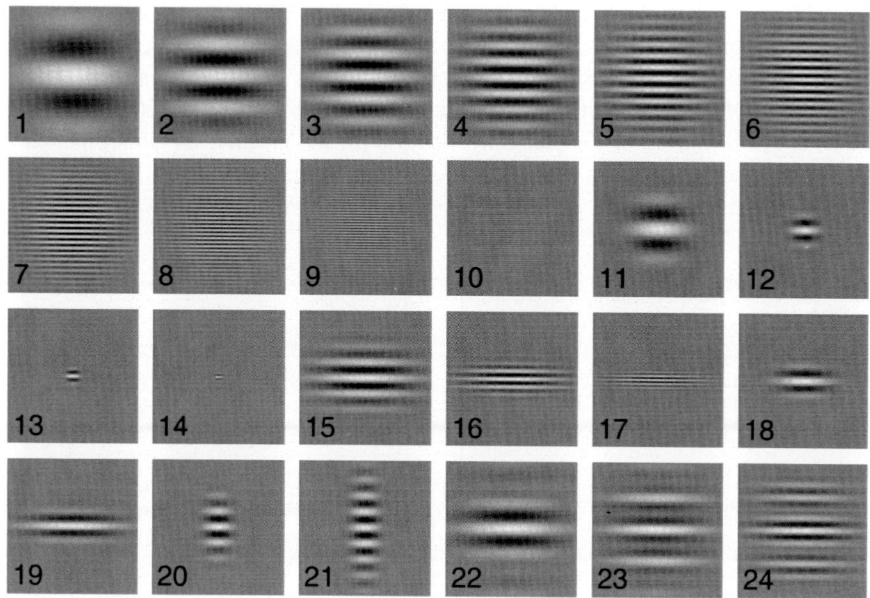

FIGURE 5.6 ModelFest stimuli

Some stimuli from the ModelFest set (the index numbers were not present in the stimuli). Figure from [18].

Fig. 5.6 shows some example gratings used in the ModelFest experiment to collect contrast sensitivity data.

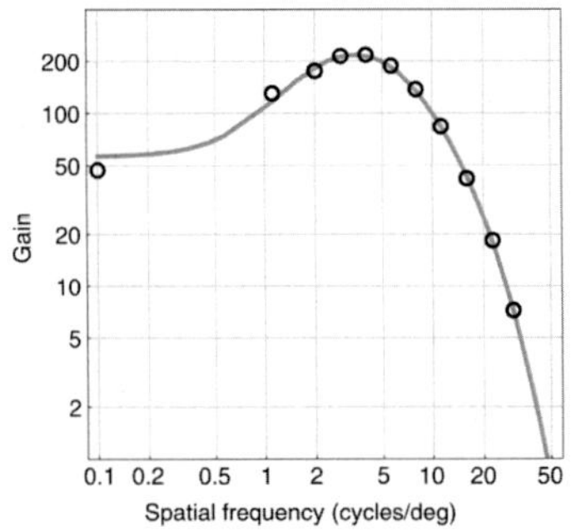

FIGURE 5.7 An example CSF

An example CSF that provides an excellent fit to the ModelFest data. This CSF is modelled as the difference between two hyperbolic secants. Figure from [18].

Fig. 5.7 shows a CSF that provides an excellent fit to the psychophysical data.

All CSFs have a bandpass nature, with a peak somewhere between 1 and 10 cycles per degree and falling off at lower and higher spatial frequencies [20]. This means that a CSF is the Fourier transform of a spatial convolution filter that has to consist of a high-frequency lobe minus a low-frequency lobe. This can be achieved for instance with a convolution kernel that is a difference of Gaussians (DoG), which we've seen has been used extensively to model the RFs of neurons in the retina and the LGN, although in practice better fits to the data are obtained with kernels that are not a DoG, see [18]. Indeed in Chapter 8 we will use for the problem of gamut mapping convolution kernels that can't be expressed in DoG form.

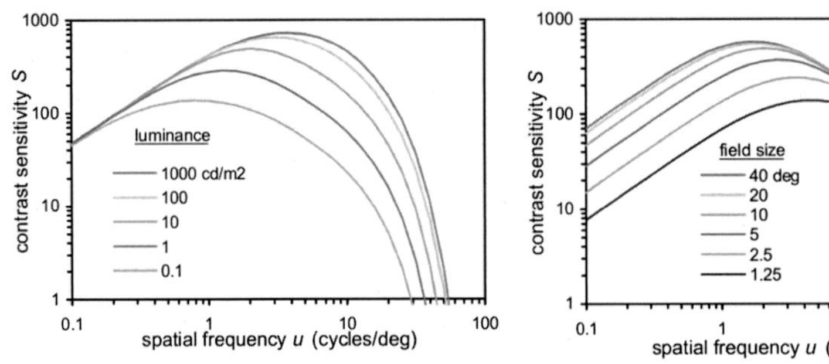

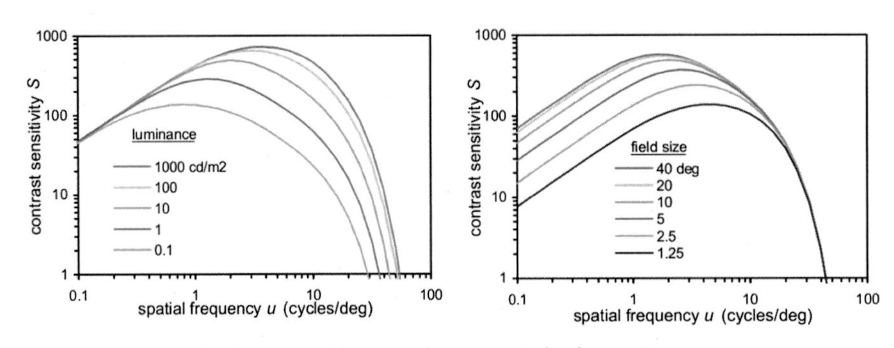

FIGURE 5.8 Barten CSF as a function of luminance and viewing angle

Barten CSF as a function of luminance (left) and viewing angle (right). Figure from [20].

In [21], Barten remarks that the CSF depends strongly on luminance, viewing angle, surround, and several other aspects like the quality of the eye lens or the sensitivity of the photoreceptors. After proposing a very elaborate model taking all these issues into account, that was shown to match very well varied psychophysical data coming from different experimental set-ups, he introduced in [20] a simplified model that has become very popular and is now known as the Barten CSF. This CSF depends only on the luminance and viewing angle.

Fig. 5.8 shows how the Barten CSF varies with luminance (left) and viewing angle (right). We can see how, as the luminance decreases, the sensitivity is reduced and the peak spatial frequency moves towards lower values: this corresponds to our inability to read fine print when the lights are dim.

We must remark, because it will be relevant later, that while the experimental datasets used to validate the Barten CSF had different luminance ranges, the maximum luminance among all of the data sets was 500 cd/m^2.

Finally, Barten introduces a correction factor for the CSF, a single value $f \leq 1$ that depends on the luminance of the surround and that has to multiply the CSF. The reason for this correction is that we adapt to the surround and if it has a much higher or much lower luminance than the object then our contrast sensitivity will be greatly reduced: the object will tend to appear uniformly dark on a bright surround or viceversa. Fig. 5.9 shows the correction factor for the Barten CSF as a function of surround luminance.

In most experiments the surround luminance is not too different from that of the object, but the effect of the surround on the perception of brightness is remarkable and we will discuss this in the following section.

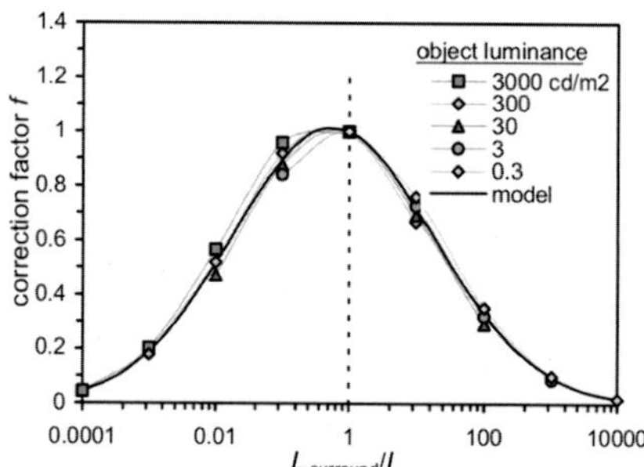

FIGURE 5.9 Correction factor for the CSF as a function of surround luminance

Correction factor for the CSF as a function of surround luminance. Figure from [20].

5.6 Effect of the surround on brightness perception: the system gamma

In a landmark work [16], Bartleson and Breneman studied brightness perception of natural images (black and white photographs) in different surround conditions: photographic prints seen in an illuminated environment, projected transparencies seen in a dark room. They used 10 images, all of them typical photographic scenes of complex luminance and spatial characteristics, which was a radical departure from the artificial and very simple image stimuli traditionally used in all brightness perception experiments.

They collected psychophysical data from 17 observers using two experimental methods. In the first one, magnitude estimation, the experimenter selected an arbitrary reference region in the image and assigned a scale value to it; the observer was asked to judge the appearance of several other regions in the same image and assign to them appropriate scale values, consistent with the scale of the reference. In the second experimental method, inter-ocular brightness matching, the observers saw with one eye a central (uniform) field in an extensive illuminated surround, and with the other eye the test image; the observer was pointed to some arbitrary area in the test image and had to adjust the luminance of the central field until their brightnesses matched. The images were shown at various levels of illuminance and different surround conditions.

Their results show, firstly, that the models of Weber–Fechner and Stevens are both unable to adequately fit the data, since in log-log coordinates the brightness vs luminance plots are significantly nonlinear.

Secondly, that the curves for relative brightness (brightness relative to that of an image reference "white") simply shift to the right when the illumination of the pictures increases while the surround remains constant.

And finally, that the shape of the relative brightness curve is indeed affected by changes in the surround. In a later work [22], Bartleson proposed a power law type of model that was able to predict the experimental data quite closely:

$$L(Y) = k_s(Y/Y_W - a_s)_s^{\gamma} - 16, \tag{5.21}$$

where L is the lightness or relative brightness, Y is the luminance, Y_W is the luminance of a reference white, and k_s, a_s, γ_s are constants that depend on the luminance of the surround. The key element here is the exponent of the power law, γ_s, which was found to be 0.33, 0.41 and 0.50 for dark, dim and light surrounds respectively.

In other words, our adaptation to the surround modifies our brightness perception function, which is modelled as a power law nonlinearity corresponds to reducing the exponent (and hence increasing the slope of the function) as the luminance of the surround decreases.

A key consequence of Bartleson and Breneman's findings is that, if we want to display an image in a dark surround but having the same appearance as if the observer was seeing the image in a light surround (i.e. the typical situation found in cinema) then we should raise the luminance of the image to a power of $1.5 = 0.50/0.33$. And if we want to display an image in a dim surround but having the same appearance as if the observer was seeing the image in a light surround (i.e. the typical situation found in TV viewing) then we should raise the luminance of the image to a power of $1.24 = 0.41/0.33$.

This exponent needed to compensate for the effect of the surround is called the *system gamma*, and describes the relationship between an image captured by a camera sensor and the image displayed on a screen. When a camera captures an image, values on the sensor are proportional to light intensity but then a nonlinearity is applied to the image signal (for reasons that we will discuss shortly, in Section 5.9). This nonlinearity used to be a power law, the process of applying the power law is known as gamma correction and the exponent is called the "encoding gamma". Displays, on the other hand, produce light whose intensity can also be expressed as a power law function of the image value, and the exponent is called the "decoding gamma".

The system gamma is the product of the encoding and the decoding gammas. If we did not adapt to the surround luminance, the encoding and decoding gammas would cancel each other out and the value for the system gamma would simply be 1.0. However, TV and display engineers have known for over 70 years that the system gamma must be set to a value of greater than 1. In the 1940s, the recommended system gamma for photographic projections, typically viewed with a dark surround, was 1.6, but between 1 and 1.1 for printed images, typically viewed with a light surround. Today, the recommended system gamma for a cinema display (typically viewed with a dark surround) is 1.5, and the system gamma for an office display (typically viewed with a light surround) is 1.1, which are in line with the results of Bartleson and Breneman.

5.7 System gamma as a function of dynamic range

As we just saw, the system gamma has traditionally been considered simply a function of the surround. But in [23] Liu and Fairchild found that, for a fixed surround, the optimum system gamma varies with each image, and in [24] Singnoo and Finlayson show that the preferred system gamma is correlated with the system gamma value that achieves the greatest histogram equalisation of the luminance image. Furthermore, Borer and Cotton found in [25] that, under a constant surround, the optimum system gamma is a linear function of the logarithm of the peak luminance of the display.

In [26] Kane and Bertalmío set out to model what was the dependency of the preferred system gamma on the dynamic range of the image and also on the dynamic range of the display.

Our hypothesis was that the preferred system gamma is the one that best equalises the brightness distribution. The model has three simple steps:

1. The image N is a normalised (i.e. in the range $[0, 1]$) and linear (i.e. with values proportional to scene luminance) grayscale image that is raised to the power of system gamma (γ_{sys}) in order to produce the image I that is displayed on a given screen

$$I(x) = N(x)^{\gamma_{sys}}. \tag{5.22}$$

2. Second, we estimate the perceived brightness B of the image by raising the image I to the power of γ_{psy}, which we shall term the psychological gamma and is essentially the exponent in a Stevens' law model for brightness:

$$B(x) = I(x)^{\gamma_{psy}}. \tag{5.23}$$

The value of the psychological gamma can be obtained from the literature (e.g. $\gamma_{psy} = 0.33$) or estimated from our own experimental data, as will be discussed below. Eq. (5.23) also implies that

$$B(x) = N(x)^{\gamma_{psy} \cdot \gamma_{psy}} \tag{5.24}$$

3. From the brightness image B we compute its histogram h and from h its normalised cumulative histogram H. The degree of histogram equalisation (F) is computed as one minus the root-mean-square (RMS) difference between H and the identity line,

$$F = 1 - \frac{1}{n+1}\sqrt{\sum_{k=0}^{n}\left(H(\frac{k}{n}) - \frac{k}{n}\right)^2}, \tag{5.25}$$

because if an image has a perfectly equalised histogram then its cumulative histogram is the identity line.

In this model the predicted value of system gamma will depend on two factors: the value of psychological gamma and the luminance distribution of the image.

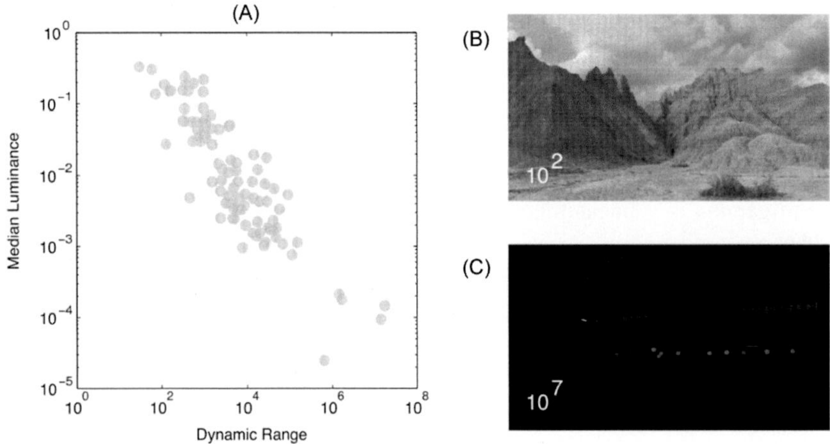

FIGURE 5.10 Image statistics from an HDR database

Image statistics from an HDR database. Original images from [27]. Figure from [26].

To investigate a subject's preference for system gamma (which we term the preferred system gamma), we select images from the high dynamic range image survey by Fairchild [27]. This database contains images with dynamic ranges from two to seven orders of magnitude, and thus images that will require different levels of system gamma to achieve histogram equalisation, for the following reason.

Using calibration data for absolute luminance values in the original scene, obtained with a photometer, we can compute the dynamic range of each image, and in

Fig. 5.10(A) we plot the normalised median luminance of each image against the dynamic range of each image. The results demonstrate that the median luminance of an image is highly correlated with the dynamic range of an image. Thus as the dynamic range of an image increases, the image becomes increasingly dark, and in turn, an increasingly compressive nonlinearity is required for histogram equalisation. This is illustrated in the same figure using an image with a low dynamic range (Fig. 5.10(B)) and an image with a high dynamic range (Fig. 5.10(C)).

To investigate viewing-condition dependency we perform our experiments with two background conditions (black and white) and three monitors with different dynamic range capabilities: a cathode ray tube (CRT) display, a liquid crystal display (LCD), and an organic light-emitting diode (OLED) display.

On each trial subjects viewed a centrally presented image and were asked to rate the quality of the image on a sliding scale. The scale was marked Terrible, Satisfactory, and Excellent at the left, middle, and right extremes, respectively. These values correspond to 0, 5, and 9 on the presented figures. During each run, subjects were presented with a variety of images displayed with different values of system gamma, that varied between 2^{-4} and 2^2 at half log-two intervals.

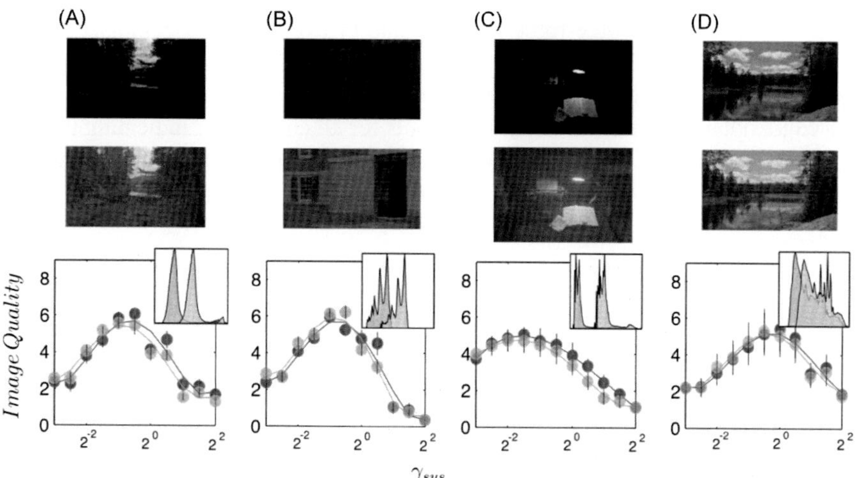

FIGURE 5.11 System gamma selection to optimise image quality

Top row: images presented with a system gamma of 1. Middle row: images presented with the preferred system gamma obtained in the experiment for the white background condition. Bottom row: image quality as a function of system gamma. (A) Amikeus Beaver Damn (5); (B) Hancock Kitchen Outside (39); (C) Lab Typewriter (46); (D) Peak Lake (72). Figure from [26].

In the first experiment we used the CRT display and the same four images as in the study by Liu and Fairchild [23]. The results are shown in Fig. 5.11. Each column shows the results for one base image. The first row shows the original image presented with a system gamma of 1 (assuming a decoding gamma of 1/2.2 in the reader's display or printer). The second row shows the images presented with the preferred system gamma chosen by the average subject. In the third row we plot image quality scores as a function of system gamma. Blue (dark gray in print version) dots denote data from the black background condition and green (light gray in print version) dots data from the white background condition.

The results show that the image quality functions tend to be shifted toward the left in the light background condition, indicating that subjects prefer a lower, more compressive system gamma to be applied to the image when a white background is used. We estimate the preferred system gamma by fitting a fourth-order polynomial to the data (solid lines in the figure) and identifying the value of system gamma that corresponds to the peak of the function. Consistent with the original Liu and Fairchild study, we find the preferred system gamma is higher when a dark background is used.

Moreover, we find the ratio between the preferred system gammas in the black and white background conditions to be 1.2, close to the value of 1.14 obtained by Liu and Fairchild, both significantly lower than the value of 1.5 obtained by Bartleson and Breneman [16]. We can also see how the estimated value of system gamma is image dependent, with higher dynamic range images (like image 46) having a lower system gamma than images of lower dynamic range (like image 72).

We can search via brute force for the value of psychological gamma γ_{psy} that maximises the Pearson's correlation between flatness F (Eq. (5.25)) and image quality for all values of system gamma applied to a given image. Encouragingly the search procedure obtains high correlation coefficients for all conditions, indicating that image quality scores are linearly related to the degree of equalisation in the brightness histogram.

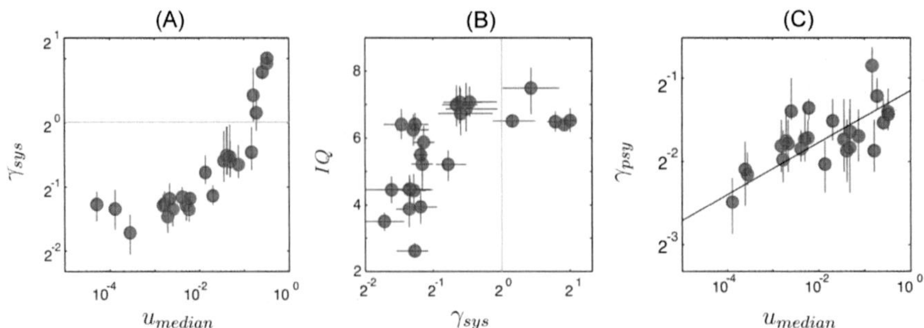

FIGURE 5.12 Results from the second experiment

Results from the second experiment, see text. Figure from [26].

In experiment two we consider just the black background condition but 26 images. The results are summarised in Fig. 5.12, which shows that:

- Lower values of system gamma γ_{sys} are required for images with a lower median luminance u_{median}, and this finding is consistent with the notion that the preferred system gamma is, at least in part, determined by the need for histogram equalisation. Let us remark that u_{median} is the median luminance of the image after it has been normalised to between 0 and 1.

- The lower the preferred system gamma, the worse the overall image appearance.
- Again we search via brute force for the value of psychological gamma γ_{psy} that maximises the Pearson's correlation between flatness F (Eq. (5.25)) and image quality for all values of system gamma applied to a given image. We find that the estimated γ_{psy} is positively correlated with the median luminance u_{median}, with a linear relationship in log-log scale: $log_2(\gamma_{psy}) = alog_2(u_{median}) + b$.

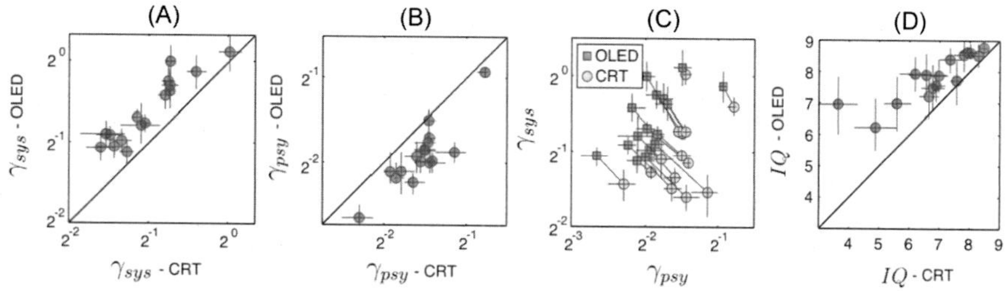

FIGURE 5.13 Results from the third experiment

Results from the third experiment, see text. In panel (C), CRT data shown with yellow circles (light gray in print version) and OLED data shown with green squares (mid gray in print version). Figure from [26].

In experiment three we perform the tests on a CRT and an OLED display: both devices have the same peak luminance of 100 cd/m^2, but the dynamic range of the OLED screen is considerably higher than that of the CRT. Results are shown in Fig. 5.13, where we can see that

- The preferred system gamma is always higher when viewed on an OLED display.
- In turn, the estimated values of psychological gamma are lower on the OLED display than on the CRT.
- If we plot the preferred system gamma against the estimated values of psychological gammas on the CRT and OLED monitors, we find that changes in the estimated preferred system gamma are matched by an equal and opposite change in the estimated psychological gamma (i.e., the changes fall upon a line with a slope of -1). Thus under the assumption that the estimated value of psychological gamma accurately reflects brightness perception, we can conclude that the perceived distribution of brightness values will be the same on the two monitor types when the preferred system gamma is used.
- Finally, in Fig. 5.13(D) we demonstrate that the maximum image quality scores obtained on the OLED display are greater than those obtained on the CRT display; however, this effect appears to converge for high image quality scores. Thus the advantage of using an OLED display over a CRT display appears to be confined to images with a high dynamic range, which seems reasonable.

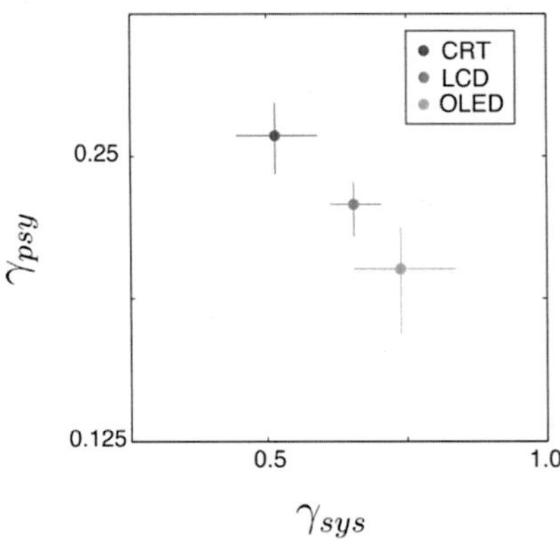

FIGURE 5.14 Estimated psychological gamma as a function of system gamma

Estimated psychological gamma as a function of system gamma, for three display types (only for the black background condition). Figure from [26].

We repeated the experiment using an LCD monitor, which has a similar peak luminance as the CRT and OLED displays and a dynamic range that is in between that of the other two screen types. Fig. 5.14 shows the results, where we can see that values for the LCD fall in between those for the CRT and OLED monitors. This strengthens the argument that dynamic range plays a role in determining the brightness function and in turn the preferred system gamma.

5.7.1 Some important considerations on brightness perception

In this study, a model of brightness perception is inferred from the data, but not directly measured. Encouragingly, the average value of psychological gamma obtained in the black background condition on the CRT (when averaged across images) is 0.31, which is close to the value of 1/3 common to many colour models but somewhat below those obtained by Stevens and Stevens [28], who estimated values that span from 0.33 (for the dark adapted eye), to 0.49 (adaption to 10,000-ft lamberts). The value of psychological gamma obtained is correlated with the normalised median luminance of the tested image, and this may reflect an adaptation to the luminance distribution of the stimulus, and/or the adapting luminance level, or alternatively some unknown aesthetic criteria, unrelated to any perceptual adaptation.

The importance of dynamic range on determining the system gamma had not been directly investigated previously; however, it may have affected the estimates produced by other studies. For instance, we saw that the work by Bartleson and Breneman computed the compensation ratio for photographic prints (viewed with a light background) against photographic projections (viewed with a dark background). The dynamic range of prints is less than two orders of magnitude, while the dynamic range of projections is at least two orders of magnitude. Our data would predict that the preferred system gamma would be lower for lower dynamic range displays, and thus the high compensation ratio of 1.5 obtained by Bartleson and Breneman relative to those found by this study (1.21) and others (1.14 in [23]) could potentially be accounted for by the different dynamic range of the two media.

Finally, we must remark that while brightness perception has been investigated using both simple and real-world stimuli, unfortunately there is no clear correspondence between the two sets of results (see [22] for a detailed discussion on this topic). In laboratory studies small test stimuli are typically viewed upon a uniform background and the relationship between the background luminance and brightness perception is investigated. In these studies the surround luminance of a monitor is fixed and is typically dark. Thus when the background luminance is altered, there is a large change in the mean luminance of the central visual field. In contrast, when real-world scenes are used, the surround luminance of a display is adjusted and this manipulation primarily affects the peripheral vision field. Brightness perception estimated using real-world scenes tends to show a relatively small modulation when the surround luminance changes, compared to that observed using simple stimuli, where the function may change more dramatically. Importantly, the ratio of exponents estimated using natural scenes predicts the ratio of preferred system gamma in the black and white background conditions, while the exponents derived using simple stimuli do not correspond well.

5.8 Brightness, lightness and HDR images

The term brightness, defined by the CIE as the "attribute of a visual sensation according to which an area appears to emit more or less light" is commonly used for images on a self-luminous display; lightness, on the other hand, is defined by the CIE as "the brightness of an area judged relative to the brightness of a similarly illuminated area that appears to be white", and originally referred to reflective surfaces [4].

The CIE lightness scale L^*, as we mentioned before, follows Stevens' law with a 0.33 exponent and was derived to match perceptual data on a black-to-white scale of reflective surfaces: $L^* = 116(Y/Y_W)^{0.33} - 16$, where Y_W is the luminance of a white perfectly diffuse surface (i.e. no specularities, glossiness, transparencies etc.) If the illumination changes (for instance if we dim the lights) both Y and Y_W would change in the same proportion so the ratio Y/Y_W would remain constant, and so would L^*. But in the case of self-luminous displays it isn't clear what the value for Y_W should be. Given that for reflective surfaces the highest possible lightness value is 100, making an analogy we could use for Y_W the peak luminance of the display, but then if Y_W varies this would not affect Y, and therefore the ratio Y/Y_W would change as well. As a result, a lightness scale is ambiguous for self-luminous displays [4].

But we wish to note that for stimuli in which there is no clear change in the overall illuminance across a scene, judgements of *brightness*, *lightness* or local *brightness-contrast* (all supra-threshold judgements) cannot be distinguished [29–31], so often the term *brightness* could be interchanged with the term *lightness*.

We have seen how it's quite common in natural scenes that the luminance has a dynamic range of 10,000:1 or more, while the dynamic range of surface reflectances is in the order of 30:1. In [32] Radonjic et al. investigate the mapping that the visual system is doing between scene luminance and surface lightness, i.e. they studied lightness perception for HDR images.

Their experimental set-up consisted of an HDR display where 5×5 grayscale checkerboard images were presented. The center square was the test stimulus, and the surrounding squares had luminances over a 10,000:1 dynamic range. Observers had to match the lightness of the center square to the lightness of a reflective surface (a neutral Munsell palette) presented elsewhere. They carried out different experiments varying the luminance distribution and DR of the checkerboards, i.e. the context of the test stimulus.

The authors were able to fit the lightness perception data with a modified Naka-Rushton equation,

$$R = \frac{(g(L-c))^n}{(g(L-c))^n + 1}, \tag{5.26}$$

where R is the visual response corresponding to lightness, L is the stimulus luminance, and g, c, n are parameter values that vary with each experiment.

The results are summarised in Fig. 5.15.

Fig. 5.15(A) plots the response function for three DR conditions ($\sim 30:1$ (black), $\sim 1,000:1$ (gray) and $\sim 10,000:1$ (blue (dark gray in print version))) where the peak luminance was the same. We can see how the slope of the response function becomes shallower as the DR increases, and in this way the response range can be allocated to match the luminances in the checkerboard context. The parameter n, that controls the slope of the response function, has to change in order to fit the data.

Fig. 5.15(B) shows how the response for a 1,000:1 context changes when all checkerboard luminances (including the test stimulus) are scaled by the same multiplicative factor. The effect is that the response is simply shifted, by an amount that is the multiplicative factor.

Fig. 5.15(C) plots the result of the same luminance scaling experiment but now for a 30:1 context. Again, the response is shifted by the scaling factor.

Finally, Fig. 5.15(D) shows what happens to the response when the minimum and maximum luminances of the context are kept fixed but the luminance distribution changes. We can see how the curve changes so that a larger portion of the response range is allocated to the stimuli that occur most often.

All these behaviours are clearly consistent with the efficient coding hypothesis, the general notion that adaptation to the statistics of the stimulus allows to optimise the use of the response range that is available. Some other interesting conclusions are that the visual system is able to maintain a lightness scale over more than 3 log units of luminance, much larger than what is needed to represent variations in scene reflectance, and that models based on luminance ratios or Weber contrast can't account for the data.

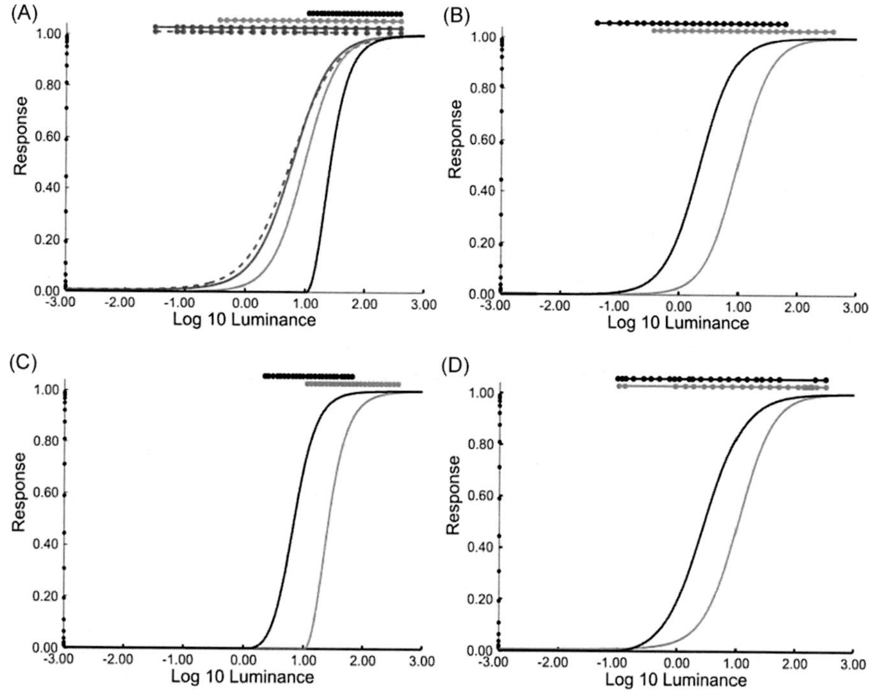

FIGURE 5.15 Lightness functions adapt to context luminance

Lightness functions adapt to context luminance. Figure from [32].

In short, the most important message here is that the brightness function depends on the image itself, shifting with the mean and changing the slope with the DR. This is the same behaviour observed for neural responses as we commented in Chapter 4, and is key for tone mapping applications, as we will see in Chapter 9.

5.9 Perceptual linearisation: gamma, log, PQ and HLG

We have seen that the brightness function has a compressive nature, meaning that its slope is higher for darker values, and gets progressively lower as the luminance increases. As a consequence, a small change in luminance in a dark region will be more noticeable than the same amount of change in luminance in a bright region.

Camera sensors transform light into numerical values that are proportional to the light intensity. The camera signal has to be quantised into a certain number N of bits in order to provide a digital output, and the quantisation process introduces errors. The maximum quantisation error is half the width of a quantisation interval, $1/2^{N+1}$ for a normalised signal, and obviously the higher N is, the smaller the quantisation errors will be.

When the signal is shown on a display, the errors will appear as luminance fluctuations. If a fluctuation is larger than a JND, i.e. if the quantisation error ΔI_Q divided by the luminance I is higher than the Weber fraction $k = \Delta I / I$, then it becomes visible. We can select N so that quantisation errors are below the Weber threshold. An insufficient bit depth N produces visual artifacts known as "banding", with smooth transitions in the image being represented due to poor quantisation as piecewise uniform regions or bands, see Fig. 5.16.

FIGURE 5.16 Quantisation with too few levels produces banding artifacts

Left: original image. Right: after quantisation with a low bit depth.

If the camera signal were just a scaled version of the scene luminance, then the quantisation errors would be more visible on the darkest parts of the image, due to the compressive form of the brightness function. Therefore, the optimum number of bits would be set by the darkest image levels, even though less bits would be enough for the brightest parts of the image. We can see that this type of coding would be inefficient.

What is actually done in practice is to apply to the camera signal, *before quantisation*, a nonlinear transform that mimics a brightness function: this process is called perceptual linearisation. In this way, the camera signal becomes proportional to perceived brightness, and now a fluctuation of some amount in a dark region will have the same visibility as a fluctuation of the same amount in a bright region.

With perceptual linearisation the quantisation intervals can become larger without introducing visible artifacts in the low luminance values, and therefore the number of bits N can be smaller than if we simply quantised a signal that were proportional to light intensity.

The nonlinear function that is applied in the camera before quantisation is called in the industry the Opto-Electrical Transfer Function (OETF). We'll briefly discuss the most popular ones: gamma correction, for SDR content, and logarithmic encoding, PQ and HLG for HDR image sequences.

5.9.1 Gamma correction

Gamma correction is simply a power law transform, except for low luminances where it's linear so as to avoid having an infinite derivative at luminance zero. This is the traditional nonlinearity applied for encoding SDR images. The exponent or "gamma", as specified in the industry standard BT.709, has a value of 0.45 but in fact the linear portion of the lower part of the curve makes the final gamma correction function be closer to a power law of exponent 0.5 [33], i.e. a square root transform: therefore, gamma correction complies with the DeVries-Rose law of brightness perception. Nonetheless we should note that camera makers often modify slightly the gamma value and it is a topic of research how to accurately estimate gamma from a given picture, e.g. see [34] and references therein.

It's standard to encode the gamma corrected image using 8 bits per channel, despite the fact that this bit depth is barely enough for SDR images and produces banding in the lowlights. For this reason, 10 bits are used in TV production, although transmission is done with 8 bits.

5.9.2 Logarithmic encoding

Although gamma correction has been the most used encoding technique, it fails when working with high dynamic range (HDR) imaging, due to quantisation issues. For high luminance values Weber–Fechner's law is a very good model, which as we have seen yields a logarithmic form for the brightness perception function. For this reason, cinema cameras perform perceptual linearisation by replacing gamma correction with a logarithmic function whose general form, common to the most popular log-encoding approaches [35,36], can be expressed as:

$$I_{out} = c \log_{10} (a \cdot A \cdot I_{lin} + b) + d, \tag{5.27}$$

where I_{out} is the encoding output, I_{lin} is the linear image captured by the sensor and demosaicked, and the parameters a, b, c, and d are constant real values (varying for different camera manufacturers and camera settings).

5.9.3 PQ: Perceptual Quantisation

The Perceptual Quantisation (PQ) approach, proposed by Miller et al. [37], is actually a family of curves. Each curve depends on the peak luminance L_W of the display where the image is going to be presented. Given L_W, the PQ function is derived in the following way.

We saw earlier how the Barten CSF is a good fit of experimental data on contrast sensitivity as a function of spatial frequency. For any given luminance level L_i the Barten formula gives us a CSF, which has a peak since the CSF is bandpass. The inverse of that peak sensitivity is therefore the minimum contrast c_i that can be detected on an image of luminance L_i: recall how experiments for determining the CSF consist in presenting to the observer a grating of contrast c and mean L over a uniform background of level L.

Let's find now the value of luminance L_{i-1} that is one JND below L_i. Per the definition of Michelson contrast

$$c_i = \frac{L_i - L_{i-1}}{L_i + L_{i-1}}, \tag{5.28}$$

and therefore

$$L_{i-1} = L_i \frac{1 - c_i}{1 + c_i}. \tag{5.29}$$

Alternatively, the luminance L_{i+1} that is one JND above L_i can be computed as:

$$L_{i+1} = L_i \frac{1 + c_i}{1 - c_i}. \tag{5.30}$$

Let the perceptual linearisation transform V be a function $V(L)$ in the range [0, 1]. Given N bits for encoding the transform, the total number of quantisation steps is $S = 2^N$. If we assign to L_i the quantised code value V_i, perceptual linearisation would mean assigning to L_{i-1} the code value that is a single quantisation step below V_i, i.e. $V_{i-1} = V_i - 1/S$. We can proceed iteratively in this manner, for instance starting with L_S as the peak luminance L_W and assigning to it the code value $V_S = 1$, next finding L_{S-1} that will have value $V_{S-1} = 1 - 1/S$, next L_{S-2} with corresponding value $V_{S-2} = V_{S-1} - 1/S$, etc. until we reach V_1, at which point the transform $V(L)$ has been completely specified.

It isn't necessary to take steps of one JND, we could use instead a fraction f of the contrast c_i at each iteration which allows to play with the total achievable range (see Miller et al. [37] for details). For instance with $f = 0.9$ and $N = 12$, the PQ curve can represent values from 10^{-6} to 10, 000 cd/m^2. In fact the PQ encoding function for 10, 000 cd/m^2 is the one proposed in [37] as the better option overall, for its balance between quantisation performance and dynamic range capability; but we must point out that, to the best of our knowledge, the Barten CSF on which PQ is based was not validated on experimental data of more than 500 cd/m^2 [21].

By building the perceptual linearisation function in this way, it is explicitly ensured that quantisation errors will not be visible, as they will always be at the JND threshold. A similar approach based on tracking CSF peaks was previously used in [38].

Tests with challenging synthetic images as well as with more representative natural HDR images showed that PQ consistently requires less bits than gamma correction to avoid banding artifacts, and for natural images 10 bits appear to be sufficient.

The perceptual linearisation function corresponding to PQ and computed as explained above can be expressed in the form of a Naka-Rushton equation:

$$V(L) = k_1 \left(\frac{c_1 + c_2 (\frac{L}{k_0})^{m_1}}{1 + c_3 (\frac{L}{k_0})^{m_1}} \right)^{m_2}. \tag{5.31}$$

An important point to make is the following: by construction, *the PQ encoding curve is not a brightness perception function*. It can't be, because the curve is built assuming maximum sensitivity at each and every luminance level, from black to white. This would be equivalent to a brightness function with crispening (maximum sensitivity) not just near the background level, but *with crispening at all luminance levels*, which can't possibly be reconciled with the psychophysical data on brightness perception and crispening of Section 5.4.

Furthermore, the PQ encoding curve is too steep to predict the brightness perception data that can be adequately modelled with Stevens' law or even with Weber–Fechner's law. This might seem surprising at first glance, because we saw how the Weber–Fechner function was obtained by integrating JNDs, and the PQ curve is also created through the integration of JNDs. But there is a very important difference: in Weber–Fechner there is a single, constant adaptation luminance, which is a reasonable approximation to brightness perception and light adaptation in many scenarios, while for PQ the adaptation luminance is allowed to vary locally at each pixel and over the whole luminance range, and this is not a model of how we perceive brightness in any scenario, not even for synthetic stimuli.

The fact that the PQ curve does not model brightness perception has important consequences for encoding the luminance of HDR images, and for the image quality metrics that are based on PQ, as we'll discuss below, but it also impacts colour representation and display metrology in HDR since PQ is the basis of a popular colour space for HDR content called IC_TC_P, which we will touch upon in Chapter 6.

5.9.4 HLG: Hybrid Log-Gamma

The Hybrid Log-Gamma (HLG) transfer function [25] is a single curve based on classic brightness perception models and designed to be backward compatible with SDR displays.

As discussed in the case of the logarithmic encoding, for high luminance values the Weber fraction is constant and therefore Weber–Fechner's law is a good approximation to brightness perception. This gives a logarithmic form for the HLG curve for the upper part of the luminance range.

For low luminance values we saw that the Weber fraction is no longer constant but depends on luminance, and the DeVries-Rose law provides a better model of brightness perception, which gives a square root form for the lower part of the HLG curve.

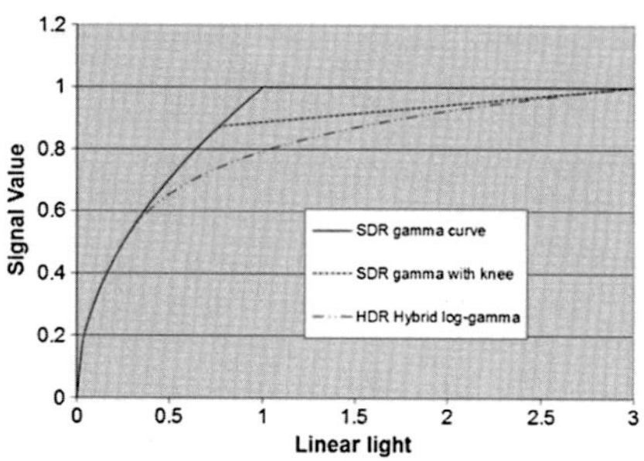

FIGURE 5.17 HLG and gamma correction

HLG and gamma correction. Figure from [25].

The HLG function is expressed as:

$$HLG(I) = \begin{cases} \sqrt{3I}, & 0 \leq I \leq \frac{1}{12} \\ a \cdot log(12I - b) + c, & \frac{1}{12} < I \leq 1 \end{cases} \tag{5.32}$$

where I is proportional to relative light intensity in a camera colour channel (R, G or B).

This has a key added advantage which is to make the signal compatible with regular SDR displays, that expect a gamma-corrected signal. We recall how the standard exponent for gamma correction is effectively 0.5, so an HDR signal encoded with HLG can be seen as it is on any legacy SDR display: for the lower part of the luminance range, corresponding to diffuse surfaces and where most of the image information tends to be, the HDR signal encoded with HLG and the SDR signal encoded with gamma correction will be almost identical. See Fig. 5.17. The highlights and brightest regions (corresponding to the logarithmic part of the HLG curve) will appear overexposed on a SDR display receiving the HDR signal, as expected.

Borer and Cotton [25] conducted validation experiments to assess the visibility of banding artifacts on highly critical 10-bit horizontal "shallow ramps", where adjacent regions varied by just one quantisation level, presented on a HDR display. Their results show that, for this type of material, the performance of HLG is marginally below that of PQ but both provide the same degree of visual impairment, described as "imperceptible" or "perceptible but not annoying" at their very worst. The authors conclude that in practice it's highly unlikely that any banding be visible on natural scenes.

5.9.5 Comparing PQ and HLG

PQ and HLG have been standardised for HDR television [39] and all current HDR devices (cameras, displays, production equipment) either support the two approaches or at least one of them. Both PQ and HLG have been proposed by companies, PQ by Dolby in the US and HLG by public broadcasters BBC in the UK and NHK in Japan. There are simple algorithms to convert from PQ to HLG and from HLG to PQ, and if the converted image is displayed at the same brightness as the original they should look identical [40].

As explained above, the motivation behind PQ was to guarantee that quantisation does not introduce visible artifacts. For this, the perceptual linearisation curve was built to withstand the worst possible case at all possible luminance levels: this is ensured by tracking the peak of the Barten CSF. We say that it's a worst case scenario because:

- The images used for determining the CSF are synthetic stimuli consisting of gratings on top of a uniform background; in this setting, thresholds can be considerably smaller than if we used natural images, where the influence of the background and surround produces a phenomenon termed *contrast masking* that reduces the visibility of fluctuations.

- The Barten CSF has a correction function for the surround, that makes the CSF considerably smaller if the surround luminance is very different from the average image luminance, but PQ ignores this factor.

Furthermore, and as discussed above in Section 5.9.3, this worst case scenario is not realistic, it corresponds to a situation that can't happen in practice, that of a brightness function with maximum sensitivity (crispening) at each and every luminance level.

In consequence, the PQ encoding function might be too conservative for HDR encoding of regular HDR content, meaning that in natural HDR images other curves might provide the same visual quality with less bits, and conversely that other curves could yield more visual quality with the same number of bits. We will discuss a step in this direction in Chapter 10, Section 10.3, where we use for perceptual linearisation a transfer curve coming from a tone mapping method based on natural image statistics and concepts of adaptation and efficient representation, detailed in Chapter 9.

There are more issues of PQ for cinema and TV production [25,40], like its high slope for lowlights which can enhance the noise, the need to specify the display already at the camera, and, especially, the fact that it is a display-referred approach (based on absolute luminance levels from the display) instead of the scene-referred approach (based on relative levels from the scene, like HLG) preferred by high-end cinema productions, in great part because this is essential when there is substantial visual effects work.

The fact that PQ is display referred has a very important consequence. Because the surround luminance has an impact on brightness perception, the brightness and contrast settings of an HDR display should be adjusted according to the viewing environment. But this is not possible when the HDR content comes with PQ encoding that enforces absolute luminance levels from the display, regardless of the surround and assuming a very dim environment luminance of 5 cd/m^2. As a result, HDR viewing with PQ encoding requires, for a high quality visual experience, that the viewing environment is quite dark, otherwise the images will look no different than in a SDR display. These are the results of the comparisons reported in a very interesting study [41], whose authors conclude that HDR (with PQ) is *"too dim for daytime"*.

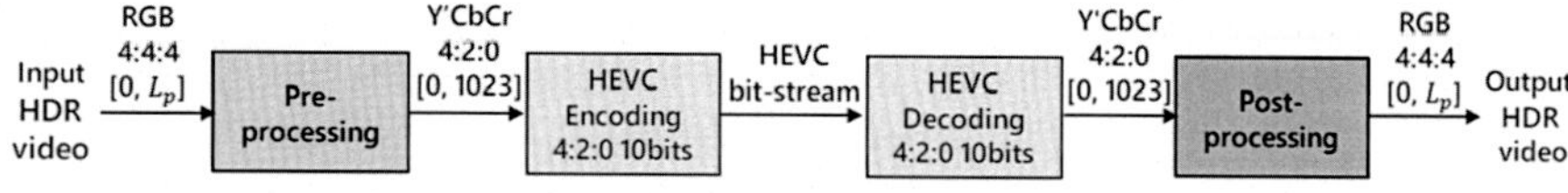

FIGURE 5.18 Overview diagram of HDR video coding

Overview diagram of HDR video coding. Figure from [42].

HDR video is always compressed, and a perceptual linearisation transfer function (TF) is an essential part of the process. Fig. 5.18 depicts an overview of the HDR video coding method. The HDR video input uses linear light intensity captured by a camera sensor. The format is RGB 4:4:4, and the range of the pixel values is from 0 to the peak luminance L_p, measured in cd/m^2. The pre-processing stage applies a TF like PQ or HLG, which essentially transforms the HDR image content into regular SDR image content. Then the image is converted to the opponent colour space Y'CbCr and the chroma channels are subsampled to 4:2:0 in 10 bits. The HEVC encoding block performs lossy compression, then the signal is transmitted, then it's uncompressed in the HEVC decoding block, and in the final post-processing block the signal is transformed back to RGB and the inverse of the TF is applied to it to produce the output HDR video.

The goal is of course to be able to reduce the transmission bitrate, at the same time ensuring that the output HDR video looks like the input HDR video as much as possible. The visibility of differences between input and output HDR videos can be assessed in two ways:

1. Ideally, performing subjective evaluation experiments, where observers compare the output to the input and rate the result by giving a score. These experiments are quite cumbersome, as they require a proper room with specialised equipment, time and human resources to arrange and perform the tests, etc.
2. Therefore, objective quality metrics are preferred in practice. Metrics take as input two images, the original HDR content (i.e. without compression) and the output HDR content that has gone through the HDR video coding pipeline (i.e. the HDR image reconstructed from the compressed signal), and provide a numerical output that is supposed to match the score that a human observer would give. For this purpose, metrics are based on models of visual perception, and validated through subjective evaluation experiments.

The two most popular metrics for HDR coding are arguably HDR-VDP2 [43] and HDR-VQM [44]. They both have a perceptual linearisation stage that emulates brightness perception and that is based on integrating thresholds coming from a CSF (using Barten's formula or one of its variants), exactly as PQ does. A number of works (see [45] and references therein) have found that both HDR-VDP2 and HDR-VQM have a good correlation with observers' scores, and therefore that they are good metrics. By applying these metrics to HDR video coded with PQ and comparing the numerical value with the result obtained for the same video encoded with HLG, both metrics yield a higher score for PQ and therefore according to them PQ is better for HDR video compression.

But in two recent works [46,45] we have put both these results into question.

First, in [46] we point out that experiments that benchmarked objective metrics for HDR coding and concluded that HDR-VDP2 and HDR-VQM were at the top of the ranking, were performed on images encoded with PQ. We encoded images with HLG and validate the performance of a number of metrics by conducting a subjective evaluation experiment. We found that the ranking of metrics changes drastically with that which was obtained when validating the metrics on PQ-encoded images: HDR-VDP2 and HDR-VQM are the best metrics according to [47], but in our tests HDR-VDP2 ranks third and HDR-VQM next to last.

Next, in [45] we argue that while previous studies have found a number of metrics that appear to correlate well with the preference of observers, those conclusions were based on experiments that do not correspond to the practical use of a current day HDR professional production scenario. So we carried out subjective evaluation experiments for PQ and HLG with a state-of-the-art HDR reference monitor used in broadcasting and post-production, and where all the observers participating in the tests are video experts. Again we found that the ranking of HDR metrics is now substantially different from what was reported earlier, with the performance of HDR-VDP2 and HDR-VQM not so good even for PQ-encoded images: a simple standard SDR metric like VIF [48], applied directly to PQ or HLG encoded signals, is shown to provide better results than those of HDR metrics.

Finally, a subjective evaluation commissioned by the European Broadcasting Union shows that for 1000 cd/m^2 displays the performance of HLG is at least equivalent or better than PQ in general [49].

References

[1] Shapley R, Enroth-Cugell C. Visual adaptation and retinal gain controls. Progress in Retinal Research 1984;3:263–346.

[2] Cornsweet TN, Pinsker H. Luminance discrimination of brief flashes under various conditions of adaptation. The Journal of Physiology 1965;176(2):294–310.

[3] Stevens SS. To honor Fechner and repeal his law. Science 1961;133(3446):80–6.

[4] Carter R. Gray scale and achromatic color difference. JOSA A 1993;10(6):1380–91.

[5] Billock VA, Tsou BH. To honor Fechner and obey Stevens: relationships between psychophysical and neural nonlinearities. Psychological Bulletin 2011 137(1):1.

[6] Heinemann EG. The relation of apparent brightness to the threshold for differences in luminance. Journal of Experimental Psychology 1961;61(5):389.

[7] Takasaki H. Lightness change of grays induced by change in reflectance of gray background. JOSA 1966;56(4):504–9.

[8] Kane D, Bertalmío M. A reevaluation of Whittle (1986, 1992) reveals the link between detection thresholds, discrimination thresholds, and brightness perception. Journal of Vision 2019;19(1).

[9] Whittle P. Increments and decrements: luminance discrimination. Vision Research 1986;26(10):1677–91.

[10] Whittle P. Brightness, discriminability and the "crispening effect". Vision Research 1992;32(8):1493–507.

[11] Blackwell HR. Contrast thresholds of the human eye. JOSA 1946;36(11):624–43.

[12] Belaıd N, Martens JB. Grey scale, the "crispening effect", and perceptual linearization. Signal Processing 1998;70(3):231–45.

[13] Moroney N. Background and the perception of lightness. 9th congress of the international colour association, vol. 4421. International Society for Optics and Photonics; 2002. p. 571–5.

[14] Carter R. Suprathreshold gray scale is implied by thresholds. Applied Optics 2018;57(29):8751–6.

[15] Kingdom F, Moulden B. A model for contrast discrimination with incremental and decremental test patches. Vision Research 1991;31(5):851–8.

[16] Bartleson C, Breneman E. Brightness perception in complex fields. Josa 1967;57(7):953–7.

[17] Campbell FW, Robson J. Application of Fourier analysis to the visibility of gratings. The Journal of Physiology 1968;197(3):551–66.

[18] Watson AB, Ahumada AJ. A standard model for foveal detection of spatial contrast. Journal of Vision 2005;5(9).

[19] Pelli DG, Bex P. Measuring contrast sensitivity. Vision Research 2013;90:10–4.

[20] Barten PG. Formula for the contrast sensitivity of the human eye. Image quality and system performance, vol. 5294. International Society for Optics and Photonics; 2003. p. 231–9.

[21] Barten PG. Contrast sensitivity of the human eye and its effects on image quality, vol. 21. Bellingham, WA: Spie Optical Engineering Press; 1999.
[22] Bartleson C. Optimum image tone reproduction. Journal of the SMPTE 1975;84(8):613–8.
[23] Liu C, Fairchild MD. Re-measuring and modeling perceived image contrast under different levels of surround illumination. In: Color and imaging conference, vol. 2007. Society for Imaging Science and Technology; 2007. p. 66–70.
[24] Singnoo J, Finlayson GD. Understanding the gamma adjustment of images. Color and imaging conference, vol. 2010. Society for Imaging Science and Technology; 2010. p. 134–9.
[25] Borer T, Cotton A. A display-independent high dynamic range television system. SMPTE Motion Imaging Journal 2016;125(4):50–6.
[26] Kane D, Bertalmío M. System gamma as a function of image–and monitor–dynamic range. Journal of Vision 2016;16(6).
[27] Fairchild MD. The hdr photographic survey. Color and imaging conference, vol. 2007. Society for Imaging Science and Technology; 2007. p. 233–8.
[28] Stevens J, Stevens SS. Brightness function: effects of adaptation. JOSA 1963;53(3):375–85.
[29] Arend LE, Goldstein R. Simultaneous constancy, lightness, and brightness. JOSA A 1987;4(12):2281–5.
[30] Arend LE, Spehar B. Lightness, brightness, and brightness contrast: 2. Reflectance variation. Perception & Psychophysics 1993;54(4):457–68.
[31] Blakeslee B, Reetz D, McCourt ME. Coming to terms with lightness and brightness: effects of stimulus configuration and instructions on brightness and lightness judgments. Journal of Vision 2008;8(11).
[32] Radonjić A, Allred SR, Gilchrist AL, Brainard DH. The dynamic range of human lightness perception. Current Biology 2011;21(22):1931–6.
[33] Poynton C. Digital video and HD: algorithms and interfaces. Elsevier; 2012.
[34] Vazquez-Corral J, Bertalmío M. Simultaneous blind gamma estimation. IEEE Signal Processing Letters 2015;22(9):1316–20.
[35] Brendel H. ALEXA Log C Curve – Usage in VFX. Tech. rep., ARRI; 2009.
[36] Sony Corporation. S-Log White Paper, S-Log within Digital Intermediate workflow designed for cinema release. Tech. rep., SONY; 2009.
[37] Miller S, Nezamabadi M, Daly S. Perceptual signal coding for more efficient usage of bit codes. SMPTE Motion Imaging Journal 2013;122(4):52–9.
[38] Mantiuk R, Daly SJ, Myszkowski K, Seidel HP. Predicting visible differences in high dynamic range images: model and its calibration. Human vision and electronic imaging x, vol. 5666. International Society for Optics and Photonics; 2005. p. 204–15.
[39] ITU-R. Image parameter values for high dynamic range television for use in production and international programme exchage. Rec BT2100 2016.
[40] Borer T, Cotton A, Pindoria M, Thompson S. Approaches to high dynamic range video. In: 2016 digital media industry & academic forum (DMIAF). IEEE; 2016. p. 71–6.
[41] https://www.hdtvtest.co.uk/news/4k-vs-201604104279.htm.
[42] Sugito Y, Cyriac P, Kane D, Bertalmío M. Improved high dynamic range video coding with a nonlinearity based on natural image statistics. In: 3rd international conference on signal processing (ICOSP 2017); 2017.
[43] Mantiuk R, Kim KJ, Rempel AG, Heidrich W. Hdr-vdp-2: a calibrated visual metric for visibility and quality predictions in all luminance conditions. ACM transactions on graphics (TOG), vol. 30. ACM; 2011. p. 40.
[44] Narwaria M, Da Silva MP, Le Callet P. Hdr-vqm: an objective quality measure for high dynamic range video. Signal Processing. Image Communication 2015;35:46–60.
[45] Sugito Y, Bertalmío M. Practical use suggests a re-evaluation of hdr objective quality metrics. In: Eleventh international conference on quality of multimedia experience (QoMEX 2019); 2019.
[46] Sugito Y, Bertalmío M. Performance evaluation of objective quality metrics on hlg-based hdr image coding. In: 2018 IEEE global conference on signal and information processing (GlobalSIP). IEEE; 2018. p. 96–100.

[47] Hanhart P, Bernardo MV, Pereira M, Pinheiro AM, Ebrahimi T. Benchmarking of objective quality metrics for hdr image quality assessment. EURASIP Journal on Image and Video Processing 2015;2015(1):39.
[48] Sheikh HR, Bovik AC. Image information and visual quality. 2004 IEEE international conference on acoustics, speech, and signal processing, vol. 3. IEEE; 2004. p. iii–709.
[49] Union EB. TR038, subjective evaluation of HLG for HDR and SDR distribution. tech. rep., EBU; 2017.

CHAPTER

Colour representation and colour gamuts

6

Colour is a perceptual quality, not a physical property of light.

There are models that for simple stimuli in controlled environments can predict very accurately the colour appearance of objects, as well as the magnitude of their colour differences. These models were developed and validated for SDR images, and their extension to the HDR case is not straightforward.

For the general case of natural images in arbitrary viewing conditions, there are many perceptual phenomena that come into play and no comprehensive vision model that is capable of handling them all in an effective way.

As a result, the colour appearance problem remains very much open, and this affects all aspects of colour representation and processing.

6.1 Light and colour

Light is defined as radiation with wavelengths within the *visible spectrum* of 380 nm to 740 nm. We are not able to see radiation outside this band, for example ultraviolet radiation (wavelength of 10 nm to 400 nm) or FM radio (wavelengths near 1 m).

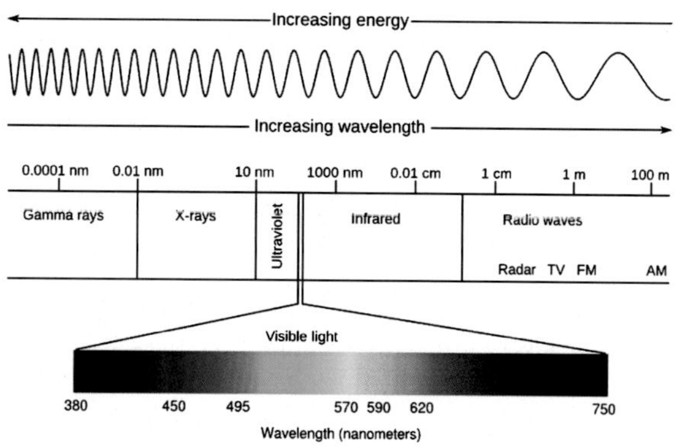

FIGURE 6.1

Electromagnetic spectrum and visible light.

Fig. 6.1 shows the full spectrum of radiation emitted by the sun, with a detail of the visible light spectrum. In this diagram, and as it's the usual practice, wavelengths

Vision Models for High Dynamic Range and Wide Colour Gamut Imaging. https://doi.org/10.1016/B978-0-12-813894-6.00011-9

in the visible spectrum are represented with colours, going from blue to red with increasing wavelength. This sort of colour-coding of wavelengths comes from the fact that, when we look at a monochromatic (i.e. single wavelength) light that is in isolation, if the light has a short wavelength we perceive it as blue, if it has middle-length wavelength we see it as green, if it has long wavelength it appears to us as red, etc.

But we must stress that *light in itself is not coloured*, colour is a perceptual quality and not a physical quantity. In particular: colour is not determined by wavelength. Monochromatic long wavelength light appears to be red when in isolation, but it can appear to be yellow in some specific surround, or orange, or pink, or even gray. See Fig. 6.2 for some examples.

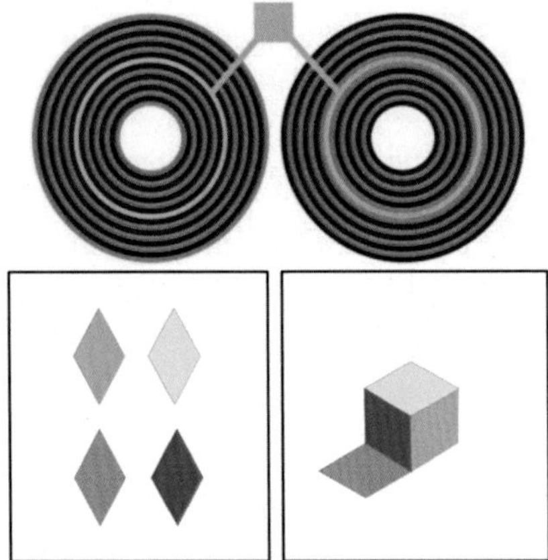

FIGURE 6.2 Wavelength does not determine colour

Top: the inner rings are identical, yet they appear to us as having different colours (from [1]). Bottom left: diamond-shaped flat surfaces of different colours. Bottom right: after arranging the surfaces into a cube, now they appear to have the same colour.

This is actually not a limitation but a key property of the visual system that provides a fundamental ecological advantage: because if colour was determined just by wavelength, objects would constantly change colour. Thanks to the influence of the context, colour perception is related to the reflectance properties of materials and we can see an object as having the same colour whether we look at it under the sun or indoors under fluorescent lighting, even though the wavelength composition of the light this object reflects is very much different in both cases.

A colour is usually characterised by three attributes:

- its hue, what we normally call “colour”: yellow, red, and so on;
- its saturation, which refers to how “pure” it is as opposed to “how mixed with white” it is: for instance, the colour red (as in blood-red) is more saturated than the colour pink (red mixed with white);
- its brightness/lightness, which expresses the intensity with which we perceive the colour.

The colour appearance of a non-emitting object depends on four factors: the spectral distribution of the illuminant, the reflectance of the object, the visual response functions of the observer, and "some as yet incompletely formulated aspect of the scene within which the object is presented" [2].

In the simple case of a flat, opaque object surrounded by a middle gray background and viewed in daylight, the colour perception problem is *solved*: we can predict quite accurately the hue, lightness and saturation of the object's perceived colour with the tools of colourimetry, as we shall see in the following section. Colourimetry is the branch of colour science concerned with specifying numerically the colour of a physically defined visual stimulus [3].

If one or more of the above assumptions is removed, though, then the problem is very much *open*, which is what happens in the general case, for natural images. But still in this situation, colourimetry is essential. It doesn't tell us what colour a visual stimulus will have, but it does tell us that if two stimuli have the same colourimetric specification (a triplet of numbers), and they are viewed in the same conditions, then they will appear identical. For instance, this is why it makes sense to use colourimetric specifications in filmmaking: while colourimetry can't say how a given object will appear in the screen, it does say that it will appear as decided in postproduction if viewed in the same conditions (and this is why movie postproduction is performed in conditions that emulate those of the cinema theatre or home viewing).

6.2 Trichromacy and colour matching

Trichromacy is a fundamental property of human colour vision. Simply put, it means that we can generate any colour by mixing three given colours, merely adjusting the amount of each. In his excellent account of the origins of colour science, Mollon [4] explains how this was a known fact and how it was already applied in the 18th century, for printing in full colour using only three types of coloured ink.

Trichromacy is due to our having three types of cone photoreceptors in the retina, therefore we must remark that it is *not* a property of light but a property of our visual system. This was not known in the 18th century: light sensation was supposed to be transmitted directly along the nerves, so trichromacy was thought to be a physical characteristic of the light. Thomas Young was the first to explain, in 1801, that the variable associated with colour in light is the wavelength and, since it varies continuously, trichromacy must be imposed by the visual system and hence there must be three kind of receptors in the eye.

The principle of trichromacy can be stated also in terms of neural responses: any isolated light stimulus is encoded by the visual system into three values, the cone responses. Two important consequences are, first, that we can have physically different light stimuli producing the same triplet of cone responses, in which case the lights will look the same colour, and second, that if two lights have the same colour within a shared context they will also look identical to one another if the shared context is changed, because if the neural responses are the same at the first stage of vision, they are interchangeable [5].

Based on the seminal work of Maxwell, in the late 1920's Wright and Guild conducted experiments in which they asked observers to colour-match a given monochromatic light by varying the intensity of a set of red, green and blue monochromatic lights [6], called the primaries. For each monochromatic test light of wavelength λ, the experiment recorded the (average, over all observers) amounts of each primary needed to match the test: $\bar{r}(\lambda)$ for the red (mid gray in print version), $\bar{g}(\lambda)$ for the green (light gray in print version), and $\bar{b}(\lambda)$ for the blue (dark gray in print version). These are the colour matching functions (CMFs) for a standard observer.

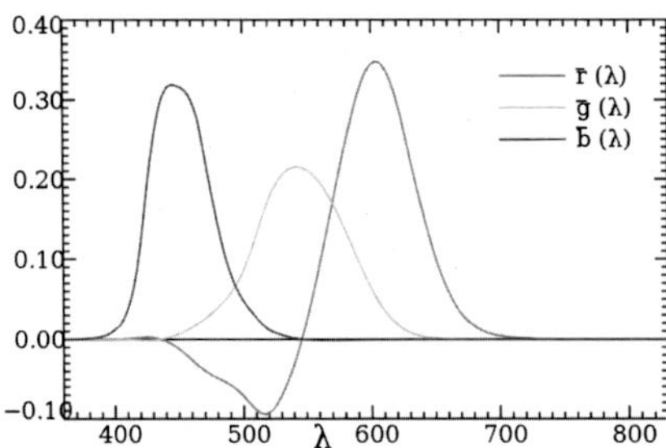

FIGURE 6.3 Colour matching functions

Colour matching functions. Figure from [7].

We can see in Fig. 6.3 that the $\bar{r}(\lambda)$ function clearly has some negative values (the other two functions have negative values as well). These correspond to cases when the test colour can't be matched as it is, unless a certain amount of primary colour (red) is added to it: after that, the green and blue lights can be adjusted so as to match the combination of test light and red.

The CMFs were derived for monochromatic test lights, but there are some linearity laws (Grassman's laws, from the mid 19th century) that we can apply [3], stating that:

- If light A matches light B and light C matches light D, then the additive mixture $A + C$ matches the additive mixture $B + D$.
- If light A matches light B then αA matches αB, with α some positive scaling factor by which the power of the light is reduced or increased.

Given a monochromatic light E_{λ_i} of wavelength λ_i, the CMFs tell us that its colour is the same as that of the additive mixture $\bar{r}(\lambda_i)\mathbf{R} + \bar{g}(\lambda_i)\mathbf{G} + \bar{b}(\lambda_i)\mathbf{B}$, where **R**, **G**, **B** are the primaries.

Any given light stimulus Q can be expressed as a sum of monochromatic lights Q_{λ_i}, each of which is just the monochromatic light E_{λ_i} scaled by the factor $Q(\lambda_i)$. Therefore, each Q_{λ_i} matches in colour the additive mixture $\bar{r}(\lambda_i)Q(\lambda_i)\mathbf{R} + \bar{g}(\lambda_i)Q(\lambda_i)\mathbf{G} + \bar{b}(\lambda_i)Q(\lambda_i)\mathbf{B}$.

From the two linearity laws above, we can see that Q has the same colour as the additive mixture $R\mathbf{R} + G\mathbf{G} + B\mathbf{B}$, where

$$R = \int_{380}^{740} \bar{r}(\lambda)Q(\lambda)d\lambda$$
$$G = \int_{380}^{740} \bar{g}(\lambda)Q(\lambda)d\lambda$$
$$B = \int_{380}^{740} \bar{b}(\lambda)Q(\lambda)d\lambda \tag{6.1}$$

This is a fundamental result, that says that we can represent the colour of any light stimulus Q by the triplet or tristimulus value (R, G, B).

Furthermore, it's easy to show that two physically different visual stimuli (with different power distributions) can produce the same tristimulus (R, G, B), in which case they will be seen as being the same colour. These stimuli are called metamers.

Another consequence of the linearity principles is that if light Q is associated to tristumulus (R, G, B), then light αQ (with α a positive scaling factor) will be associated to tristimulus $(\alpha R, \alpha G, \alpha B)$.

Scaling the intensity of the stimulus by α simply scales the corresponding tristimulus value by the same factor, but the chromatic information remains the same. For this reason, and since the sum $R + G + B$ corresponds to the intensity of the stimulus, it's usual to decouple luminance from chromaticity through normalisation of the tristimulus by the intensity

$$r = \frac{R}{R+G+B}$$
$$g = \frac{G}{R+G+B}$$
$$b = \frac{B}{R+G+B} \tag{6.2}$$

As a result, $r + g + b = 1$ and hence having r, g is enough to represent the chromatic information (as b can always be recovered from r, g). Fig. 6.4 presents an rg chromaticity diagram, i.e. a planar representation of the r and g components of all colours.

The boundary horseshoe-shaped curve corresponds to monochromatic lights, and therefore any point $(r_{\lambda_i}, g_{\lambda_i})$ along that curve is simply the value of the colour matching functions: $(\bar{r}(\lambda_i), \bar{g}(\lambda_i))$. Notice how for short wavelengths the curve goes into the half-plane where $r < 0$, because for those wavelengths $\bar{r} < 0$ as we saw in Fig. 6.3.

From the linearity principles, if some light stimulus Q is an additive mixture of monochromatic lights E_1 and E_2, then the chromaticity coordinates (r_Q, g_Q) of Q will be in the segment that connects the coordinates (r_1, g_1) and (r_2, g_2) of E_1 and E_2 respectively. And since any light stimulus can be decomposed into monochromatic lights, any light stimulus is represented in this diagram *inside* the tongue-shaped region: there are no colours outside it.

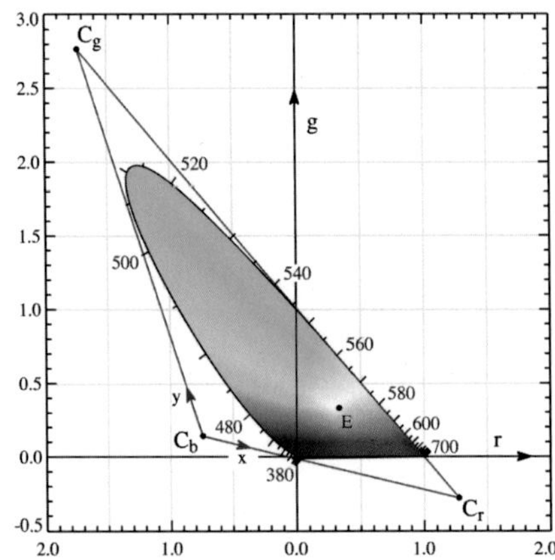

FIGURE 6.4 The rg chromaticity diagram

The rg chromaticity diagram. Figure from [8].

For white light $R = G = B$ and therefore the chromaticity of white is the point $(\frac{1}{3}, \frac{1}{3})$.

6.3 The first standard colour spaces

Recapping, a colour sensation can be described with three parameters; given a test colour, we call its *tristimulus* values the amounts of three colours (primaries in some additive colour model) that are needed to match that test colour. If two single, isolated coloured lights have different spectral distributions but the same tristimulus values, then they will be perceived as being of the same colour.

A *colour space* is a method that associates colours with tristimulus values. Therefore, it is described by three primaries and their corresponding colour matching functions. Given their relationship to tristimulus values, colour spaces are three-dimensional: each colour can be represented as a point in a three-dimensional plot.

In 1931, the International Commission on Illumination (or CIE, for its French name) amalgamated Wright and Guild's data [9] and proposed two sets of colour matching functions for a standard observer, known as CIE RGB and CIE XYZ; this standard for colourimetry is still today one of the most used methods for specifying colours in the industry. The CIE RGB colour matching functions are the functions $\bar{r}(\lambda)$, $\bar{g}(\lambda)$, $\bar{b}(\lambda)$ mentioned earlier. The tristimulus values (R, G, B) for a light $E(\lambda)$ are computed from these functions as stated in Eq. (6.1).

For each wavelength λ, one of the three functions is negative. This posed a problem, since the calculators of the time were manually operated and hence errors were quite common in the computation of the tristimulus values [10]. That's why the CIE XYZ colour matching functions $\bar{x}(\lambda)$, $\bar{y}(\lambda)$, $\bar{z}(\lambda)$ were also introduced alongside the

CIE RGB ones. From the CIE XYZ functions, the (X, Y, Z) tristimulus values for a light source with spectral distribution $E(\lambda)$ can be computed as:

$$X = \int_{380}^{740} \bar{x}(\lambda)E(\lambda)d\lambda$$
$$Y = \int_{380}^{740} \bar{y}(\lambda)E(\lambda)d\lambda$$
$$Z = \int_{380}^{740} \bar{z}(\lambda)E(\lambda)d\lambda. \tag{6.3}$$

The colour matching functions $\bar{x}(\lambda)$, $\bar{y}(\lambda)$, $\bar{z}(\lambda)$ are obtained as a linear combination of $\bar{r}(\lambda)$, $\bar{g}(\lambda)$, $\bar{b}(\lambda)$ by imposing certain criteria, chiefly among them:

- $\bar{x}(\lambda)$, $\bar{y}(\lambda)$, $\bar{z}(\lambda)$ must always be positive;
- $\bar{y}(\lambda)$ is identical to the standard luminosity function $V(\lambda)$, which is a dimensionless function describing the sensitivity to light as a function of wavelength; therefore, $Y = \int \bar{y}(\lambda)E(\lambda)d\lambda$ would correspond to the luminance of the colour stimulus;
- $\bar{x}(\lambda)$, $\bar{y}(\lambda)$, $\bar{z}(\lambda)$ are normalised so that they produce equal tristimulus values $X = Y = Z$ for a white light, i.e. a light with a uniform (flat) spectrum.

Because the CMFs of CIE XYZ are a linear transformation of the CMFs of CIE RGB, this means that we can go from one colour space to the other with a linear, invertible transform:

$$\begin{bmatrix} X \\ Y \\ Z \end{bmatrix} = \mathbf{M} \begin{bmatrix} R \\ G \\ B \end{bmatrix} \tag{6.4}$$

$$\begin{bmatrix} R \\ G \\ B \end{bmatrix} = \mathbf{M}^{-1} \begin{bmatrix} X \\ Y \\ Z \end{bmatrix} \tag{6.5}$$

where $\mathbf{M}$ is a 3×3 matrix.

An important point we must stress is the following. It can be shown that for any set of physically realisable primaries there are wavelengths λ for which the colour matching values are negative. Since $\bar{x}(\lambda)$, $\bar{y}(\lambda)$, $\bar{z}(\lambda)$ are always positive, this implies that their primaries can never be physically realisable. This is why the primaries for CIE XYZ are called *virtual primaries.*

We now define the values x, y, z:

$$x = \frac{X}{X+Y+Z}$$
$$y = \frac{Y}{X+Y+Z}$$

$$z = \frac{Z}{X + Y + Z}. \tag{6.6}$$

It is easy to see that, for lights E_1 and $E_2 = \alpha E_1$, these values are identical: $x_1 = x_2$, $y_1 = y_2$, $z_1 = z_2$. This is why x, y, z are called the *chromaticity coordinates*, because they don't change if the light stimulus only varies its intensity. We will now see that for x, y, z we have the same properties that we mentioned for r, g, b in the previous section.

By construction $x + y + z = 1$, so $z = 1 - x - y$ and all the information of the chromaticity coordinates is contained in the pair (x, y). Therefore, all the possible chromaticities can be represented in a 2D plane, the plane with axes x and y, and this is called the CIE xy chromaticity diagram; see Fig. 6.5.

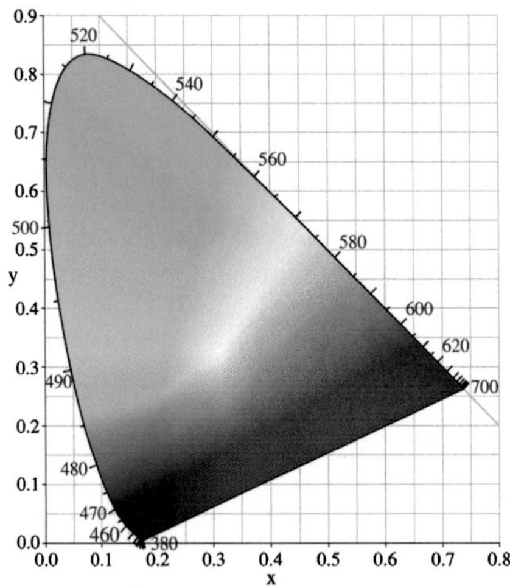

FIGURE 6.5 CIE xy chromaticity diagram

CIE xy chromaticity diagram. Figure from [7].

This tongue-shaped region represents all the differents chromaticities that can be perceived by a standard observer; it can be seen as the result of performing this operation: slicing the XYZ volume with the plane $X + Y + Z = 1$, then projecting the resulting plane onto the XY plane. See Fig. 6.6.

It is worth remarking that the triplet of values formed by chromaticity (x, y) and luminance Y perfectly describes a colour, and from (x, y, Y) we can obtain (X, Y, Z).

The upper boundary of the chromaticity diagram is a horseshoe-shaped curve corresponding to monochromatic colours: this curve is called the spectrum locus [10]. The lower boundary is the purple line, and corresponds to mixtures of lights from the extrema of the spectrum.

If monochromatic lights E_1 and E_2 have coordinates (x_1, y_1) and (x_2, y_2) (that will lie on the spectrum locus because the lights are monochromatic), the mixture $E_3 = E_1 + E_2$ will have coordinates (x_3, y_3) located in the segment joining (x_1, y_1)

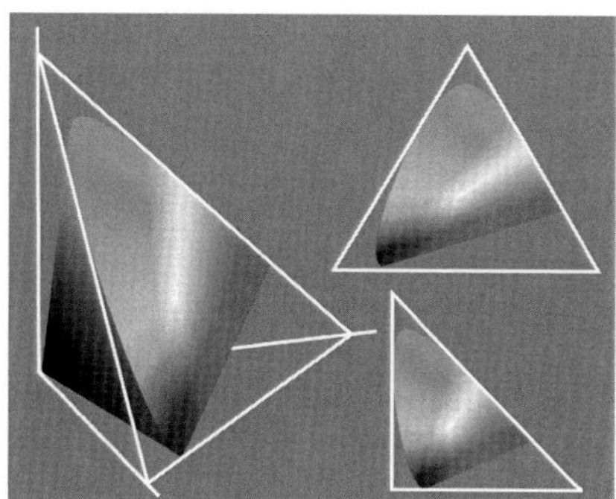

FIGURE 6.6 XYZ **volume and** xy **diagram**

Left: XYZ volume. Top right: after slicing volume with plane $X+Y+Z=1$. Bottom right: after projecting plane onto XY plane.

and (x_2, y_2). Therefore, the tongue-shaped region delimited by the spectrum locus and the purple line represents all the possible chromaticities that we can perceive, as mentioned above.

Helmholtz showed that each monochromatic light had a complementary, i.e. the mix of both lights yields white [4]. Monochromatic lights with wavelengths in the range between red and yellow–green have monochromatic complementaries with wavelengths in the range between blue–green and violets. The complementary of green is not a monochromatic light but purple, a mixture of blue and red light from the two ends of the visible spectrum.

Perfect white (i.e. light with a completely uniform power spectrum) has coordinates $x = y = \frac{1}{3}$, so as we mix a monochromatic light with white, its chromaticity coordinates move inwards and the saturation of the colours is reduced. A pure monochromatic light has 100% saturation while white has 0% saturation. But in practice, white lights never have a completely flat spectrum. The CIE has defined a set of standard illuminants: A for incandescent light, B for sunlight, C for average daylight, D for phases of daylight, E is the equal-energy illuminant, while illuminants F represent fluorescent lamps of various compositions [11]. The illuminants in the D series are defined simply by denoting the temperature in Kelvin degrees of the black-body radiator whose power spectrum is closer to that of the illuminant. A black-body radiator is an object that does not reflect light and emits radiation, and the power spectrum of this radiation is uniquely described by the temperature of the object. This is why in photography it is common to express the tonality of an illuminant by its *colour temperature*: a bluish white will have a high colour temperature, whereas a reddish-white will have a lower colour temperature. For instance, CIE illuminant D65 corresponds to the phase of daylight with a power spectrum close to that of a black-body radiator at 6500°K, and D60 to 6000°K. These two are the common illuminants used in cinema and TV standards.

Another very important consequence of the linearity property stated before is that any system that uses three primaries to represent colours will only be capable of representing the chromaticities lying inside the triangle determined by the chromaticities of the primaries. Furthermore, because of the convex shape of the chromaticity diagram, any such triangle will be fully contained in the diagram, leaving out chromaticity points. Therefore, for any trichromatic system there will always be colours that the system is not capable of representing.

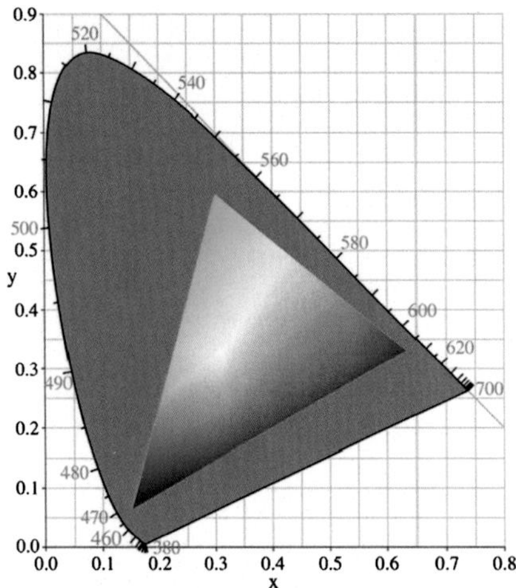

FIGURE 6.7 Chromaticities of a display

Chromaticities of a display. Image from [12].

For instance, Fig. 6.7 shows the chromaticity diagram for a display, where the vertices correspond to the chromaticities of the red (mid gray in print version), green (light gray in print version) and blue (dark gray in print version) light-emitting elements of the device. A display with more saturated primaries will have associated a larger triangle in the chromaticity diagram, and therefore will be able to reproduce more colours. In Section 6.6 we will return to the issue of characterising the range of colours achievable by a device, called its colour gamut.

6.4 Perceptually uniform colour spaces

The CIE XYZ colour space (and consequently the CIE xy chromaticity diagram as well) suffers from several limitations in terms of the perception of colours:

- the distance between two points in XYZ space or in the xy diagram is not proportional to the perceived difference between the colours corresponding to the points;
- a mixture of two lights in equal proportions will have chromaticity coordinates that do not lie exactly at the middle of the segment joining the chromaticities of the original two lights.

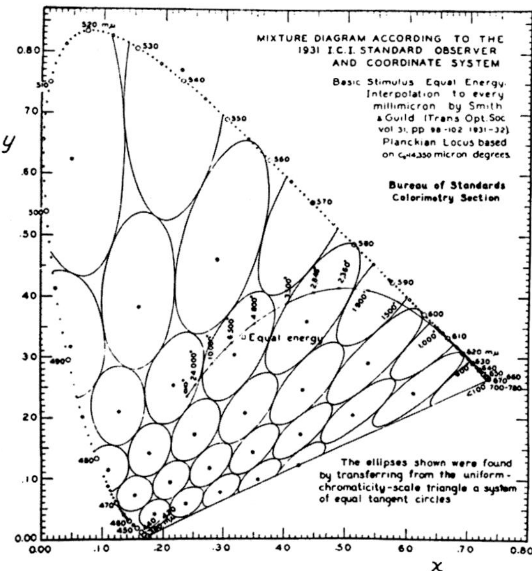

FIGURE 6.8 MacAdam ellipses

Ellipses representing the chromaticities of circles of equal size and constant perceptual distances from their center points. Figure from [13].

This can be observed in Fig. 6.8, where perceptual differences of the same magnitude are represented as ellipses in the CIE xy diagram, whereas if the CIE XYZ space were *uniform* we should have circles instead.

But in colour reproduction systems, perceptual uniformity is a very useful property because it allows us to define error tolerances, and therefore much work has been devoted to the developing of uniform colour spaces. Research was carried out independently in two lines: finding a uniform lightness scale and devising a uniform chromaticity diagram for colours of constant lightness [10].

In 1976 the CIE introduced the CIE 1976 $L^*a^*b^*$ colour space, abbreviated CIELAB. It was designed to be perceptually uniform, with the channel L^* representing lightness (perceived luminance), opponent channel a^* representing *red–green* response and opponent channel b^* representing *yellow–blue* response, where the chromaticity coordinates are chosen so that the Euclidean distance between two points in CIELAB space is proportional to the perceptual difference between the colours corresponding to those points.

First of all, CIELAB emulates adaptation to the ambient illuminant by performing a normalisation with respect to the tristimulus of a reference white. This is a crude approximation to the colour constancy property of the visual system, directly based on the von Kries' law [10]: if (R, G, B) and (R', G', B') represent the colours of the same object after full adaptation to two different lights, then these values are related by the formula

$$R' = k_r R, \quad G' = k_g G, \quad B' = k_b B, \tag{6.7}$$

where the coefficients k_r, k_g, k_b are one for each type of receptor and they are independent of wavelength. The von Kries' law provides a good approximation of the experimental facts as long as the change in chromatic adaptation is moderate and the luminance of test stimulus and surround are almost constant.

The coefficients can be computed by reasoning in the following way. Let (R_o, G_o, B_o) be the perceived colour of light source A when viewed under light B; after we adapt to A it becomes achromatic, therefore the values (R'_o, G'_o, B'_o) must be equal, $R'_o = G'_o = B'_o$, and we adjust k_r, k_g, k_b so that this holds.

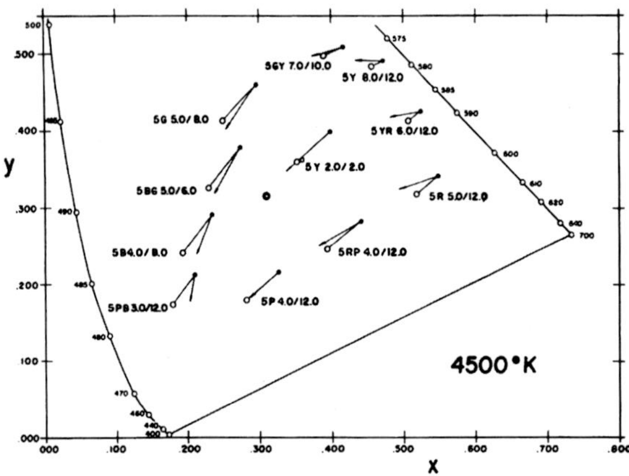

FIGURE 6.9 von Kries emulates colour constancy

Chromaticity points for 11 objects under CIE illuminant C (circles) and under a fluorescent lamp of colour temperature 4500°K (dots). Dots to arrow heads: adaptive colour shift computed with von Kries' law. Figure from [2].

Fig. 6.9 shows the chromaticity points for 11 objects under CIE illuminant C (circles) and under a fluorescent lamp of colour temperature 4500°K (dots). The segments going from circles to dots represent the colourimetric shift. It goes *away* from the blue region of the diagram, which is consistent with the fluorescent lamp having more energy in the blue region of the spectrum, hence adaptation to this light source makes objects look more reddish-yellow. The dots to arrow heads represent the adaptive colour shift computed with von Kries' law. The distance between circle and corresponding arrow head represents the resultant colour shift. We can see then that chromatic adaptation basically tries to counteract the colourimetric shift, resulting in an approximate constancy of colour perception. In some cases (like in the object with Munsell notation $5P4/12$ in the figure) the colour constancy is almost perfect, while in others it can be rather limited. For instance, the object $5Y8/12$ has a net shift towards green, which explains the greenish appearance given to butter by a cool-white fluorescent lamp. Worthey [14] explains how these limitations in colour constancy arise from the overlap of the cone response functions, when for von Kries' to produce perfect constancy these functions should be narrow and non-overlapping.

In short, von Kries' coefficient law models the adaptation process that produces colour constancy and makes us perceive illuminants as being approximately white; it is a very simple way to modify the chromaticity coordinates so that, in many situations, they correspond more closely to the *perception* of colour.

After applying the chromatic adaptation transform of von Kries to the (X, Y, Z) tristimulus, CIELAB applies a power-law nonlinearity of power $\frac{1}{3}$ to estimate the lightness. We recall from Chapter 5 that, under some experimental conditions, lightness is approximately proportional to the luminance raised to the power of $\frac{1}{3}$ [3].

Finally, the colour opponent signals a^* and b^* are derived. The theory of colour opponency was proposed by Hering in the late 19th century. He observed that there isn't any hue that can be described as a combination of red and green, nor as a combination of yellow and blue. Therefore, he postulated that there are two opponent axes for colour perception, one being the *red–green* axis and the other one the *yellow–blue* axis, and any colour can be represented by two values, which are the proportions of the colour on each opponent channel. For instance, *orange* could be expressed as 1 in both axes, as it is a mixture of *red* and *yellow*, while *purple* could be expressed as 1 in the *red–green* axis and -1 in the *yellow–blue* axis, since it is a mixture of *red* and *blue*.

If we recall from Chapters 2 and 3 that L-cones had their peak sensitivity at red, M-cones at green and S-cones at blue, then at first glance it would seem that the psychophysical theory of colour opponency has a direct biological correlate in the cone opponent signals originated in the retina, where $L - M$ would correspond to *red–green* and $S - (L + M)$ to *blue–yellow*. But the truth is that there isn't a straightforward correspondence: linear combination of cone responses is not enough to predict colour appearance from neurophysiological data, cone signals are combined in a non-linear way; please see [15] for a very interesting discussion on this topic.

The differences for the colour-opponent channels a^*, b^* are computed after applying the lightness perception nonlinearity on the normalised tristimulus values:

$$L^* = 116 f\left(\frac{Y}{Y_n}\right) - 16 \tag{6.8}$$

$$\begin{aligned} a^* &= 500\left(f\left(\frac{X}{X_n}\right) - f\left(\frac{Y}{Y_n}\right)\right) \\ b^* &= 200\left(f\left(\frac{Y}{Y_n}\right) - f\left(\frac{Z}{Z_n}\right)\right) \end{aligned} \tag{6.9}$$

$$f(x)s = \begin{cases} x^{\frac{1}{3}}, & \text{if } x > \left(\frac{6}{29}\right)^3 \\ \frac{1}{3}\left(\frac{29}{6}\right)^2 x + \frac{4}{29}, & \text{otherwise,} \end{cases} \tag{6.10}$$

where (X_n, Y_n, Z_n) are the tristimulus values of the reference white, the illuminant to which we are adapted.

In CIELAB the chromaticity coordinates (a^*, b^*) can be positive or negative: $a^* > 0$ indicates redness, $a^* < 0$ greenness, $b^* > 0$ yellowness and $b^* < 0$ blueness (see Fig. 6.10). For this reason it is often more convenient to express CIELAB colours in cylindrical coordinates $L^*C^*h^*$, where

- $C^* = \sqrt{a^{*2} + b^{*2}}$, the radius from the origin, is the *chroma*, which can be defined as the degree of colourfulness with respect to a white colour of the same brightness: decreasing C^* the colours become muted and approach gray; clearly, saturation S and chroma C^* are related, and in CIELAB saturation is defined as chroma over lightness, $S = C^*/L^*$.
- $h^* = arctan\frac{b^*}{a^*}$, the angle from the positive a^* axis, is the *hue*: a hue angle of $h* = 0°$ corresponds to red (mid gray in print version), $h* = 60°$ corresponds to yellow (light gray in print version), $h* = 120°$ corresponds to green (gray in print version), etc.

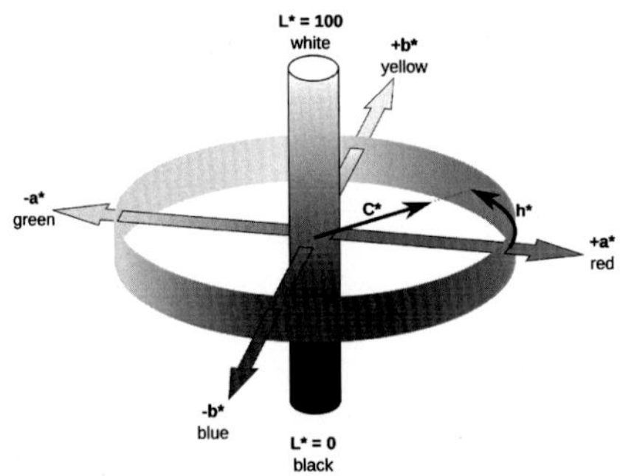

FIGURE 6.10

CIELAB colour space in both Cartesian and cylindrical coordinates.

Fairchild [16] points out that CIELAB represents rather well perceptual data on uniform scales for lightness, hue and chroma coming from the Munsell system, obtained from matches of diffuse reflective surfaces seen under standard illuminants. Plots of constant chroma in the a^*b^* plane should be circular, and plots of constant hue should be segments emanating radially from the point $(0, 0)$, and this is approximately the case for the Munsell data. But when using data from experiments done on emissive displays, that can achieve higher chroma, then the representation in CIELAB departs from the observed results, and constant hue segments become curved. In some parts of the colour space (mainly around blue) CIELAB suffers from *cross-contamination* [17]: changing only one attribute (such as hue) produces changes in the perception of another attribute (such as saturation). This is a consequence of the deficiencies of the system with respect to the correlates for hue [10]: the correlate for hue is the angle $arctan\frac{b^*}{a^*}$, and therefore constant hue should correspond to planes passing through the L^* axis, but what is observed experimentally are curved surfaces instead of planes. These surfaces depart more from the intended planes near the negative b^* axis, hence the problems around blue.

A colour space with a very good prediction of constant perceived hue is IPT, proposed by Ebner and Farichild [18]. Here the (X, Y, Z) tristimulus is first converted to cone tristimulus values (L, M, S) by means of a linear transform (multiplication by a 3×3 matrix), then a power law nonlinearity is applied to the cone responses, and a final linear transform yields the (I, P, T) tristimulus where I corresponds to lightness, P to $red\text{–}green$ opponency and T corresponds to $yellow\text{–}blue$ opponency.

6.5 HDR colour spaces

Both CIELAB and IPT, while very successful, were designed to match perceptual data on experiments performed using reflective surfaces, whose dynamic range is fairly limited, in the order of 30:1; therefore, their direct application to HDR imaging seems questionable. To overcome these limitations, a number of HDR colour spaces have been introduced.

In [19] Fairchild and Wyble proposed HDR versions of CIELAB and IPT, simply called hdr-CIELAB and hdr-IPT. In each case, the power law nonlinearity for lightness perception was replaced by a Naka-Rushton equation (alternatively called Michaelis–Menten equation), which we have seen in Chapters 4 and 5 is an adequate model of cone responses; the parameters of each HDR colour space were optimised so that they matched as closely as possible the original colour spaces in the SDR domain (i.e. for luminances below perfect diffuse white). A subsequent study by Fairchild and Chen in [20] shows that CIELAB and IPT maintain a reasonable performance, and remain approximately uniform, up to luminances that are 8 times higher than the diffuse white, while hdr-CIELAB and hdr-IPT both under-predict lightness in the range above diffuse white. They conclude that more experiments are necessary and that probably a different sigmoidal function, other than the Naka-Rushton equation, should be found to define a proper HDR colour space.

The IC_TC_P colour space [21], developed by Dolby, has a similar approach as hdr-IPT in that it can also be seen as an extension of IPT where the power law nonlinearity of the original has been replaced, in this case by the PQ encoding curve that we discussed in Chapter 5. In order to calculate IC_TC_P, the RGB tristimulus is first converted to LMS by a linear transformation, then the estimated cone responses are passed through the PQ encoding function, and finally another linear transform is applied to generate the lightness value I, the $yellow\text{–}blue$ opponency C_T and the $red\text{–}green$ opponency C_P.

Another HDR colour space that uses the PQ encoding function as the lightness nonlinearity is $J_za_zb_z$ [22]. Starting from XYZ tristimulus, it applies a linear conversion to LMS, then the PQ encoding function, then a linear transform to generate lightness J_z and chrominances a_z, b_z, followed by an adjustment of J_z for highlights.

We saw in Chapter 5 that the PQ encoding curve is not a model of brightness perception, it's a nonlinearity developed as a way to ensure that quantisation errors are below the threshold of visibility. For this reason, a common view in the colour science community is that $ICTCP$ and $J_z a_z b_z$, being based on PQ, can't be properly called perceptually uniform colour spaces, since their lightness function does not model lightness perception; they should be seen as encoding spaces.

6.6 Colour gamuts

The CIE defines colour gamut in this way: *"a range of colours achievable on a given colour reproduction medium (or present in an image on that medium) under a given set of viewing conditions – it is a volume in colour space."*

It is important to notice then that, strictly, devices do not have a unique colour gamut, since there are external factors such as the surround illumination that influence the way colours produced by the device are perceived. In practice, though, this consideration is ignored and for any given device a unique gamut is characterised, either by analytical methods that model the device's properties, or by measuring the positions in colour space of a sparse set of test colours from which a gamut boundary descriptor is estimated, see [17] for details.

Each device can produce luminances up to a maximum value Y_w, called peak luminance. For example the peak luminance of TV sets has traditionally been in the order of 100 cd/m^2. Normally the value of Y is scaled so that $Y_w = 1$ or $Y_w = 100$ in arbitrary units.

In the CIE XYZ colour space, Y is computed as $Y = 0.18R + 0.81G + 0.01B$ and we see that B has very little influence in the luminance. Therefore, for $Y_w = 1$ a colour of luminance up to $Y = 0.01$ can have any chromaticity and in particular it can be purely blue (i.e. it can have $R = G = 0$), but a colour of luminance higher than 0.01 *must* have some component of red or green because the blue component alone would not be enough to produce the intended luminance; as a result, as the luminance increases the blue colours become less saturated. On the other hand, we can have a pure red colour up to $Y = 0.18$ and pure green up to $Y = 0.81$. As Y increases, the possible colours reduce more and more their saturation, until when $Y = 1 = Y_w$ the colour can only be white. Beyond Y_w there is no gamut, as Y_w is the peak luminance of the display.

Most displays are three-primaries, RGB devices; there are also displays with RGB plus white, and displays with more than three primaries, but we won't be discussing them here. As a consequence, the 3D plot of the colour gamut of an RGB display, in xyY coordinates, must be a curved prism. It has a triangular base at $Y = 0$, like the one shown in Fig. 6.7, where the vertices of the triangle correspond to the chromaticities of the primaries of the display.

With increasing Y the boundaries of the gamut bend more and more, until they finally meet in a single point at the cusp $(\frac{1}{3}, \frac{1}{3}, Y_w)$, which is the point corresponding to the peak luminance. See Fig. 6.11 (left).

The same concepts apply when plotting the 3D gamut in CIELAB coordinates: the luminance limits the saturation that a colour can have so the walls of the gamut curve

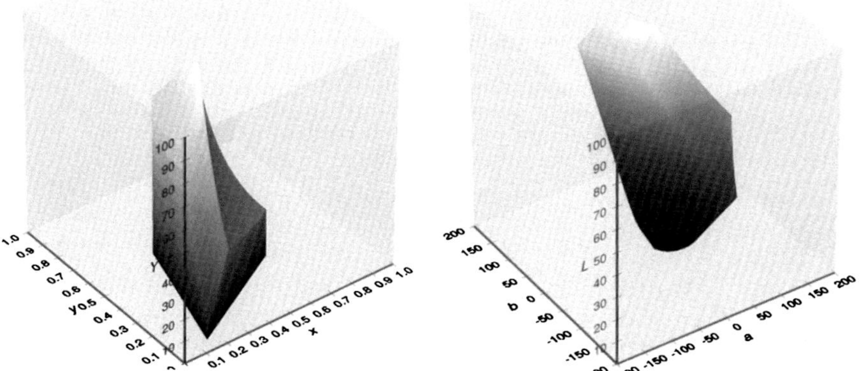

FIGURE 6.11 The colour gamut of an RGB display device

The colour gamut of an RGB display device. Left: in xyY coordinates, figure from [23]. Right: in CIELAB colour space, figure from [24].

as Y increases, and when $Y = Y_w$ the only possible colour is white so the gamut ends at a peak. With CIELAB, the base of the gamut is not a triangle as in xyY because with zero luminance we have $L^* = a^* = b^* = 0$, so the base of the gamut is also a point. See Fig. 6.11 (right).

Different display technologies have different colour gamuts, and in order to compare them the most common metric in the display industry has traditionally been to compute the areas of the triangles formed by the primaries in some chromaticity diagram like CIE xy [25]. This is a 2D representation, but it makes sense because the peak luminance of display devices has been around 100 cd/m^2 for many years and therefore the main difference among two gamuts could be characterised by the chromaticities of the primaries; put another way, if two devices that have the same peak luminance also have the same primaries, their gamuts should be the same.

In fact, these triangles are routinely referred to as "the colour gamut of the device", even though gamuts per definition should be in 3D. The standard colour gamuts in the industry *are in 2D*, and they are characterised by the chromaticity coordinates for the primaries and the reference white point. Fig. 6.12 shows, from left to right, the standard gamuts for cinema and TV: Rec. 709 with D_{65} white point, the gamut for TV and most displays; DCI-P3 with D_{60} white point, the gamut for cinema, a 25% larger than Rec. 709 (actually the D_{60} white point of DCI-P3 is very controversial, and many voices in the industry call for a D_{65} white point as in the standard TV gamut); and finally Rec. 2020 for ultra-high definition television (UHDTV), the largest standard gamut, designed to encompass the others and most object colours as well [25]. Display manufacturers adapt the colourimetry of their devices to the corresponding standard, and gamut comparisons and size estimates have traditionally been done in 2D.

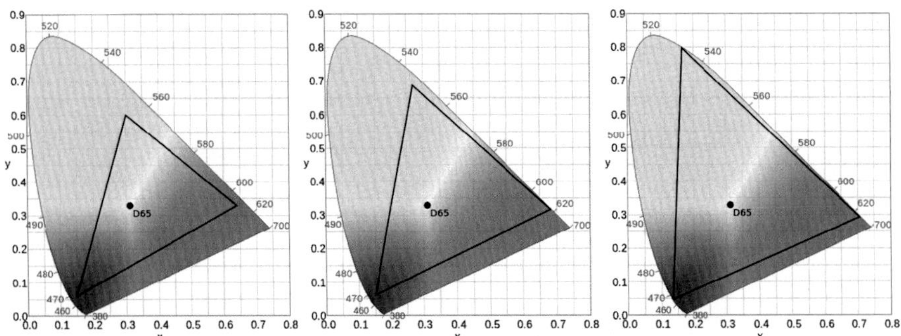

FIGURE 6.12 Standard colour gamuts in cinema and TV

Standard colour gamuts in cinema and TV. From left to right: Rec. 709 for TV (image from [26]), DCI-P3 for cinema (image from [27]) and Rec. 2020 for UHDTV (image from [28]).

6.7 Brighter displays have larger gamuts

From the analysis above it's easy to see that a display with $Y_w = 100$ cd/m^2 imposes more limitations on the chromaticities of the colours it can reproduce than a display with $Y_w = 1000$ cd/m^2. The consequence is that, as the peak luminance increases, the volume of the colour gamut increases as well, and therefore for the same set of primaries a brighter display can represent more colours, more vivid, than a dimmer display.

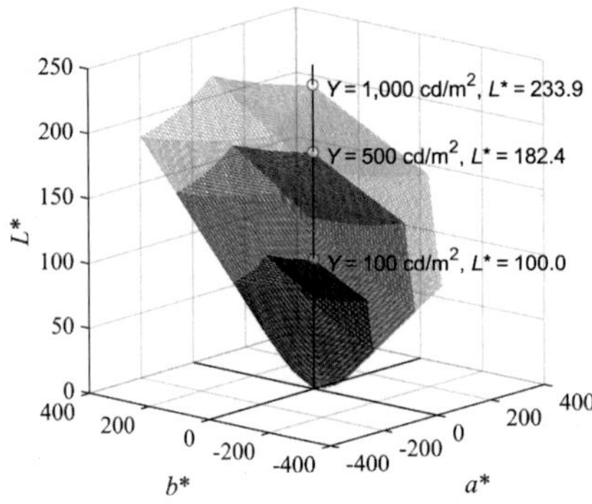

FIGURE 6.13 Colour gamut increases with peak luminance

Colour gamut increases with peak luminance. Adapted from [25].

This is clearly shown in Fig. 6.13, that depicts three colour gamuts with the same primaries but different peak luminance. As Y_w increases the gamut becomes taller but also wider, so the radius from the origin is larger, and we recall that this radius $C^* = \sqrt{a^{*2} + b^{*2}}$ is the chroma, the degree of colourfulness. Hence bright displays can reproduce colourful colours that a regular display can't.

As a result HDR screens, with $Y_w = 1000$ cd/m^2 and higher, not only can display contrast, details and highlights that are beyond the reach of SDR devices, they can also produce colours that we can't see in a SDR display, like a bright blue sky of 10 cd/m^2.

This poses a problem for display metrology, as device gamuts can no longer be characterised by the 2D triangle of the chromaticities of the primaries. Some manufacturers propose to use the CIELAB volume of the colour gamut, often called "colour volume", as a measure of the gamut size and as a value to compare different devices. But in [25] Masaoka points out a very important limitation of this view: if we scale the peak luminance Y_w by a factor k, then the height L^* of the colour gamut is scaled by $k^{1/3}$, and therefore the volume of the colour gamut is scaled by the cube of that, i.e. k, and as a result the colour volume is simply proportional to the peak luminance. And then, as Fig. 6.14 shows, a device with the small Rec. 709 gamut but a high peak luminance of $Y_w = 1000$ cd/m^2 has the same colour volume as a $Y_w = 400$ cd/m^2 Rec. 2020 display, which by design is capable of reproducing the colours of most objects, as mentioned earlier. Therefore the CIELAB colour volume does not seem to be a good measure of the colour capabilities of a display.

It has been proposed instead that the colour gamut for HDR displays be measured as the colour volume in IC_TC_P space. But here the problem is that IC_TC_P overestimates the colour volume. This is due to the fact that the PQ curve that serves as the basis of IC_TC_P does not model brightness perception: it is a much steeper nonlinearity than actual perception models as those of CIELAB, hdr-CIELAB or hdr-IPT, and this extra steepness produces an exaggerated volume.

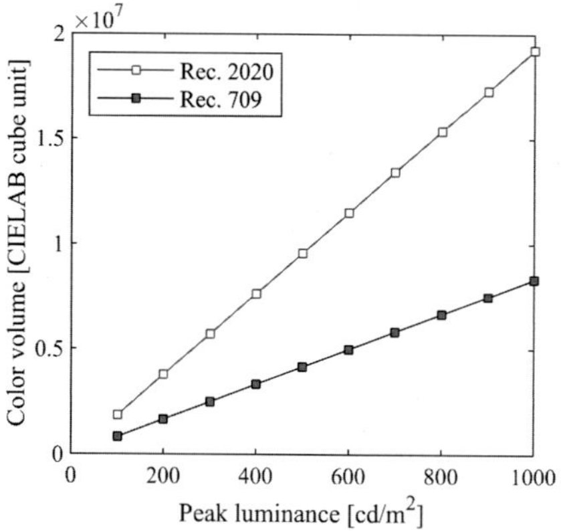

FIGURE 6.14 Colour volume is proportional to peak luminance

Colour volume is proportional to peak luminance. Adapted from [25].

Masaoka concludes that it's reasonable to use the xy chromaticity area to evaluate SDR display gamuts, and that a single-valued colour volume metric is not really practical for HDR displays: he proposes to use instead combinations of conventional metrics, like the chromaticity gamut area, peak and black luminance.

6.8 3D gamuts are not the whole picture

We would also add that, regardless of the colour space that is used, the colour volume of an HDR display can be a misleading measure for other reasons as well. To begin with, the average picture level or average luminance is quite similar in a SDR movie as in its HDR version, it's only on the very bright highlights and the very dark shadows that the HDR image is clearly different. Therefore, even if the peak luminance of the HDR display is 10 times larger than that of an SDR display, in practice most of the extra 3D gamut of the HDR display is not going to be used.

Another very important point to consider is that most HDR monitors have a physical limit on the average luminance they can display and this limit is well below the peak luminance; so for example a display with peak luminance of 1000 cd/m^2 can reach that value for highlights of small area, but if it has to show a fully white image at maximum value then all the pixels will be at around 600 cd/m^2 or even less, otherwise the energy consumption and heat dissipation requirements would be too high.

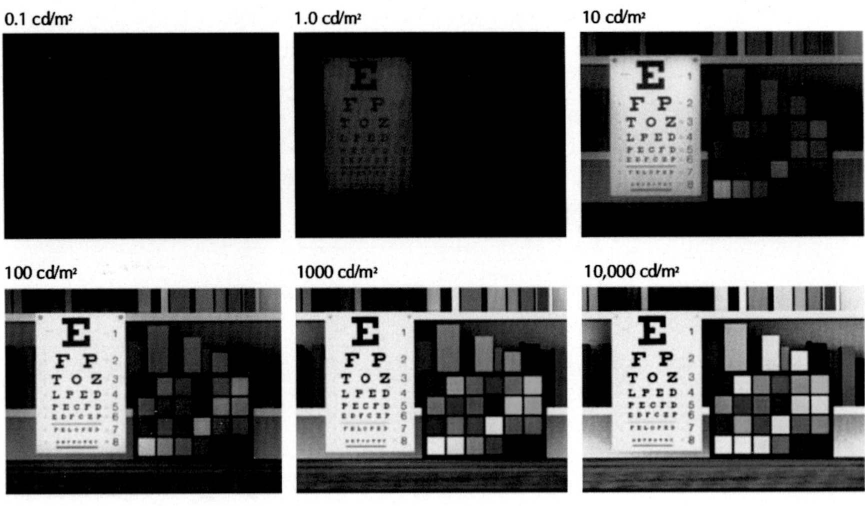

FIGURE 6.15

Hunt and Stevens effects. Image from [29].

The viewing conditions have a strong impact on the way we perceive colours. For instance, saturation decreases when the illumination decreases (the Hunt effect), and contrast also decreases with diminishing illumination (the Stevens effect); see Fig. 6.15. We mentioned at the start of Section 6.6 how, strictly, the definition of a colour gamut involves the viewing conditions, but that in practice they are ignored. This oversimplification was probably not too problematic for SDR displays with a Rec.709 gamut, but for HDR/WCG devices we really need to consider the environment in order to assess the colour capabilities of a display. This requires not a colour space but a *colour appearance model*. For given colourimetry under specified reference viewing conditions, a colour appearance model predicts the colourimetry required under the specified test viewing conditions for producing the same colour appearance [10].

The CIE has proposed several colour appearance models, the most recent of which is CIECAM02, published in 2002. The two major pieces of CIECAM02 are a chromatic adaptation transform and equations for computing correlates of perceptual attributes, such as brightness, lightness, chroma, saturation, colourfulness and hue [30]. The chromatic adaptation transform considers chromaticity variations in the adopted white point and was derived based on experimental data from corresponding colours data sets. It is followed by a non-linear compression stage, based on physiological data, before computing perceptual attributes correlates. The perceptual attribute correlates were derived considering data sets such as the Munsell Book of Colour. CIECAM02 is constrained to be invertible in closed form and to take into account a sub-set of colour appearance phenomena.

CIECAM02 was developed for (and validated with) SDR images. For HDR images there is a colour appearance model proposed by Kuang et al. [31] in 2007 called iCAM06. It decomposes the image into base and detail layers, applying on the base layer a von Kries normalisation (to approximate colour constancy) and tone compression with the Naka-Rushton equation (to emulate photoreceptor responses), while on the detail layer a power law transform is applied to predict Stevens effect. After this the processed base and detail layers are combined, followed by transforms that predict Hunt's effect and the influence of the surround (as in the system gamma experiments discussed in Chapter 5). The performance of iCAM06 on HDR images was evaluated by psychophysical tests in which observers had to compare the outputs of the colour appearance model with the results of the most successful tone mapping operators at the time; the iCAM06 model ranked first among all methods.

And finally, the image content itself has an impact on the colour gamut. At the beginning of the chapter we saw how wavelength does not determine colour, because the context of an object influences how we see it, as illustrated in Fig. 6.2. But the context does not only alter perception, it also expands to a great extent the gamut of colours that we can see [5]: for instance the colours brown, maroon or gray are never perceived in isolation, we see them only when a light is viewed next to other light stimuli.

We are not aware of the existence of effective models that take all the above aspects into account.

6.9 Brightness, hue and saturation are not independent

An image that has been created to be displayed in a wide gamut, say DCI-P3, can't be directly shown in a screen with a smaller gamut, say Rec. 709, because the image colours outside the small gamut will not be reproduced properly and visual artifacts like hue shifts and loss of detail may occur. Conversely, if we were to display a reduced gamut image in a wide gamut device, without any conversion, the image appearance would not improve in any way and we would not be making use of the enhanced capabilities of the wide gamut display. In order to prevent these issues, it is necessary to establish correspondences between colours in both gamuts, a mapping in colour space, that preserves as much as possible the intended look of the images, and this is the gamut mapping problem, that we will see in detail in Chapter 8.

From the analysis in previous sections, two displays with similar peak luminance but different xy chromaticity area will not have the same ability to reproduce colours: the device with the largest chromaticity area can display colours that are too saturated for the other device, beyond its reach. Therefore, gamut mapping entails modifying the saturation of colours so as to account for the capabilities of the target device, but at the same time brightness and hue should remain approximately the same, in order to preserve the image look. This is a very challenging problem, for a number of reasons. Chief among them is the fact that colour perception is a very complicated process which is far from being fully understood (and influenced by factors external to the display system, like the viewing conditions). As an example, we will briefly discuss here how saturation, brightness and hue are interrelated, and therefore changing one will affect the others.

6.9.1 Saturated objects appear brighter

We saw in Section 6.3 how the luminance Y was defined in terms of the standard luminosity function $V(\lambda)$, which is a dimension-less function describing the sensitivity to light as a function of wavelength; therefore, $Y = \int V(\lambda)E(\lambda)d\lambda$ corresponds to the luminance of the light stimulus of spectral radiance $E(\lambda)$. In all perceptual colour spaces brightness, or lightness, is a nonlinear function of the luminance, e.g. a power law in CIELAB or the Naka-Rushton equation in hdr-IPT.

But the function $V(\lambda)$ is based on results obtained with a certain experimental procedure called flicker photometry, and it is not valid for matching the brightness of colours with different chromaticity [32]. In fact, the perceived brightness of a colour is not determined by its luminance, it's affected by the saturation and the hue as well. This is known as the Helmholtz–Kohlrausch effect, illustrated in Fig. 6.16. All the rectangles in this figure have the same luminance, the same Y value. In the top we can see that the brightness depends on hue, and in the bottom we see that brightness depends on saturation.

As a consequence, if we were to modify the saturation of a colour while preserving all other attributes, its brightness would appear to change, which has implications for the gamut mapping problem, as we will see in Chapter 8. There are cortical correlates for saturation, and the interaction between achromatic and chromatic signals that produces brightness perception has been shown to happen at V1 [34,35], not before, with some neural mechanisms having been proposed [36].

There are some psychophysical models for brightness as a function of luminance and hue, e.g. [37,38], but the review in [39] found that none of the state of the art

FIGURE 6.16 Helmholtz–Kohlrausch effect

Illustration of the Helmholtz–Kohlrausch effect, from [33].

models were able to fully predict perceived brightness, despite the fact that all of them included the Helmholtz–Kohlrausch effect.

On the other hand, the perception of colour saturation in natural scenes can be predicted quite well from the most popular saturation measures, derived from standard colour spaces [40].

6.9.2 Increasing the brightness or saturation of a reflective object can make it appear fluorescent

Another important aspect of the interrelation between saturation and brightness, particularly relevant when mastering images for HDR displays, is the following.

In colour science, the Rösch-MacAdam colour solid denotes the volume of all the colours that can be generated by reflective surfaces under an illuminant of given spectral distribution and power. The colour solid is a subset of the larger volume of all possible colours, and outside the colour solid there are the colour points corresponding to light sources, which can be at the same time very saturated and very bright.

Fig. 6.17 illustrates these ideas: the left image shows an example colour solid in CIELAB colour space and for standard illuminant D_{65}, while on the right we can see the spectral locus for monochromatic stimuli. The points in-between the spectral locus and the colour solid correspond to light-emitting objects.

When mastering a movie for HDR, colourists have much more room to increase saturation and brightness. Rising the brightness of an object is equivalent to moving up its corresponding point in CIELAB, and increasing the saturation means moving the point horizontally, away from the lightness axis. So if either of these displacements is too large, the resulting point, which started inside the colour solid, ends up outside of it and as a result the colour-corrected object may now look fluorescent, appearing to be light-emitting. This is related to the concept of G_0 of Evans, recounted by Fairchild in [16]: for a given chromaticity, its G_0 value is the luminance at which

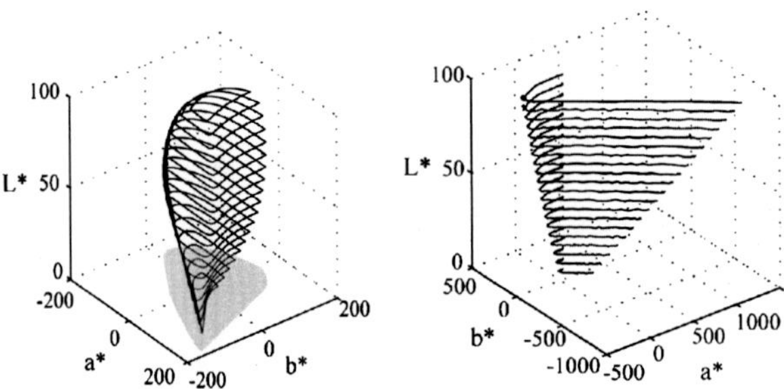

FIGURE 6.17 The colour solid and the spectral locus

Left: colour solid in CIELAB for illuminant D_{65}. Right: spectral locus for monochromatic stimuli. Image from [41].

the appearance of the colour transitions from being that of a reflective surface to being that of a light source.

References

[1] Monnier P. Standard definitions of chromatic induction fail to describe induction with s-cone patterned backgrounds. Vision Research 2008;48(27):2708–14.

[2] Judd D. Color appearance. In: Contributions to color science, vol. 545. National Bureau of Standards (USA); 1979. p. 539–64.

[3] Wyszecki G, Stiles WS. Color science: concepts and methods, quantitative data and formulas. John Wiley & Sons; 1982.

[4] Mollon J. The origins of modern color science. In: The science of color. Oxford: Optical Society of America; 2003. p. 1–39.

[5] Shevell SK, Kingdom FA. Color in complex scenes. Annual Review of Psychology 2008;59:143–66.

[6] Wright W. A re-determination of the trichromatic coefficients of the spectral colours. Transactions of the Optical Society 1929;30(4):141. http://stacks.iop.org/1475-4878/30/i=4/a=301.

[7] http://en.wikipedia.org/wiki/Color_matching_function#Color_matching_functions.

[8] https://commons.wikimedia.org/w/index.php?curid=195285.

[9] Fairman H, Brill M, Hemmendinger H. How the CIE 1931 color-matching functions were derived from Wright-Guild data. Color Research and Application 1997;22(1):11–23. https://doi.org/10.1002/(SICI)1520-6378(199702)22:1<11::AID-COL4>3.0.CO;2-7.

[10] Sharma G. Color fundamentals for digital imaging. In: Digital color imaging handbook. CRC Press; 2003.

[11] http://en.wikipedia.org/wiki/Standard_illuminant.

[12] http://en.wikipedia.org/wiki/Color_gamut.

[13] Judd D. Estimation of chromaticity differences and nearest color temperature on the standard 1931 ICI colorimetric coordinate system. In: Contributions to color science, vol. 545. National Bureau of Standards (USA); 1979. p. 207–12.

[14] Worthey J. Limitations of color constancy. Journal of the Optical Society of America A 1985;2(7):1014–26.

[15] Shevell SK, Martin PR. Color opponency: tutorial. JOSA A 2017;34(7):1099–108.

[16] Fairchild MD. Color appearance models. John Wiley & Sons; 2013.

[17] Morovič J. Color gamut mapping. John Wiley & Sons; 2008.

[18] Ebner F, Fairchild MD. Development and testing of a color space (ipt) with improved hue uniformity. In: Color and imaging conference. Society for Imaging Science and Technology; 1998. p. 8–13.
[19] Fairchild MD, Wyble DR. hdr-cielab and hdr-ipt: Simple models for describing the color of high-dynamic-range and wide-color-gamut images. In: Color and imaging conference. Society for Imaging Science and Technology; 2010. p. 322–6.
[20] Fairchild MD, Chen PH. Brightness, lightness, and specifying color in high-dynamic-range scenes and images. Image quality and system performance VIII, vol. 7867. International Society for Optics and Photonics; 2011. p. 78670O.
[21] Union IT. Rec BT.2100-1, "Image parameter values for high dynamic range television for use in production and international programme exchange". Tech. rep., ITU; 2017.
[22] Safdar M, Cui G, Kim YJ, Luo MR. Perceptually uniform color space for image signals including high dynamic range and wide gamut. Optics Express 2017;25(13):15131–51.
[23] https://en.wikipedia.org/wiki/CIE_1931_color_space.
[24] https://en.wikipedia.org/wiki/CIELAB_color_space.
[25] Masaoka K. Cielab-metric color volume for hdr displays. In: International display workshop (IDW'17); 2017.
[26] https://commons.wikimedia.org/w/index.php?curid=21864671.
[27] https://commons.wikimedia.org/w/index.php?curid=46686806.
[28] https://commons.wikimedia.org/w/index.php?curid=21864661.
[29] Fairchild M. Color appearance models: CIECAM02 and beyond. Tutorial notes, IS&T/SID 12th color imaging conference; 2004.
[30] Moroney N, Fairchild M, Hunt R, Li C, Luo M, Newman T. The CIECAM02 color appearance model. In: Proceedings of color and imaging conference. Society for Imaging Science and Technology; 2002.
[31] Kuang J, Johnson GM, Fairchild MD. icam06: a refined image appearance model for hdr image rendering. Journal of Visual Communication and Image Representation 2007;18(5):406–14.
[32] Fairchild MD, Pirrotta E. Predicting the lightness of chromatic object colors using cielab. Color Research & Application 1991;16(6):385–93.
[33] Elliot AJ, Fairchild MD, Franklin A. Handbook of color psychology. Cambridge University Press; 2015.
[34] Hanazawa A, Komatsu H, Murakami I. Neural selectivity for hue and saturation of colour in the primary visual cortex of the monkey. European Journal of Neuroscience 2000;12(5):1753–63.
[35] Xing D, Ouni A, Chen S, Sahmoud H, Gordon J, Shapley R. Brightness–color interactions in human early visual cortex. Journal of Neuroscience 2015;35(5):2226–32.
[36] Billock VA, Tsou BH. Sensory recoding via neural synchronization: integrating hue and luminance into chromatic brightness and saturation. JOSA A 2005;22(10):2289–98.
[37] Ayama M, Ikeda M. Brightness-to-luminance ratio of colored light in the entire chromaticity diagram. Color Research & Application: Endorsed by Inter-Society Color Council, The Colour Group (Great Britain), Canadian Society for Color, Color Science Association of Japan, Dutch Society for the Study of Color, The Swedish Colour Centre Foundation, Colour Society of Australia, Centre Français de la Couleur 1998;23(5):274–87.
[38] Pridmore RW. Chroma, chromatic luminance, and luminous reflectance. Part ii: Related models of chroma, colorfulness, and brightness. Color Research & Application 2009;34(1):55–67.
[39] Withouck M, Smet KA, Ryckaert WR, Pointer MR, Deconinck G, Koenderink J, et al. Brightness perception of unrelated self-luminous colors. JOSA A 2013;30(6):1248–55.
[40] Schiller F, Gegenfurtner KR. Perception of saturation in natural scenes. JOSA A 2016;33(3):A194–206.
[41] Perales E, Martínez-Verdú FM, Linhares JMM, Nascimento SMC. Number of discernible colors for color-deficient observers estimated from the macadam limits. JOSA A 2010;27(10):2106–14.

CHAPTER 7

Histogram equalisation and vision models

In this chapter we show how an image processing method for colour and contrast enhancement, that performs local histogram equalisation (LHE), is linked to neural activity models and the Retinex theory of colour vision. The common thread behind all these subjects is efficient representation.

The LHE method can be extended to reproduce an important visual perception phenomenon which is assimilation, a type of visual induction effect. The traditional view has been that assimilation must be a cortical process, but we show that it can start already in the retina.

The LHE method is based on minimising an energy functional through an iterative process. If the functional is regularised, the minimisation can be achieved in a single step by convolving the input with a kernel, which has important implications for the performance of algorithms based on LHE, as we will see later in the book.

7.1 From image processing to neural models

7.1.1 Histogram equalisation

Histogram equalisation is a classical, very basic image processing technique dating at least to the early 1970s (see [1] and references therein), aiming at enhancing the contrast and improving the appearance of images by way of re-distributing their levels uniformly across the available range. In this sense, an image would be optimal if its histogram were flat or "equalised", meaning that all the range is used and all levels are represented by the same amount of pixels. Therefore, when an image has a flat histogram its cumulative histogram is simply a ramp, and this allows for a very straightforward computation for the histogram equalisation procedure: assuming we are working on a graylevel image in the range [0,1], we have to substitute each level g in the original image by the value of its normalised cumulative histogram, $H(g)$. The solution is computed very fast using a look-up table (LUT). An example result can be seen in Fig. 7.1; notice that, while the range has been expanded and the resulting image has a more even histogram, it's not perfectly uniform.

The example in Fig. 7.1 corresponds to an underexposed image, and Fig. 7.2 shows another histogram equalisation example, in this case of an overexposed image. The results in both figures show clear artifacts, and it's well known that histogram equalisation often does not improve the appearance of images but makes them worse.

Vision Models for High Dynamic Range and Wide Colour Gamut Imaging. https://doi.org/10.1016/B978-0-12-813894-6.00012-0

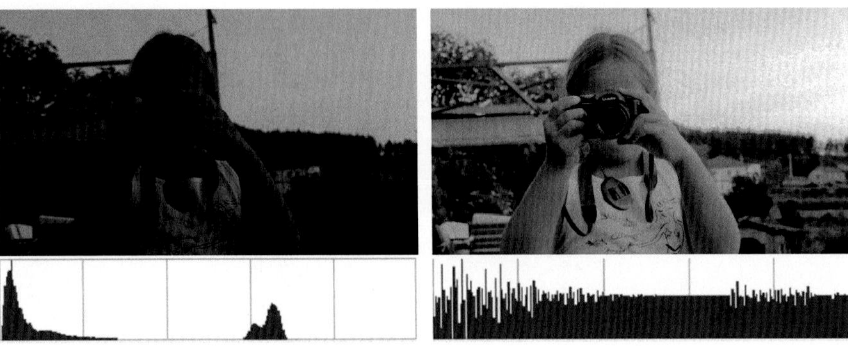

FIGURE 7.1 Histogram equalisation

Left: original image and associated histogram. Right: after histogram equalisation.

This is aggravated by the fact that the equalisation procedure is a one-shot technique, that only produces a final result, without any "in-between", so if the resulting image shows any type of unpleasant artifact there is nothing to do about it.

FIGURE 7.2 Histogram equalisation

Left: original image. Right: after histogram equalisation.

This issue was addressed by Sapiro and Caselles in [2], who proved that the minimisation of the energy functional

$$E(I) = 2\sum_x (I(x) - \frac{1}{2})^2 - \frac{1}{AB}\sum_x \sum_y |I(x) - I(y)| \tag{7.1}$$

produces an image I with a flat histogram. The range of I is $[0, 1]$, x, y are pixels and A, B are the image dimensions. While the result of histogram equalisation is very often unsatisfactory and can't be altered, the method in [2] starts with an input image I_0 and applies to it step after step of the minimisation of Eq. (7.1), letting the user decide when to stop. If the user lets the minimisation run to convergence, she'll get the same result as with a LUT, but otherwise a better result can be obtained if the iterative procedure stops before the appearance of severe artifacts. The squared differences in the first term of Eq. (7.1) and the absolute differences in the second one are required to ensure that the minimisation yields an image with equalised histogram, see [2] for details.

The energy in Eq. (7.1) can be interpreted as the difference between two positive and competing terms,

$$E(I) = D(I) - C(I). \tag{7.2}$$

The term D measures the dispersion around the average value of $\frac{1}{2}$, as in the *gray world* hypothesis for colour constancy which states that in a sufficiently varied scene the average colour will be perceived as gray (an observation made by Judd [3,4] and formalised by Buchsbaum [5]) and therefore the illuminant colour can be estimated from the colour average of the scene; this implies that the minimisation of $E(I)$ will make the image mean tend to $\frac{1}{2}$, so that $D(I)$ is small, and corresponding to the case where the illuminant is white.

The term C measures the contrast as the sum of the absolute value of the pixel differences; because of the negative sign in front of this term, minimising $E(I)$ will increase the contrast.

7.1.2 Perceptually-based contrast enhancement

The abovementioned measure of contrast is global, not local, i.e. the differences are computed regardless of the spatial locations of the pixels. This is not consistent with how we *perceive* contrast, which is in a localised manner, at each point having neighbours exert a higher influence than far-away points. Using the concepts introduced by the popular perceptually-based colour correction method ACE [6], Bertalmío et al. propose in [7] an adapted version of the functional of Eq. (7.1) that complies with some very basic visual perception principles, namely those of locality, colour constancy and *white patch* (the latter stating that the brightest spot in the image is perceived as white, an observation that is often attributed to the Retinex theory [8] but which has a long history that dates back at least to the works of Helmholtz, as explained by [10,11]):

$$\begin{aligned} E(I) = \frac{\alpha}{2}\sum_x (I(x) - \frac{1}{2})^2 - \gamma \sum_x \sum_y w(x,y)|I(x) - I(y)| \\ + \frac{\beta}{2}\sum_x (I(x) - I_0(x))^2. \end{aligned} \tag{7.3}$$

Here w is a distance function such that its value decreases as the distance between x and y increases, I_0 is the original image and α, β and γ are positive weights (which can be chosen so as to guarantee the white patch property, see [7] for details).

The gradient descent equation for the functional in Eq. (7.3) is the following:

$$I_t(x) = -\alpha(I(x) - \frac{1}{2}) + \gamma \sum_y w(x,y) sgn(I(x) - I(y)) - \beta(I(x) - I_0(x)). \tag{7.4}$$

Starting from $I = I_0$, we iterate Eq. (7.4) until we reach a steady state, which will be the result of this algorithm.

By minimising the energy in Eq. (7.3) we are locally enhancing contrast (second term) and promoting colour constancy by discounting the illuminant (first term), while preventing the image from departing too much from its original values (third term).

We can intuitively explain how the contrast term operates by considering the following. Since $\gamma > 0$, in order to minimise $E(I)$ we need to increase $\sum_x \sum_y w(x, y)|I(x) - I(y)|$, i.e. the local contrast, a weighted sum of local pixel differences. So if a pixel has a larger value than all the pixels in its neighbourhood, then increasing the local contrast is achieved by enlarging these differences, increasing the pixel value even more. Conversely, when a pixel has a smaller value than all the pixels in its neighbourhood, then increasing the local contrast is achieved by making the pixel value even smaller. If $\gamma < 0$, then the minimisation of $E(I)$ would reduce, not increase, the contrast, as pointed out in [12].

The radius of the kernel w controls how local the contrast enhancement is: a very large radius corresponds to the global case, whereas with a small value the contrast term is computed over small pixel neighbourhoods. We could also say that the minimisation of Eq. (7.3) approximates *local* histogram equalisation, where the locality is given by the distance function w: if instead of decreasing with the distance between x and y we had that $w(x, y) = 1 \forall (x, y)$ and $\beta = 0$, then Eq. (7.3) would become Eq. (7.1) and therefore iterating Eq. (7.4) would perform *global* histogram equalisation.

The local histogram equalisation (LHE) method of [7] has several good properties:

1. It has a very good local contrast enhancement performance, producing results without halos, spurious colours or any other kind of visual artifact.
2. It can deal with both underexposed and overexposed pictures. See Figs. 7.3 and 7.4.
3. It “flattens” the histogram, approaching histogram equalisation. See Fig. 7.5.
4. It reproduces visual perception phenomena such as simultaneous contrast and the Mach Band effect. See Figs. 7.6 and 7.7.
5. It yields very good colour constancy results, being able to remove strong colour casts and to deal with non-uniform illumination (a challenging scenario for most colour constancy algorithms, as discussed in [13]). See Fig. 7.8.

7.1.3 Connection with the Retinex theory of colour vision

Regarding colour constancy, there is also a very interesting and close connection between LHE and the classical approach of Retinex.

Land makes in [8] a very clear and detailed explanation of his Retinex theory and the experiments that led to its postulation. After scores of perceptual matching tests, his conclusion was that our perception of the colour of an object had a physical correlate in what he called scaled integrated reflectance, which is defined, for each

FIGURE 7.3 Underexposed picture

Underexposed picture. Left: original. Right: result of [7].

FIGURE 7.4 Overexposed picture

Overexposed picture. Left: original. Right: result of [7].

FIGURE 7.5 Equalising the histogram

Top row: original (left), result of [7] (right). Bottom row: corresponding luminance histograms.

waveband (long, medium and short, corresponding to cone response sensitivities,) as a ratio: the integral of the radiance of the object over the waveband, divided by the integral over the same waveband of the radiance of a white object under the same scene illuminant. The scaling is a non-linear function that relates reflectance to lightness sensation.

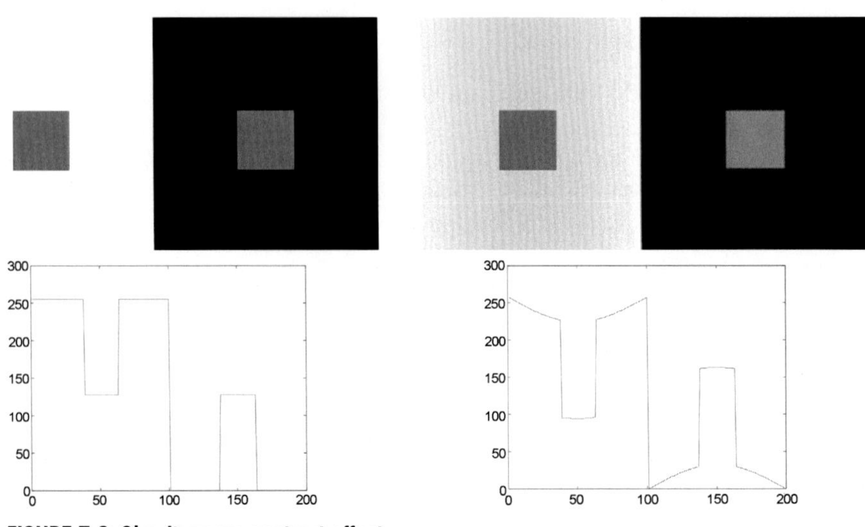

FIGURE 7.6 Simultaneous contrast effect

Simultaneous contrast effect. Left: original image and associated profile of middle row. Right: after the local histogram equalisation (LHE) method of [7]. Notice how the small squares in the center have the same gray value in the original image, but we perceive them differently: the gray square over the white background is seen darker than the gray square over the black background; this effect is reproduced by the LHE result.

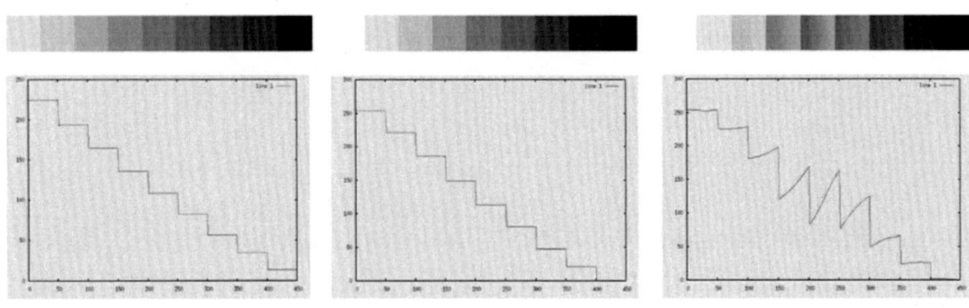

FIGURE 7.7 Mach band effect

Mach band effect. Left: original image (top) and corresponding profile of a sample row (bottom). Middle: after histogram equalisation. Right: result of applying the method in [7] to the original image; notice how this method reproduces the Mach band effect, a perceptual phenomenon by which the contrast at edges is exaggerated and we see a thin lighter strip at the lighter side of the edge and a think darker strip at the darker side of the edge.

But this implies that in order to perceive the colour of an object somehow our visual system is comparing the light coming from the object with the light coming from a reference white, and Land wondered how we are able to find this reference white *"in the unevenly light world without reference sheets of white paper"* [8]. The sensation of white will be generated by the area of maximum radiance in all three bands (this is the von Kries' model or "white-patch" assumption); this area could be

FIGURE 7.8 Colour constancy

Colour constancy. Left: original. Right: result of [7].

used as reference, but Land didn't know how our visual system could *"ascertain the reflectance of an area without in effect placing a comparison standard next to the area"* [8].

The solution he proposed consisted of comparing far-away points through paths: the ratio of the values at the two end-points of the path can be computed as the sequential product of the ratios of each point of the path with the following point. The (original) Retinex algorithm of Land and McCann [9] consists of assigning, for each point and each waveband (long, middle, short), an estimate reflectance obtained as the average of the sequential products obtained on many paths, all ending in the given point. Land thought that this was a plausible explanation of how our visual system estimates reflectances but he didn't want to venture where exactly this type of operations were being carried out, in the retina or at the cortex; therefore he chose the name "Retinex" for his approach.

The Retinex algorithm is directly applied to digital images in a straightforward way, where the pixels will be the points and the three colour channels R, G and B play the role of the wavebands. In [14] it was shown that the basic Retinex formulation always increases brightness so if applied to overexposed pictures some type of compensation or post-processing mechanism would be required; also, if the algorithm is iterated the results may improve but the convergence image is flat white, so there is some "sweet spot" of the number of iterations yielding the best output [12].

Another major source of problems is the reliance of Retinex on paths: their length, shape and number condition the results and many works have been proposed trying to optimise the selection of these variables.

In our kernel-based Retinex (KBR) formulation [12], we take all the essential elements of the Retinex theory [8] (channel independence, the ratio reset mechanism, local averages, non-linear correction) and propose an implementation that is intrinsically 2D, and therefore free of the issues associated with the 1D paths used in several Retinex variants. The results obtained with this algorithm comply with all the expected properties of Retinex (such as performing colour constancy while being unable to deal with overexposed images) but don't suffer from the usual shortcomings such as sensitivity to noise, appearance of halos, etc. In [12] it is proven that there isn't any energy that is minimised by the iterative application of the KBR algorithm, and this fact is linked to its limitations regarding overexposed pictures. Using the analysis of contrast performed in [15], we are able to determine in [12] how to modify the basic KBR equation so that it can also handle overexposed images, and the resulting, modified KBR equation turns out to be essentially the gradient descent of the energy given by Eq. (7.3).

In other words, the LHE method of [7] can be seen as an iterative application of Retinex, although in a modified version that allows to produce good results also in the case of overexposed images.

7.1.4 Connection with neuroscience

The activity of a population of neurons in the region $V1$ of the visual cortex evolves in time according to the Wilson–Cowan equations (see [16–18]). Treating $V1$ as a planar sheet of nervous tissue, the state $a(r,\phi,t)$ of a population of cells with cortical space coordinates $r \in \mathbb{R}^2$ and orientation preference $\phi \in [0,\pi)$ can be modelled with the following PDE ([16]):

$$\frac{\partial a(r,\phi,t)}{\partial t} = -\alpha a(r,\phi,t) + \mu \int_0^{\pi} \int_{\mathbb{R}^2} \omega(r,\phi \| r',\phi')\sigma(a(r',\phi',t))dr'd\phi' + h(r,\phi,t), \tag{7.5}$$

where α, μ are coupling coefficients, $h(r,\phi,t)$ is the external input (visual stimuli), $\omega(r,\phi\|r',\phi')$ is a kernel that decays with the differences $|r-r'|$, $|\phi-\phi'|$ and σ is a sigmoid function. If we ignore the orientation ϕ and assume that the input h is constant in time, it can be shown that Eq. (7.5) is very similar to the gradient descent Eq. (7.4), where neural activity a plays the role of image value I, sigmoid function σ behaves as the derivative of the absolute value function, and the visual input h is the initial image I_0.

Therefore, the LHE model of [7] is closely related to the Wilson–Cowan neural activity model, as pointed out in [7] and further discussed in [19].

Recently, we have shown in [20] that if we go back to the energy functional of LHE and properly extend it so as to consider the orientation ϕ, the gradient descent of this new functional also has a Wilson–Cowan form but now it's capable of reproducing more perceptual phenomena, like the Poggendorff illusion; see Fig. 7.9.

7.1.5 Efficient representation allows to explain these connections

In summary, the works discussed above have shown deep and perhaps surprising connections between histogram equalisation, colour and contrast perception and neural activity models.

The common theme behind these topics, and the idea that is in fact connecting them, is efficient representation. We discussed in Chapter 4 how histogram equalisation allows for efficient coding, how the retina encodes colour and contrast in a way that is approximately independent from the illumination in order to be more effi-

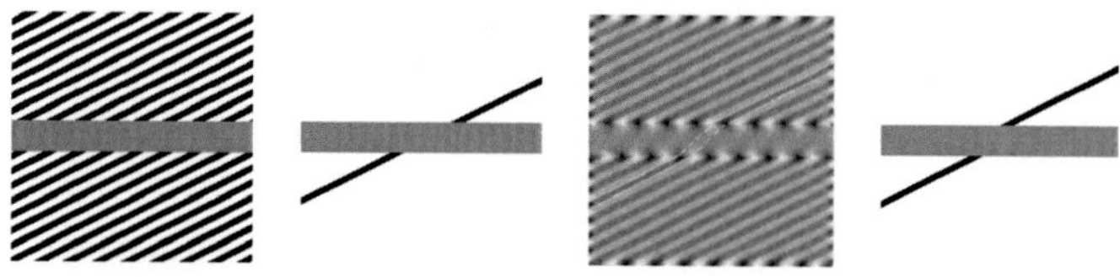

FIGURE 7.9 Poggendorff illusion

Left: original image, a variation of the Poggendorff illusion; the presence of the horizontal grey bar induces a misalignment of the background stripes. Middle left: the classical Poggendorff illusion, extracted from the previous image; the horizontal grey bar is superposed on top of the black stripe, creating the illusion of misalignment. Middle right: output of model in [20], that can be seen as an extension of the LHE model of [7] that now considers orientation in the weighting function w. Right: the perceived alignment is reconstructed and isolated after processing. Figure from [20].

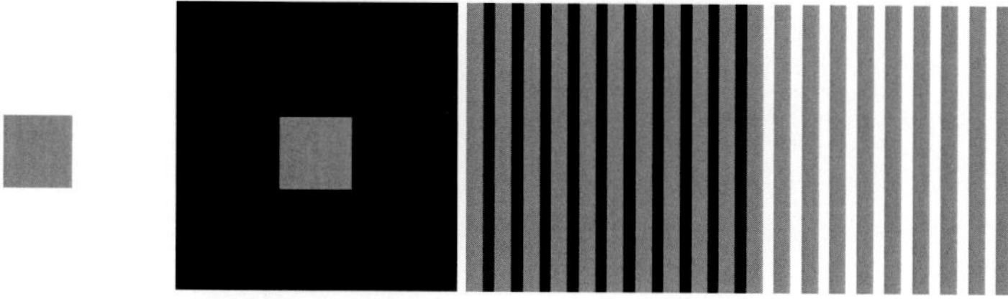

FIGURE 7.10 Achromatic induction

Left: the small center squares are equal but are perceived differently, with induced *contrast* with respect to their surround. Right: the gray bars are equal but are perceived differently, with induced *assimilation* to their surround.

cient, and how neural activity can often be explained by models that perform efficient coding.

7.2 Visual induction: contrast and assimilation

Visual induction is a perceptual phenomenon by which the appearance of an object, in terms of brightness or colour, is affected by its surround. Visual induction can be chromatic or achromatic, and in each case it can take two forms: *contrast*, when the appearance of the object shifts away from that of its surroundings (e.g. a dark object on a light background appears even darker, or a light object in a dark surround becomes even lighter), or *assimilation*, in which case the appearance of the object becomes more similar to that of its surround. See Figs. 7.10 and 7.11 for some examples.

As pointed out by [22], lightness assimilation occurs in situations of high spatial frequency while lightness contrast is associated with relatively lower spatial frequencies. In the classic experiments of Helson [23], observers had to judge the appearance of grey bars over white or black backgrounds. When the bars were very thin, the ob-

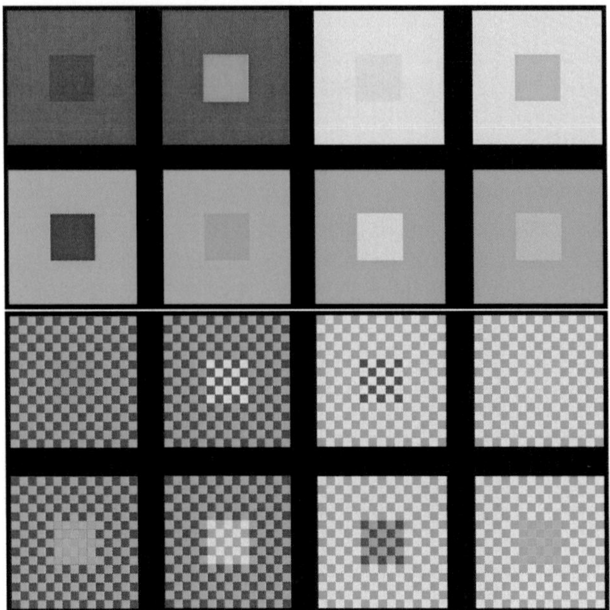

FIGURE 7.11 Chromatic induction

Top: the small center squares in each column are equal but are perceived differently, with induced *contrast* with respect to their surround. Bottom: half of the very small squares in the center checkerboard in each column are equal but are perceived differently, with induced *assimilation* to their surround. Figure from [21].

servers reported assimilation; as the bars increased in width, the assimilation effect became less pronounced, and after some point the observers started to report contrast, whose effect became increasingly more pronounced as the width of the bars turned larger. See Fig. 7.12. A similar result for the chromatic case was reported by Fach and Sharpe [24].

7.2.1 The LHE model can't reproduce assimilation

Looking at Eq. (7.4), we can see that the spatial arrangement of the image data is only taken into account by the weighting function w. But in practice w is very wide, and therefore we can expect that the local contrast enhancement procedure of [7] will always produce lightness contrast, not assimilation, since as we mentioned previously assimilation is linked to high spatial frequencies [22].

Fig. 7.13 confirms this: image (A) produces lightness assimilation, because all gray bars have the same value but they are perceived darker when surrounded by black and lighter when surrounded by white; on the other hand, the result produced by [7] in image (B) actually emulates lightness contrast rather than assimilation, as the line profiles in images (C) and (D) show.

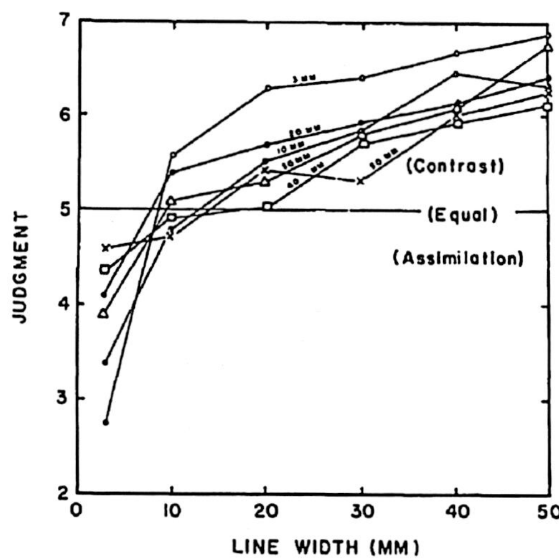

FIGURE 7.12 Induction type depends on spatial frequency

Induction type depends on spatial frequency. Low spatial frequencies induce contrast, while high spatial frequencies induce assimilation. Reprinted with permission from Helson H, Studies of anomalous contrast and assimilation. JOSA 1963; 53(1):179–184. ©The Optical Society [23].

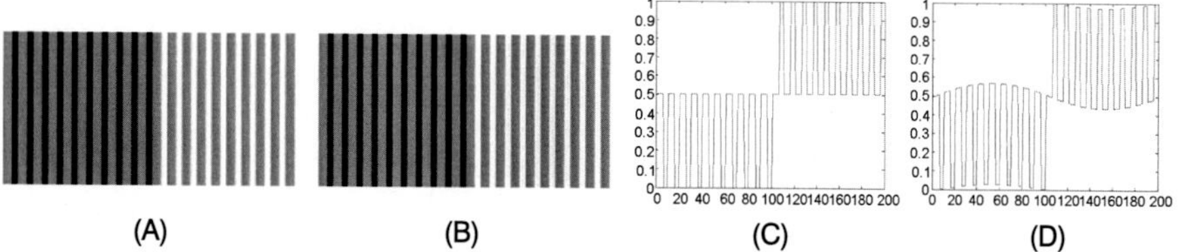

FIGURE 7.13 The LHE model can't reproduce assimilation

(A) Original image, example of lightness assimilation: the grey bars have all the same value but appear different over black and white backgrounds. (B) Result of applying the model of [7] to image (A). (C) Profile of a line from image (A). (D) Profile of a line from image (B): notice how the model of [7] actually emulates lightness contrast rather than assimilation.

7.2.2 Introducing spatial frequency into the LHE model

In order to overcome the intrinsic limitations of [7] with respect to lightness induction, we should introduce spatial frequency in the energy functional. We do so in [25], proposing a new model consisting of the following PDE, a modification of the gradient descent equation (7.4):

$$I_t(x) = -\alpha(I(x) - \mu(x)) + \gamma(1 + (\sigma(x))^c) \times \sum_y w(x, y) sgn(I(x) - I(y)) - \beta(I(x) - I_0(x)), \tag{7.6}$$

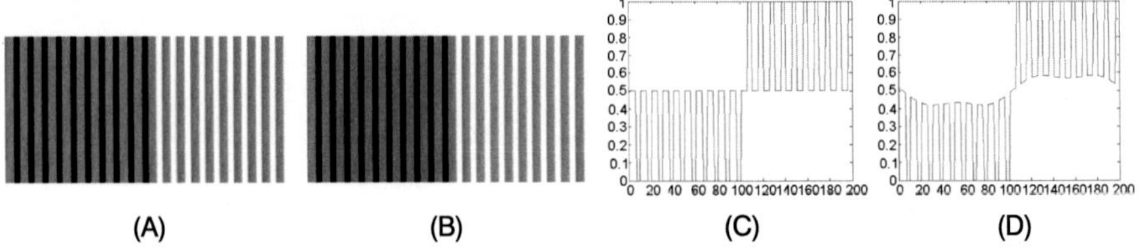

FIGURE 7.14 Predictions for assimilation example

(A) Original image, example of lightness assimilation: the grey bars have all the same value but appear different over black and white backgrounds. (B) Result of applying the model in [25] to image (A). (C) Profile of a line from image (A). (D) Profile of a line from image (B): notice how this model is capable of emulating lightness assimilation.

where $\mu(x)$ is the mean average of the original image data computed over a neighbourhood of x, $\sigma(x)$ is the standard deviation of the image data computed over a small neighbourhood of x, and the exponent c is a positive constant.

The differences with respect to Eq. (7.4) are that now the average in the first term is no longer global (the $1/2$ value of Eq. (7.4)) but local, and that the weight for the second term is no longer a constant, but it changes both spatially and with each iteration, according to the local standard deviation σ: if the neighbourhood over which it is computed is sufficiently small, the standard deviation can provide a simple estimate of spatial frequency. But also, the standard deviation is commonly used in the vision literature as an estimate of local contrast. We have this contrast $\sigma(x)$ raised to a power c, and this is also the case with other neural models where a power law is applied to the contrast, as we will briefly discuss later.

Again, this is a Wilson–Cowan type of neural activity model, where I_0 is the visual input. We take I_0 as a non-linear modification of the radiance stimulus, e.g. I_0 could be the result of applying the Naka-Rushton equation to the radiance stimuli. As we did with Eq. (7.4), we start with an image $I = I_0$ and iterate Eq. (7.6) until convergence, obtaining a result which we'll see is able to predict perceptual phenomena as well as improve the efficiency of the representation.

The model in [25] can predict lightness assimilation in the previous example of the alternating gray bars of Fig. 7.13, as we show in Fig. 7.14.

Finally, [25] argues that the proposed model of Eq. (7.6) is a good candidate for a neural model providing the contrast constancy effects described by [26] and [27]. We will now see how this new neural model, applied to signals already encoded by photoreceptors, further improves efficiency by reducing redundancy: flattening the histogram and whitening the power spectrum.

Fig. 7.15(A) shows a high dynamic range image linearly scaled to the range $[0, 1]$. Clearly this kind of mapping is useless, which is a way to explain the need for light adaptation and gain control mechanisms in our photoreceptors. Fig. 7.15(B) shows the result of applying the Naka-Rushton equation to the previous HDR image. Fig. 7.15(C) shows the result of applying the model in [25] to image 7.15(B). As we can see, the original image (whose values should be approximately proportional to

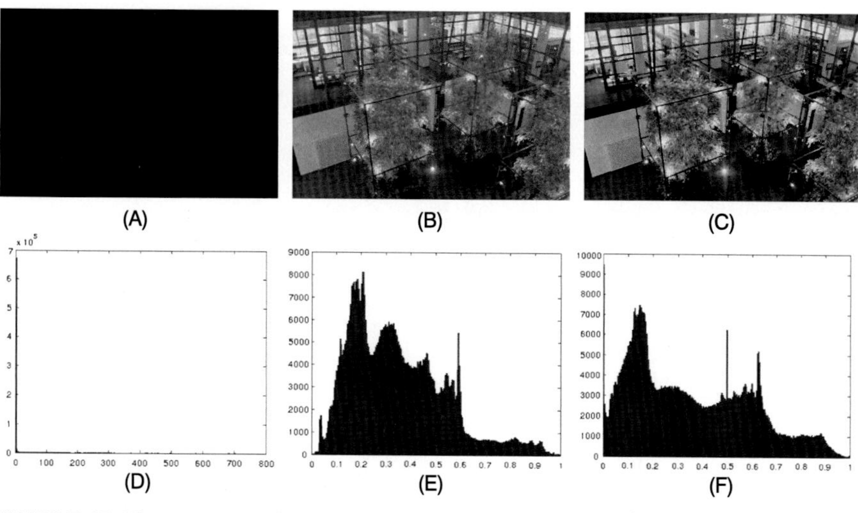

FIGURE 7.15 Histogram equalisation

(A) High dynamic range (HDR) image, linearly scaled. (B) Result of applying the Naka-Rushton equation to the HDR image. (C) Result of applying the model of [25] to image (B). (D) Histogram of (A). (E) Histogram of (B). (F) Histogram of (C). Figure from [25].

the light intensity values of the scene) has a very lopsided histogram, which is made considerably more uniform by applying the Naka-Rushton equation and even more flat if we apply the method in [25] to the Naka-Rushton output. Local contrast is clearly enhanced as well, see for instance the window frames, the book cases behind the windows, etc.

Fig. 7.16(A) shows the result of applying the Naka-Rushton equation to a high dynamic range image. Fig. 7.16(B) shows the result of applying the model of [7] to image 7.16(A) (this is roughly equivalent to the tone mapping approach proposed by [28]). Fig. 7.16(C) shows the result of applying the model in [25] to image 7.16(A). Fig. 7.16(D) compares the power spectrum of the three previous images. We can see that [25] improves spectrum whitening over the other two results. In this image the contrast enhancement is more subtle but still noticeable, especially in the interior of the tree-trunk and on the leaves and grass in the foreground.

7.2.3 Retinal lateral inhibition can explain assimilation

We have seen in Chapter 2 how lateral inhibition is produced in the retina by horizontal and amacrine cells, which send inhibitory feedback to photoreceptors and bipolar cells after pooling signals over their neighbourhoods. Lateral inhibition creates the typical center-surround structure of the receptive field (RF) of retinal ganglion cells (RGCs), with the excitatory center due to the feed-forward cells (photoreceptors

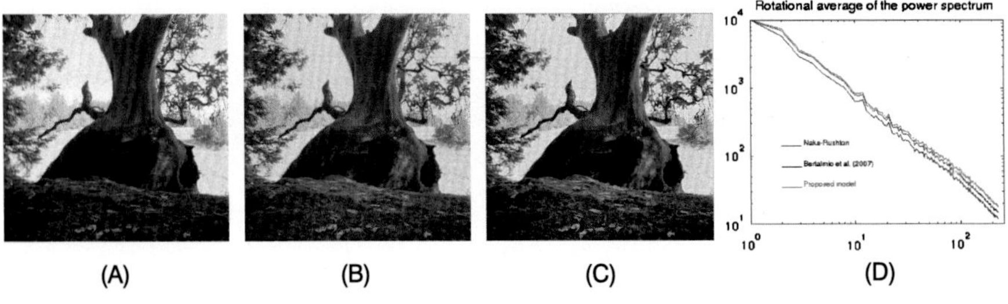

FIGURE 7.16 Spectrum whitening

(A) Result of applying the Naka-Rushton equation to a high dynamic range image. (B) Result of applying the model of [7] to image (A). (C) Result of applying the model of [25] to image (A). (D) Power spectrum of images (A) to (C). Figure from [25].

and bipolar cells) and the inhibitory surround due to interneurons (horizontal and amacrine cells); see Fig. 7.17.

We also mentioned how this center-surround organisation is a very important instance of efficient representation, as it allows to represent with less resources large uniform regions because they generate little activity (or no activity at all in the case that center and surround signals cancel each other perfectly).

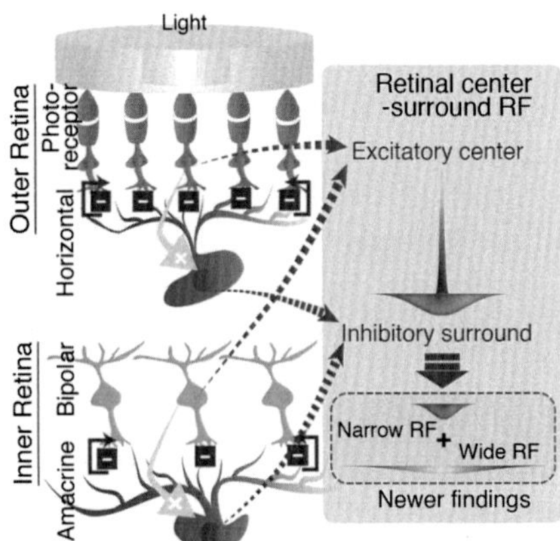

FIGURE 7.17 Lateral inhibition in the retina

Lateral inhibition in the retina occurs as the feedback from the interneurons, horizontal cells and amacrine cells, which receive excitatory inputs from photoreceptors and bipolar cells, respectively, inhibit the excited photoreceptors and bipolar cells and their neighbourhood. The typical center-surround RF structure of a ganglion cell is the combination of the excitatory center created by the feedforward cells and inhibitory surround formed by the interneurons. The study in [29] tests the effect of the wide RF component of these interneurons in addition to the classic (narrow) RF (inset). Figure from [29].

Center-surround RFs are usually modelled as a difference of Gaussians (DoG), with the standard deviation of the center Gaussian given by the RF size of the feedforward cells and the standard deviation of the surround Gaussian given by the RF size of the interneurons.

Visual perception models based on convolution with DoG filters have been shown to predict a significant number of visual illusions and other perceptual phenomena [30], including both types of visual induction, contrast and assimilation.

But neurophysiological studies that tested the DoG model using parameter values corresponding to estimated RF sizes of RGCs concluded that while induction in the form of contrast can be generated in the retina, assimilation *can not* because it would need a much longer range of interaction between image regions than what lateral inhibition can provide.

In other words, assimilation requires a much wider RF for the lateral inhibition, and thus it was assumed that assimilation had to take place at a later stage than the retina, most probably at the cortex, since in any case it's known that cortical mechanisms involving orientation, depth estimation and attention can have an impact on assimilation.

In [29] Yeonan-Kim and Bertalmío showed that, in fact, assimilation can *start* already in the retina. We took to classic retinal models, those of Wilson [31] and van Hateren [32], that have been validated by matching single-cell recording data and whose parameters have been selected based on neurophysiological evidence, and adapted them so that parasol RGCs (in the M-pathway which as we mentioned is where luminance processing largely takes place) have a surround that is now dual, with a narrow component of large amplitude and a wide component of smaller amplitude; see Fig. 7.18. This different form for the surround is based on more recent neurophysiological data showing that retinal interneurons have RFs that are much more extended than previously assumed, and RGC responses show a component that goes beyond the classical RF. For instance, for horizontal cells it has been shown that the direct dendritic connection from photoreceptors to horizontal cells creates a narrow RF component, while electric coupling among adjoining horizontal cells produces a wide RF component. The reason why this wide, extra-classical RF component had been previously ignored when modelling RGC responses might just have to do with earlier studies focusing on the more detectable center and narrow surround elements, ignoring weaker peripheral effects from the extra-classic surround.

With these updates, the retinal models of Wilson and van Hateren are now able to predict both assimilation and contrast, matching the psychophysical data from three different experiments, including the classic work by Helson [23]. We concluded that the long-range effect of visual induction required to explain assimilation phenomena does not need to involve post-retinal processing as it was previously assumed, it might start already at the interaction between photoreceptor and horizontal cell, at the very first stage of the visual processing.

Another very interesting result from [29] is that the wide RF component can provide RGCs with a more global estimate of the mean, which suggest a link to the neural mechanisms of the Retinex theory. In particular, Land proposed in [33] a Retinex algorithm that is conceptually similar to a DoG formulation, requiring the contribution

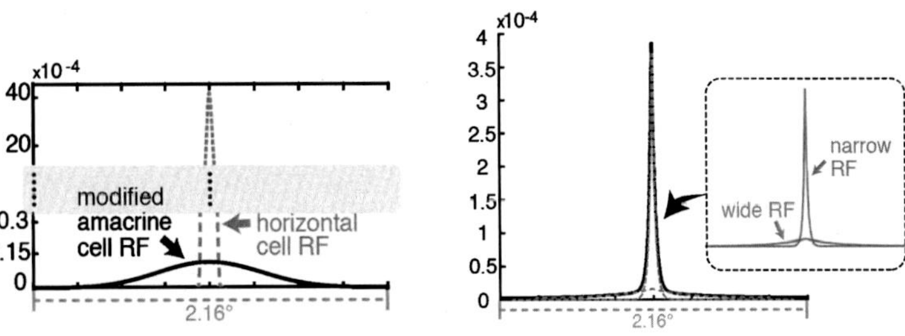

FIGURE 7.18 Dual wide-narrow surround

Dual wide-narrow surround. Left: Wilson's model. Right: van Hateren's model. Figure from [29].

from far-away regions in a way that we can now see can be accomplished by the wide-RF interneuron feedback.

Later works pursued further the center-surround paradigm for Retinex, mimicking retinal processes and using DoG filters as models of RFs of RGCs, using single or multiple scales for the surround, which again ties with the dual narrow-wide surround RF discussed in [29]. For a detailed analysis of the retinal and cortical components of Retinex algorithms please see [34].

7.2.4 The LHE model with dual wide-narrow surround produces both assimilation and contrast

The modifications introduced to the LHE model in [25] allow it to perform both assimilation and contrast, but multiplying the contrast term by a weight depending on the local standard deviation doesn't fit well with the basic postulates of Wilson and Cowan's theory.

For this reason, in [35] we proposed a different update of the LHE model, incorporating the dual wide-narrow form of the surround discussed in the previous section. This new formulation, while still based on efficient coding and having the proper elements of a Wilson–Cowan equation, is now able to qualitatively replicate psychophysical data on assimilation and contrast, both on achromatic and chromatic data.

Going back to the generalised form of the Wilson–Cowan equations used in [7]

$$I_t(x,t) = -\alpha(I(x,t) - \frac{1}{2}) + \gamma \sum_y w(x,y) sgn(I(x,t) - I(y,t)) \\ - \beta(I(x,t) - I_0(x)), \tag{7.7}$$

we replace the target global mean average $\frac{1}{2}$ with $\mu(x,t)$, which is a local and time-varying average:

$$I_t(x,t) = -\alpha(I(x,t) - \mu(x,t)) + \gamma \sum_y w(x,y) sgn(I(x,t) - I(y,t))$$
$$- \beta(I(x,t) - I_0(x)). \qquad (7.8)$$

The value for $\mu(x,t)$ is computed by convolving the image I with a kernel K obtained as a weighted sum of two Gaussians, a wider one with less weight and a narrow one with larger weight, as shown in Fig. 7.19.

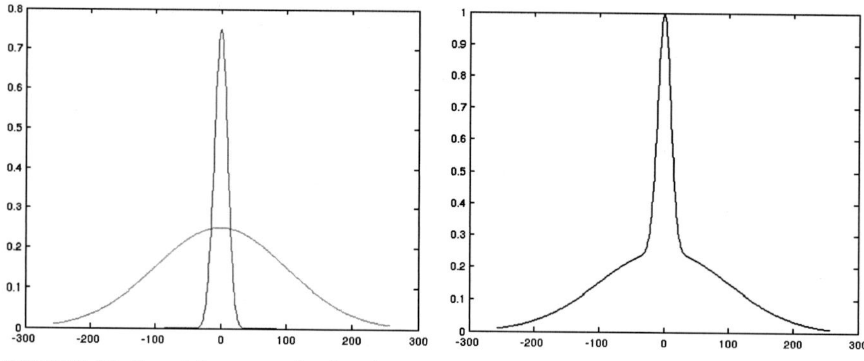

FIGURE 7.19 Kernel for computing local mean

Left: scaled Gaussians. Right: resulting kernel K for computing the local mean.

The model given by Eq. (7.8) now complies with the basic postulates of Wilson–Cowan. Again, how we apply it in practice to an input image I_0 is by running Eq. (7.8) to steady state, with $I = I_0$ as initial condition.

The results thus obtained show both assimilation and contrast, as it can be seen in Figs. 7.20, 7.21 and 7.22. By this we mean that by applying Eq. (7.8) we obtain results where regions that had the same value in the original image now have different values, and they are different in a way that is consistent with the perceptual phenomena of assimilation and contrast. For example, in Fig. 7.21 we see that Eq. (7.8) has made darker the gray bars that were on a dark background, and it has made lighter the gray bars that were on a light background, and therefore Eq. (7.8) is producing a result that mimics assimilation.

We want to point out that all these achromatic and chromatic induction results have been obtained with the same set of parameter values and kernel sizes, i.e. *the change from assimilation to contrast is only due to the spatial content of the images since the model is fixed.*

With Eq. (7.8) we can qualitatively replicate the psychophysical results of the induction experiments of both Helson [23] for achromatic stimuli and Fach and Sharpe [24] for chromatic images. As mentioned earlier, in these experiments observers were presented with stimuli consisting of gratings, identical bars over a uniform background, and reported the strongest assimilation effects for the largest spatial frequencies (thinnest bars); the strength of the effects would decrease with the increase in bar width, turning into contrast effects for wide bars. Fig. 7.23 shows plots

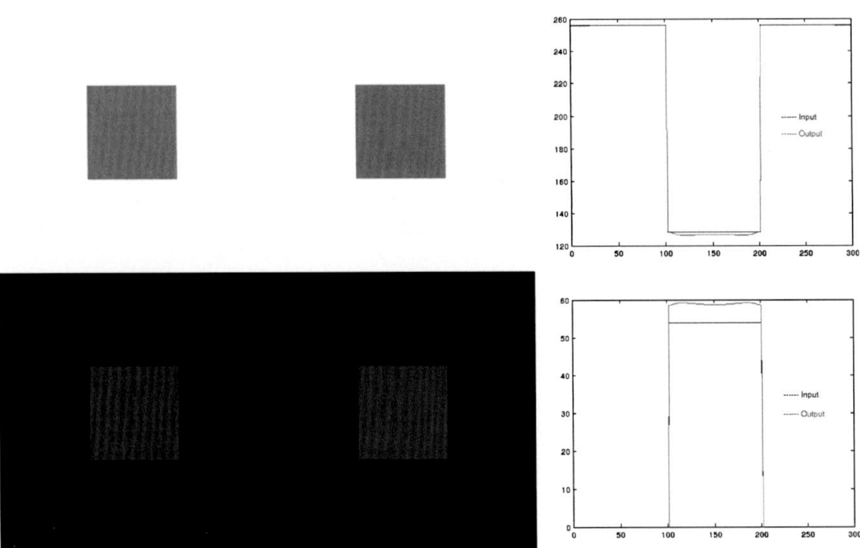

FIGURE 7.20 Achromatic induction example: contrast

Left: input. Middle: output by method proposed in [35]. Right: plot of image values of the middle row of input and output images.

of some of the psychophysical data obtained in these experiments, and these plots are matched qualitatively by the results obtained with Eq. (7.8), as it can be seen in Fig. 7.24.

7.3 Retinal processing optimises contrast coding

In Section 7.2.3 we saw that lateral inhibition in the retina had an impact on a number of perceptual phenomena, including visual induction and colour constancy, and how the properties of these phenomena depended on the sizes of the RFs of the retinal cells.

In [36] we went back to the retinal models we updated and analysed in [29], studied what were their most essential elements, and produced the simplest possible form of equations to model the retinal feedback system that are nonetheless capable of predicting a number of significant contrast perception phenomena like brightness induction (assimilation and contrast) and the band-pass form of the contrast sensitivity function.

These equations form a system of partial differential equations (PDEs), and the steady state solution V to an input signal I_0 is shown to have the following form:

$$V = K * I_0, \tag{7.9}$$

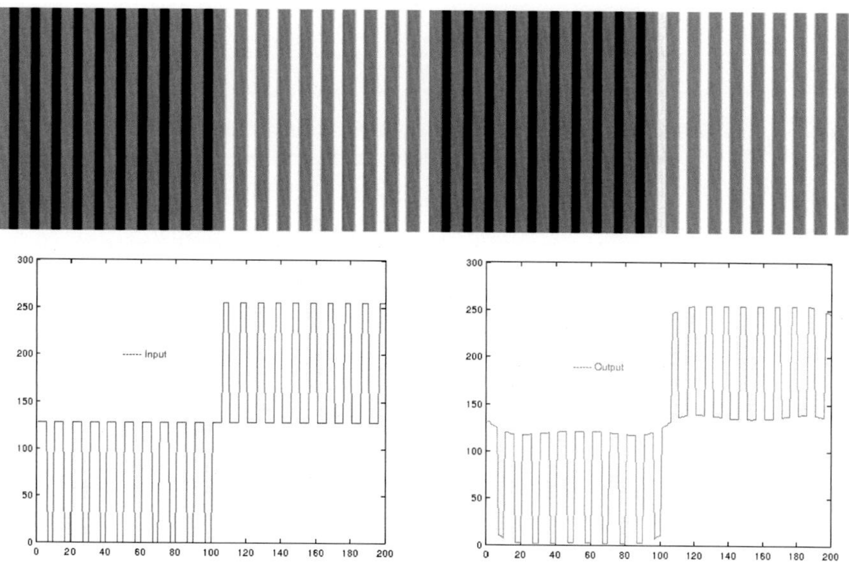

FIGURE 7.21 Achromatic induction example: assimilation

Left: input. Middle left: output by method in [35]. Middle right and right: plot of image values of the middle row of input and output images.

where K is a kernel whose shape depends on K_a and K_b, which are the RFs of the feed-forward cells and interneurons, respectively:

$$K = \mathcal{F}^{-1}\left(\frac{\mathcal{F}(K_a)}{1+\mathcal{F}(K_b)},\right) \tag{7.10}$$

with $\mathcal{F}$ representing the Fourier transform. An example kernel K is shown in Fig. 7.25. Notice that while it has a center-surround form, from Eq. (7.10) it can be seen that K is not well approximated by a difference of Gaussians. This is consistent with the experimental results on CSFs that we discussed in Chapter 5.

If we now consider the following energy functional:

$$\begin{aligned} E(V) = &\frac{\alpha}{2}\int_{\Omega}(V(x) - K_1 * I_0(x))^2 dx \\ &- \frac{\gamma}{2}\int_{\Omega^2} K_2(x,y)(V(x) - V(y))^2 dxdy, \end{aligned} \tag{7.11}$$

it can be shown that the minimum of E is achieved by the following image V:

$$V = \mathcal{F}^{-1}\left(\frac{\alpha\mathcal{F}(I_0)\mathcal{F}(K_1)}{(\alpha-\gamma)+\gamma\mathcal{F}(K_2)}\right), \tag{7.12}$$

which is equivalent to Eq. (7.9), the steady-state solution of the simplified retinal model, when $\alpha = 2$, $\gamma = 1$, $K_1 = K_a/2$, $K_2 = K_b$.

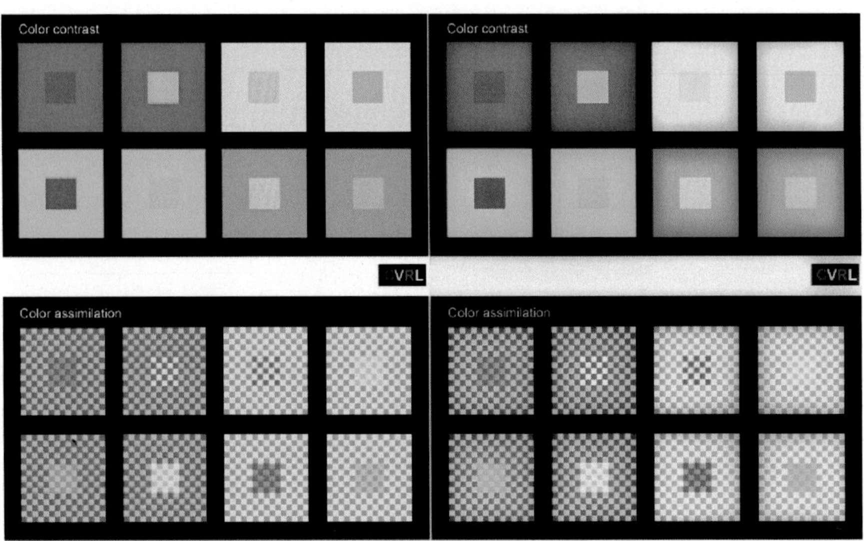

FIGURE 7.22 Chromatic induction examples

Left: input. Right: output from model in [35]. Original figure from [21].

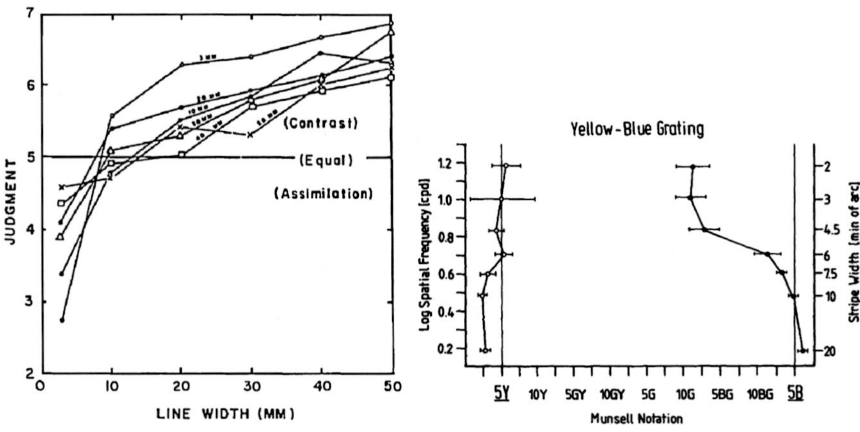

FIGURE 7.23 Psychophysical data on induction

Left: data from Helson, Studies of anomalous contrast and assimilation. JOSA 1963; 53(1):179–184. ©The Optical Society [23]. Right: data from Fach and Sharpe [24].

We can see then that the retinal model proposed in [36] produces a result V that minimises a certain energy E, and this energy has a very close relationship with the energy associated with the LHE model of [7] and that we reproduce again here for

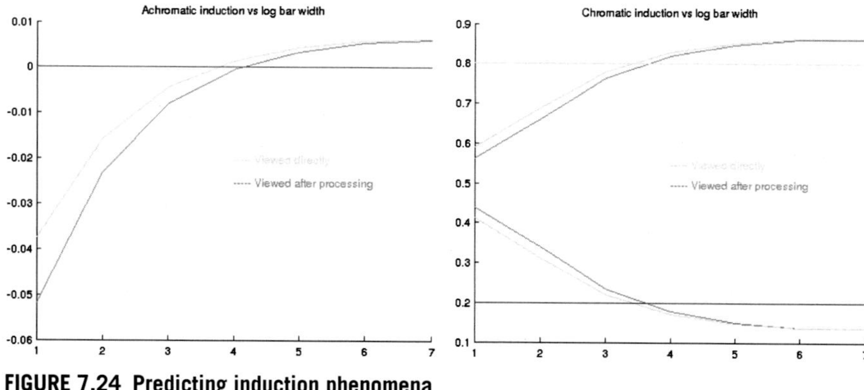

FIGURE 7.24 Predicting induction phenomena

Left: achromatic grating, assimilation below blue (dark gray in print version) line, contrast above blue line. Right: yellow–blue grating; top plots correspond to yellow (ligth gray in print version), and yellow line separates assimilation from contrast effects; bottom plots correspond to blue, and blue line separates assimilation from contrast effects. Comparing with Fig. 7.23 we can see that the plots in the current figure approximate the plots of both Helson for achromatic stimuli and Fach and Sharpe for chromatic images, which were obtained through psychophysical tests.

convenience:

$$E_{LHE}(V) = \frac{\alpha}{2}\int_{\Omega}(V(x) - \frac{1}{2})^2 - \gamma\int_{\Omega^2} w(x,y)|V(x) - V(y)| + \frac{\beta}{2}\int_{\Omega}(V(x) - I_0(x))^2. \tag{7.13}$$

The main difference between Eqs. (7.9) and (7.13) resides in the contrast term, i.e. the second term of the right-hand side. This term has the form of the weighted average of the absolute value of differences in (7.13), while in (7.9) these differences are raised to the power of two, which has the effect of *regularising* the functional $E(V)$, making it convex, and therefore its minimum can be computed with a single convolution as shown in Eq. (7.9), while E_{LHE} is non-convex and as a consequence its minimum has to be found by the iteration of the gradient descent equation.

7.4 A final reformulation of the LHE model

When looking at an image on a display, observers commonly place themselves at a distance from the screen so that the angle of view is larger for larger displays. We have seen previously how the magnitude of the induction effects depends on spatial frequency, and as a consequence, the visual induction phenomena will also change from one screen to another: the same image may show significant assimilation effects when viewed on a mobile phone, and less assimilation or even contrast when viewed in the cinema. For instance, if the usual field of view when watching a TV screen is of

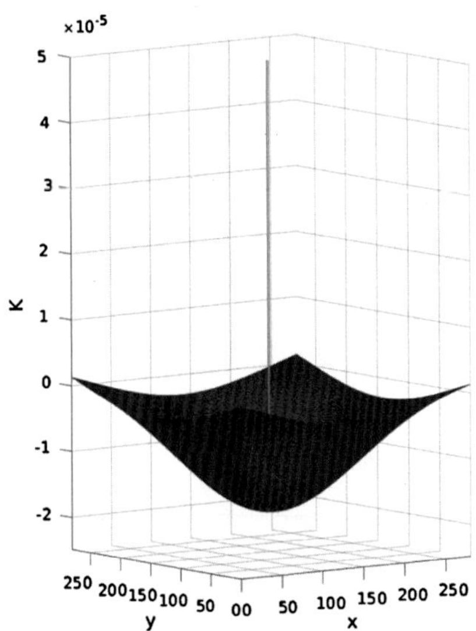

FIGURE 7.25 Kernel for steady state solution in simplified retinal model

Kernel for the steady state solution of the simplified retinal model of [36]. It has a center-surround form but it is not well approximated by a difference of Gaussians.

A degrees and when watching a mobile phone screen it is B degrees, and it happens that $A > B$; therefore, an image grating of C cycles has spatial frequency C/A on the TV that is smaller than the spatial frequency C/B of the same image on the phone. As a consequence the same image produces a larger assimilation effect when viewed on a phone than on a TV.

In order to deal with this, aiming at pre-processing images so as to compensate for the induction effect introduced by the difference in screen sizes, we presented in [37] a model for visual induction that extends the original LHE formulation of [7] by combining ideas introduced in works discussed above:

- We use a local mean computation as in [25].
- We estimate the mean by a convolution with a kernel that is the sum of two Gaussians, a wider one with small amplitude and a narrow one with larger amplitude, as in [35].
- In the contrast term of the energy functional we replace the absolute value function with an exponential to the power of two, as in the retinal model of [36].

The proposed model consists then of the following energy functional:

$$E(O) = \frac{\alpha}{2}\int_{\Omega}(O(x) - K_m * O(x))^2 dx - \frac{\gamma}{2}\int_{\Omega^2} K_c(x,y)(O(x) - O(y))^2 dxdy + \frac{\beta}{2}\int_{\Omega}(O(x) - I(x))^2, \tag{7.14}$$

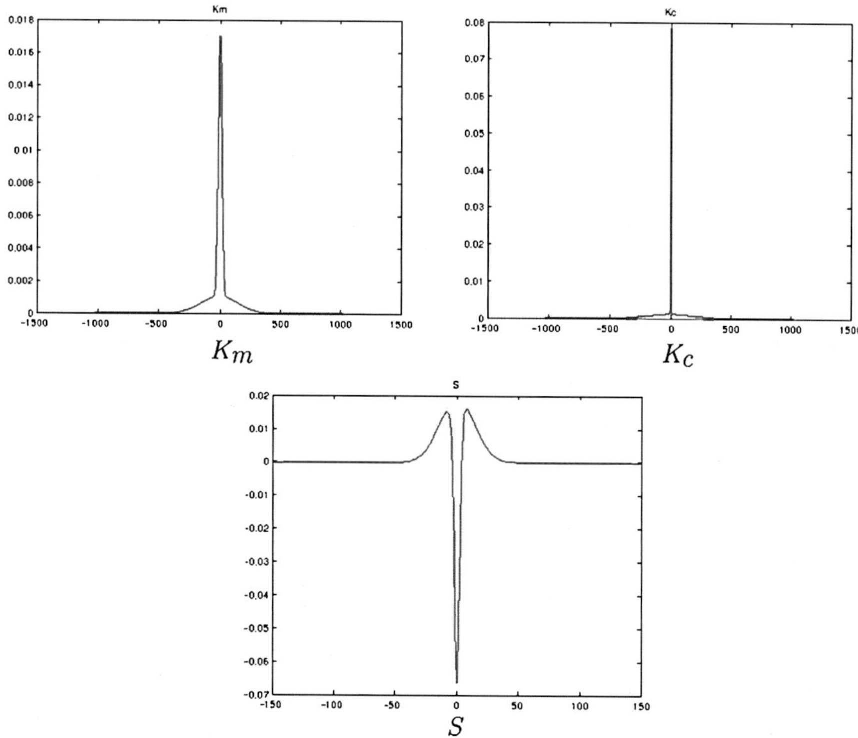

FIGURE 7.26 Kernels for improved LHE model

Kernels for the improved LHE model of [37]. K_m is the kernel for computing the local mean, K_c is the kernel used for the weighted average in the contrast term, and S is the kernel that when convolved with the input provides the output of the method.

where the first term penalises deviations from the (local) mean, the second term enhances contrast and the third term penalises deviations from the input image I.

Again, the presence of the square function in the contrast term has the effect of regularising the energy functional and its minimum can be computed directly, without the need to iterate the gradient descent:

$$O = S * I, \tag{7.15}$$

with

$$S = \mathcal{F}^{-}1\left(\frac{\beta}{\alpha+\beta-\gamma-\alpha\mathcal{F}(K_m)+\gamma\mathcal{F}(K_c)}\right). \tag{7.16}$$

So in terms of induction effects, the appearance of input image I is modelled by output image O, obtained through convolution with kernel S.

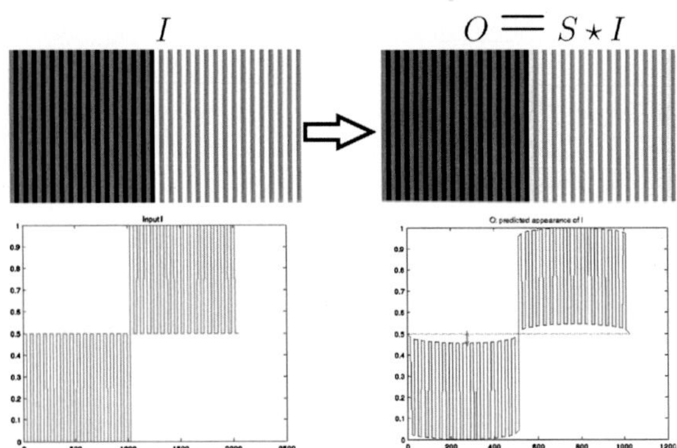

FIGURE 7.27 Improved LHE model predicts assimilation

The improved LHE model of [37] predicts assimilation. Left: input image (top) and values for the middle row of the image (bottom). Right: output. Red (mid gray in print version) arrow marks the amount of the assimilation effect.

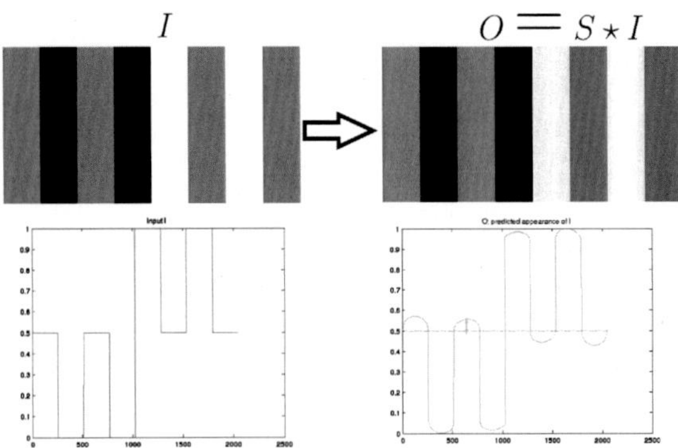

FIGURE 7.28 Improved LHE model predicts contrast

The improved LHE model of [37] predicts contrast. Left: input image (top) and values for the middle row of the image (bottom). Right: output. Green (ligth gray in print version) arrow marks the amount of the assimilation effect.

Fig. 7.26 shows the form of the kernels K_m, K_c and S (these are just 1D plots but the kernels of course are 2D, and rotationally symmetric) that yield the best results: we found a set of values for the parameters determining S that allow this model to qualitatively replicate the results from the psychophysical experiments of Helson [23] for a wide range of bar sizes.

Example outputs from this improved model are shown in Figs. 7.27 and 7.28.

A plot of the strength of the induction effect as a function of gray bar width is shown in Fig. 7.29: notice the qualitative similarity with the psychophysical data obtained by Helson, shown previously in Fig. 7.12. The red (mid gray in print version) and green (light gray in print version) arrows in Fig. 7.29 correspond to the amplitude

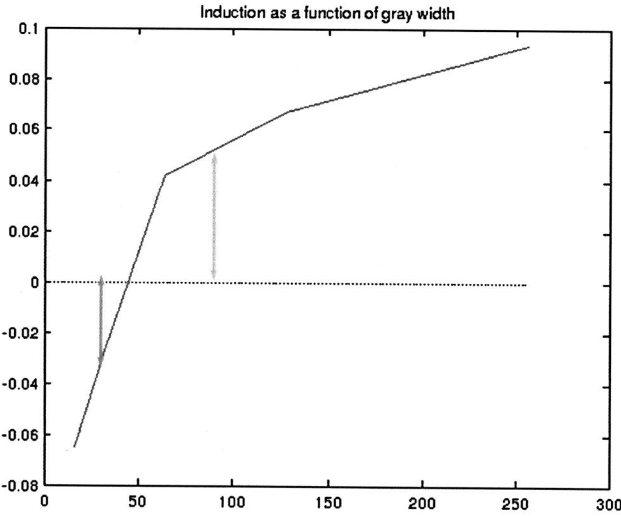

FIGURE 7.29 Improved LHE model reproduces Helson's data

The improved LHE model of [37] is able to qualitatively reproduce the psychophysical data on induction obtained by Helson [23] and shown previously in Fig. 7.12.

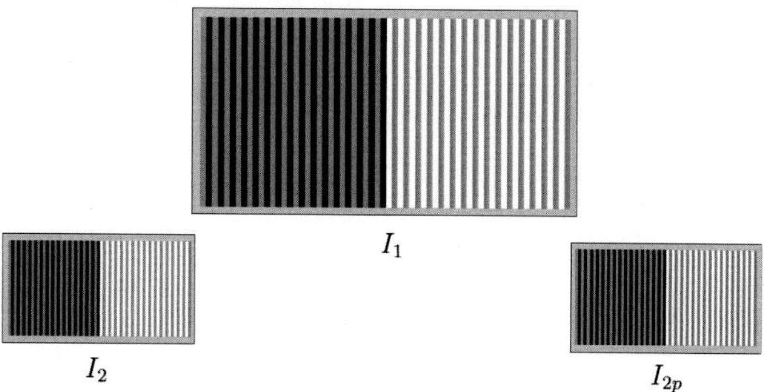

FIGURE 7.30 Pre-compensating for changes in induction

Pre-compensating for changes in induction. I_1 and I_2 are identical in values but different in size, so they look different to us due to visual induction effects. Image I_{2p} has been obtained using the method in [37], and its appearance is now quite similar to I_1.

of the assimilation and contrast effects in Figs. 7.27 and 7.28, where the arrows are marked as well.

7.4.1 Pre-correcting for changes in induction

Let I_1 and I_2 be two differently-sized instances of the same image (e.g. the same image shown in a cinema screen and in a mobile phone screen). They will differ also in terms of spatial frequency so the improved LHE model predicts different outputs for them:

$$O_1 = S * I_1 \neq S * I_2 = O_2. \tag{7.17}$$

We want to find a compensation kernel C_{12} such that if we pre-process I_2 by convolving it with C_{12} yielding I_{2p}, then the model predicts the same output for I_{2p} as for image I_1:

$$I_{2p} = C_{12} * I_2, \ O_{2p} = S * I_{2p}, \ O_{2p} = O_1. \tag{7.18}$$

This can be achieved if C_{12} has the form:

$$C_{12} = \mathcal{F}^{-1}\left(\frac{\mathcal{F}(O_1)}{\mathcal{F}(O_2)}\right). \tag{7.19}$$

As a result, we can apply this method to cinema exhibition in the following way. If an image I has been colour-graded for display 1, e.g. a TV that produces I_1, and we know it will be shown on display 2, e.g. a mobile phone screen that produces image I_2, from the screen sizes of both displays we can compute C_{12} and convolve this kernel with the image: this way, when it is displayed in 2 the induction effects will be the same as if the image was shown in display 1.

Fig. 7.30 shows an example of the potential of this method. Here image I_2 is a scaled version of I_1, and because of this the visual induction effects make them look different although their values are identical: for instance you can check how the gray bars over the white background look lighter in I_2 than in I_1, which is to be expected as the assimilation effects become stronger when the spatial frequency increases. Image I_{2p} has been obtained by convolving I_2 with C_{12} as explained above, and you can see now that the appearance of I_{2p} and I_1 are quite similar. What remains to be done is to extend the model to predict chromatic induction data, and to perform tests on natural images.

References

[1] Pratt W. Digital image processing. 4th edition. John Wiley & Sons; 2007.

[2] Sapiro G, Caselles V. Histogram modification via differential equations. Journal of Differential Equations 1997;135:238–66.

[3] Judd D. Hue, saturation, and lightness of surface colors with chromatic illumination. Journal of the Optical Society of America 1940;30(1):2–32.

[4] Judd D. The unsolved problem of color perception. In: Contributions to color science, vol. 545. National Bureau of Standards (USA); 1979. p. 516–22.

[5] Buchsbaum G. A spatial processor model for object colour perception. Journal of the Franklin Institute 1980;310:337–50.

[6] Rizzi A, Gatta C, Marini D. A new algorithm for unsupervised global and local color correction. Pattern Recognition Letters 2003;24:1663–77.

[7] Bertalmío M, Caselles V, Provenzi E, Rizzi A. Perceptual color correction through variational techniques. IEEE Transactions on Image Processing 2007;16:1058–72.

[8] Land E. The Retinex theory of color vision. Scientific American 1977;237:108–28.
[9] Land Edwin H, McCann John J. Lightness and retinex theory. JOSA 1971;61(1):1–11.
[10] Judd D. Appraisal of Land's work on two-primary color projections. In: Contributions to color science, vol. 545. National Bureau of Standards (USA); 1979. p. 471–86.
[11] Judd D. Color appearance. In: Contributions to color science, vol. 545. National Bureau of Standards (USA); 1979. p. 539–64.
[12] Bertalmío M, Caselles V, Provenzi E. Issues about Retinex theory and contrast enhancement. International Journal of Computer Vision 2009;83(1):101–19.
[13] Bertalmío M. Image processing for cinema. CRC Press, Taylor & Francis; 2014.
[14] Provenzi E, De Carli L, Rizzi A, Marini D. Mathematical definition and analysis of the Retinex algorithm. JOSA A 2005;22(12):2613–21.
[15] Palma-Amestoy R, Provenzi E, Bertalmío M, Caselles V. A perceptually inspired variational framework for color enhancement. IEEE Transactions on Pattern Analysis and Machine Intelligence 2009;31(3):458–74.
[16] Bressloff P, Cowan J, Golubitsky M, Thomas P, Wiener M. What geometric visual hallucinations tell us about the visual cortex. Neural Computation 2002;14(3):473–91.
[17] Wilson H, Cowan J. Excitatory and inhibitory interactions in localised populations of model neurons. Biophysical Journal 1972;12:1–24.
[18] Wilson H, Cowan J. A mathematical theory of the functional dynamics of cortical and thalamic nervous tissue. Biological Cybernetics 1973;13(2):55–80.
[19] Bertalmío M, Cowan J. Implementing the Retinex algorithm with Wilson–Cowan equations. Journal of Physiology, Paris 2009.
[20] Bertalmío M, Calatroni L, Franceschi V, Franceschiello B, Prandi D. A cortical-inspired model for orientation-dependent contrast perception: a link with Wilson–Cowan equations. In: International conference on scale space and variational methods in computer vision; 2019.
[21] Stockman A, Brainard DH. Color vision mechanisms. Vision and Vision Optics 2010:1–104.
[22] Shevell S. Color appearance. In: The science of color. Elsevier Science Ltd; 2003. p. 149–90.
[23] Helson H. Studies of anomalous contrast and assimilation. JOSA 1963;53(1):179–84.
[24] Fach C, Sharpe LT. Assimilative hue shifts in color depend on bar width. Perception & Psychophysics 1986;40(6):412–8.
[25] Bertalmío M. From image processing to computational neuroscience: a neural model based on histogram equalization. Frontiers in Computational Neuroscience 2014;8:71.
[26] Georgeson M, Sullivan G. Contrast constancy: deblurring in human vision by spatial frequency channels. The Journal of Physiology 1975;252(3):627–56.
[27] Martinez L, Molano-Mazón M, Wang X, Sommer F, Hirsch J. Statistical wiring of thalamic receptive fields optimizes spatial sampling of the retinal image. Neuron 2014;81(4):943–56.
[28] Ferradans S, Bertalmío M, Provenzi E, Caselles V. An analysis of visual adaptation and contrast perception for tone mapping. IEEE Transactions on Pattern Analysis and Machine Intelligence 2011.
[29] Yeonan-Kim J, Bertalmío M. Retinal lateral inhibition provides the biological basis of long-range spatial induction. PLoS ONE 2016;11(12):e0168963.
[30] Blakeslee B, McCourt ME. A multiscale spatial filtering account of the white effect, simultaneous brightness contrast and grating induction. Vision Research 1999;39(26):4361–77.
[31] Wilson HR. A neural model of foveal light adaptation and afterimage formation. Visual Neuroscience 1997;14(3):403–23.
[32] van Hateren H. A cellular and molecular model of response kinetics and adaptation in primate cones and horizontal cells. Journal of Vision 2005;5(4).
[33] Land EH. Recent advances in Retinex theory and some implications for cortical computations: color vision and the natural image. Proceedings of the National Academy of Sciences of the United States of America 1983;80(16):5163.
[34] Yeonan-Kim J, Bertalmío M. Analysis of retinal and cortical components of Retinex algorithms. Journal of Electronic Imaging 2017;26(3):031208.
[35] Bertalmío M. Correcting for induction phenomena on displays of different size. In: AIC Midterm meeting; 2015.

[36] Kim J, Batard T, Bertalmío M. Retinal processing optimizes contrast coding. Journal of Vision 2016;16(12).

[37] Bertalmío M, Batard T, Kim J. Correcting for induction phenomena on displays of different size. Journal of Vision 2016;16(12).

CHAPTER

Vision models for gamut mapping in cinema

8

Gamut mapping is the problem of transforming the colours of image or video content so as to fully exploit the colour palette of the display device where the content will be shown, while preserving the artistic intent of the original content's creator. In particular in the cinema industry, the rapid advancement in display technologies has created a pressing need to develop automatic and fast gamut mapping algorithms.

In this chapter we show how the local histogram equalisation approach presented earlier can be used to develop gamut mapping algorithms that are of low computational complexity, produce results that are free from artifacts and outperform state-of-the-art methods according to psychophysical tests.

Another contribution of our research is to highlight the limitations of existing image quality metrics when applied to the gamut mapping problem, as none of them, including two state-of-the-art deep learning metrics for image perception, are able to predict the preferences of the observers.

8.1 Introduction

We begin this chapter by recalling some basic concepts on colour.

The range of colours that a device is able to reproduce is called its colour gamut. A very common and convenient way of describing colours is to ignore their luminance component and just represent the chromatic content on a 2D plane known as the CIE xy chromaticity diagram, shown in Fig. 8.1. In this figure the tongue-shaped region corresponds to the chromaticities of all the colours a standard observer can perceive.

Most existing displays are based on the trichromacy property of human vision, creating colours by mixing three well-chosen red, green and blue primaries in different proportions. The chromaticities of these primaries determine a triangle in the CIE xy chromaticity diagram, and this triangle is the colour gamut of the display in question. Therefore, for any given three-primary display there will be many colours that we could perceive but the display is not able to generate, i.e. all the colours with chromaticities outside the triangle associated to the display. Also, devices with different sets of primaries will have different gamuts.

For this reason, in order to facilitate inter-operability a number of standard distribution gamuts have been defined, and for cinema the most relevant ones are shown

Vision Models for High Dynamic Range and Wide Colour Gamut Imaging. https://doi.org/10.1016/B978-0-12-813894-6.00013-2

in Fig. 1: DCI-P3 [1] is the standard gamut used in cinema postproduction and recommended for digital cinema projection, BT.709 [2] is used for cable and broadcast TV, DVD, Blu-Ray and streaming, and BT.2020 [3] is a very wide colour gamut for the next generation UHDTV, currently only achievable by some state-of-the-art laser projectors. Fig. 1 also shows Pointer's gamut [4], which covers all the frequently occurring real surface colours; we can see how only BT.2020 is able to completely include Pointer's gamut.

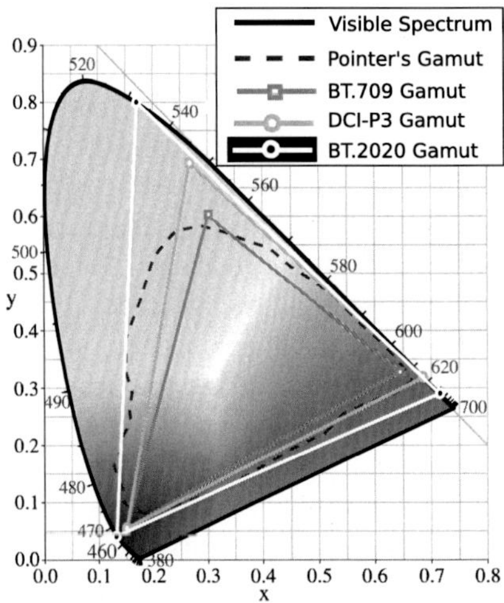

FIGURE 8.1

Gamuts on CIE xy chromaticity diagram.

The adaptation to a standard gamut implies altering the range of colours of the original content. This process is either carried out within the camera in live TV broadcasts (or low-budget movie productions), or performed off-line by expert technicians in the cinema industry. In practice, for the purpose of modifying the movie gamut, colourists at the post-production stage build 3D look-up-tables (LUTs) for each movie or specific scenes in it. These LUTs contain millions of entries and colourists only specify a few colours manually, while the rest are interpolated regardless of their spatial or temporal distribution [5]. Subsequently, the resulting movie may have false colours that were not originally present.

To tackle this problem, colourists usually perform intensive manual correction in a shot-by-shot, object-by-object basis. This process is difficult, time consuming and expensive, and therefore it makes an automated procedure called gamut mapping (GM) very desirable: GM transforms an image so that its colours better fit the target gamut.

In general, there are two types of GM procedures. First is gamut reduction (GR), in which colours are mapped from a larger source gamut to a smaller destination gamut. A common situation where GR is necessary is when a movie intended for cinema viewing is displayed on a TV [6], [7]. Second is gamut extension (GE), that involves mapping colours from a smaller source gamut to a larger destination gamut. For example, wide-gamut state-of the-art displays often receive movies that are encoded with limited-gamuts as a precaution measure against regular (or poor) display devices; therefore, we cannot exploit the full colour rendering potential of these new devices unless we use a GE procedure [5]. The process of GE is gaining importance with the introduction of new display technologies and laser projectors [8], [9]. These new displays use pure (very saturated) colour primaries which enable them to cover much wider gamuts, so now a tablet screen may have a DCI-P3 gamut for instance, while all the content it shows comes in the smaller BT.709 standard.

We must also point out that GR is required, not optional, when the gamut of the display does not cover some colours of the input image: if GR isn't performed in this case, the image on the screen may show visible artifacts. But GE is not an essential process, no visual artifact is generated by omitting to perform GE; rather, GE is considered as an enhancement operation [10].

In this chapter we will present a number of gamut mapping works done in collaboration with Syed Waqas Zamir and Javier Vazquez-Corral, and that are based on the vision models related to local histogram equalisation that we reviewed in Chapter 7.

8.2 Related work

There are a large number of gamut mapping algorithms (GMAs) that have been proposed in the literature, we refer the interested reader to the comprehensive book of Morovič [10]. In general, GMAs can be divided into two main categories: global GMAs and local GMAs. Global (also known as non-local or non-adaptive) GMAs map colours of an image to the target gamut independently, completely ignoring the spatial distribution of colours in the image. Whereas local GMAs modify pixel values by taking into account their neighbourhoods; as a result, two identical values surrounded by different neighbourhoods will be mapped to two different values.

One class of global GR algorithms (GRAs) consists of gamut clipping methods. Gamut clipping is a very common approach to perform gamut reduction where colours that lie inside the destination gamut are left untouched while those colours that fall outside are projected onto the destination gamut boundary. In order to produce reduced-gamut images, gamut clipping techniques use particular strategies and mapping directions, e.g. [11–16].

Clipping GRAs, due to their inherent behaviour, project whole out-of-gamut colour segments into single points on the target gamut, and therefore they may produce images with a visible loss of detail. To avoid this issue, another class of global GRAs called gamut compression algorithms modify all colours of an input image, both inside and outside the target gamut. Such functionality enables compression GRAs to map out-of-gamut colour segments to in-gamut colour segments (instead of single points), though the results they produce may lack in saturation, specially when there is a large difference between the source gamut and the target gamut. See for instance [17–24].

Local GRAs are also known as 'spatial' methods. Some popular approaches are:

- Frequency-based, that use a spatial filter to separate the high-frequency image detail and perform global gamut mapping on the base layer [25–28].
- Inspired in the Retinex framework, performing spatial comparisons to retain source image gradients in the reproduced images [29–32].
- Optimisation-based, where the gamut mapping is an iterative procedure that minimises some error metric [33–38].

While the majority of the published GMAs deal with the problem of gamut reduction, the case is very different for gamut extension: only a few works have been proposed in this direction. We can mention for instance the GE algorithms (GEAs) of [39–47]. One simple solution to perform gamut extension would be to take any compression GRA and use it in the reverse direction, but this way of approaching GE may yield images that are unnatural and unpleasant in appearance.

8.3 Gamut mapping with the LHE model

We recall from Chapter 7 how in [48] the authors propose a variational method for colour and contrast enhancement consisting in minimising the following energy functional, a process that performs local histogram equalisation (LHE):

$$E(I) = \frac{\alpha}{2}\sum_x\left(I(x)-\frac{1}{2}\right)^2 - \frac{\gamma}{2}\sum_x\sum_y w(x,y)|I(x)-I(y)| + \frac{\beta}{2}\sum_x (I(x)-I_0(x))^2, \tag{8.1}$$

where α, β and γ are constant weights, I is a colour channel in the range $[0, 1]$, I_0 is the original image channel, $w(x, y)$ is a normalised Gaussian kernel of standard deviation σ, and x and y are pixel locations. The constants α and β are always positive, so minimising $E(I)$ penalises the departure from the original image (third term of the functional) and from a mean value of $1/2$ (first term).

If γ is positive, minimising $E(I)$ amounts to increasing $\sum_x \sum_y w(x, y)|I(x) - I(y)|$, i.e. the local contrast. If, on the contrary, $\gamma < 0$, then the minimisation of Eq. (8.1) *reduces*, not increases, the contrast, as pointed out in [49].

Fig. 8.2, top row, shows example outputs that can be obtained with this approach. The left image is the input, the middle image is the result that minimises Eq. (8.1) for some $\gamma > 0$, and the image on the right is the result that minimises Eq. (8.1) for some $\gamma < 0$. Notice how, in the middle image, the contrast has been enhanced and the colours have become more saturated, while the opposite happens in the image on the right, where contrast has been reduced as well as the chromaticity of the pixel colours. This is corroborated in the bottom row, that plots the chromaticity diagrams for the pictures above.

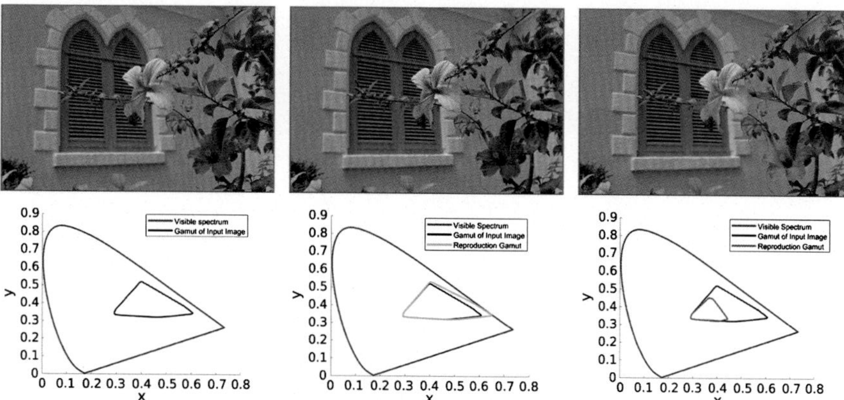

FIGURE 8.2 Example images and their corresponding gamut plots

Example images and their corresponding gamut plots. Left column: input image. Middle column: extended-gamut image by applying the LHE model with $\gamma > 0$ [48]. Right column: reduced-gamut image by applying the LHE model with $\gamma < 0$ [48]. Original image from [50].

8.3.1 Gamut mapping by applying LHE in RGB

The example of Fig. 8.2 highlights the potential for the variational method of [48] to be used for gamut mapping: with $\gamma > 0$ it is capable of producing gamut extension, and gamut reduction when $\gamma < 0$. That is the approach we followed in [51]: after replacing the value of $1/2$ in the first term of the functional with the global mean average value μ, the minimum of Eq. (8.1) for each channel $I \in \{R, G, B\}$ can be obtained by iterating

$$I^{k+1}(x) = \frac{I^k(x) + \Delta t \left(\alpha\mu + \beta I_0(x) + \frac{\gamma}{2} R_{I^k}(x)\right)}{1 + \Delta t(\alpha + \beta)}, \tag{8.2}$$

where the initial condition is $I^{k=0}(x) = I_0(x)$, $R_{I^k}(x)$ indicates the contrast function

$$R_{I^k}(x) = \frac{\sum_y w(x, y) s\left(I^k(x) - I^k(y)\right)}{\sum_y w(x, y)}, \tag{8.3}$$

and the slope function $s(\cdot)$ is a regularised approximation of the sign function, which in [48] is chosen as a polynomial of degree 7.

In detail, for gamut reduction we propose in [51] an iterative approach, where at each iteration we run Eq. (8.2) independently on each channel $I \in \{R, G, B\}$ for some particular α, β, and γ until we reach the steady state. At this point we check all pixels and those that have values that are inside the target gamut are kept untouched for all subsequent iterations, i.e. these pixels will be part of the final output. Next we decrease the value of γ and proceed to the following iteration, repeating the process until all the out-of-gamut colours come inside the destination gamut. An example of this iterative procedure is shown in Fig. 8.3, where the vivid green colour marks

FIGURE 8.3 Gradual mapping of colours

Gradual mapping of colours. Out-of-gamut colours (in green) when (A) $\gamma = 0$, (B) $\gamma = -0.22$, (C) $\gamma = -0.83$, (D) $\gamma = -3.21$. As γ decreases the number of out-of-gamut pixels is reduced. Figure from [51].

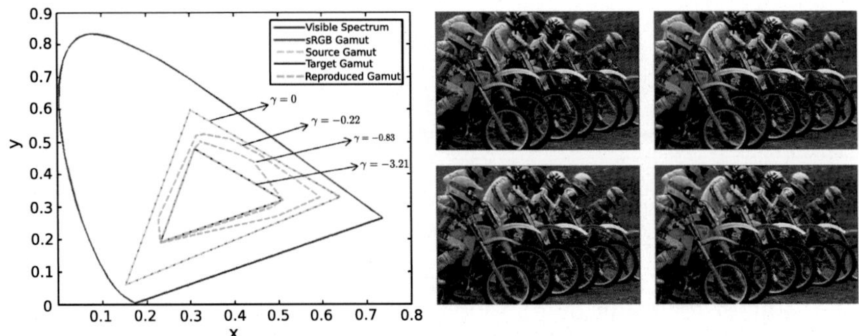

FIGURE 8.4 GR approach

GR approach. (A) Gamuts on chromaticity diagram. (B) Top left: original image. Top right: $\gamma = -0.22$. Bottom left: $\gamma = -0.83$. Bottom right: $\gamma = -3.21$. Figure from [51].

the pixels that are out-of-gamut at that iteration. Fig. 8.4 shows how the gamut is iteratively reduced towards the target.

The proposed method is shown in Fig. 8.5 to outperform the state of the art visually, with better colour and detail preservation. This is the case both for static images and for video sequences, which show no spatiotemporal artifacts.

For gamut extension the process is no longer iterative, but our GEA consists of three stages. First we slightly shift to the right the histogram of each channel, so as to prevent pixel values from going to black in the subsequent contrast enhancement step. Next we run Eq. (8.2) with positive α, β, and γ until we reach the steady state. The value of γ is selected so that the processed image has a gamut slightly larger than the destination gamut. Finally, the out-of-gamut values are mapped back inside using the GRA described previously.

While in [51] we present experimental comparisons and visual and quantitative assessments of the results of our GEA, these tests were rather limited because they were not performed in realistic conditions: the target gamut was not an actual wide gamut like the DCI-P3 used in cinema, and the validation did not involve psychophysical tests in the proper set-up (e.g. large screen, low ambient light like in a movie theater). We addressed these limitations in our following work, described in the next section.

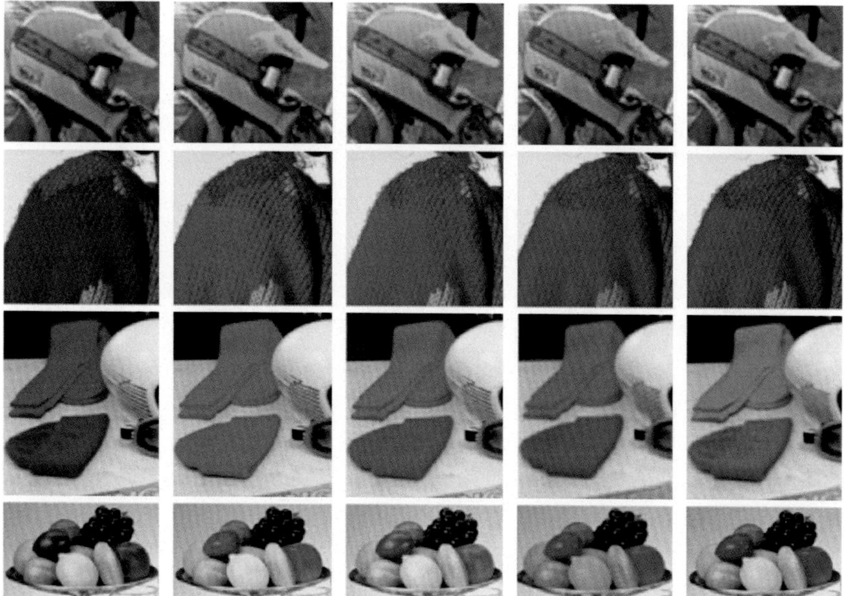

FIGURE 8.5 Detail preservation with GRAs

Detail preservation using GRAs on still images. Column 1: original cropped regions. Column 2: output of HPMINDE [12]. Column 3: output of Lau et al. [34]. Column 4: output of Alsam et al. [30]. Column 5: output of our GRA in RGB from [51]. Figure from [51].

Finally, let's point out that all the GM methods discussed in this chapter always assume that one of the gamuts is fully included in the other; when this is not the case, we can apply the extension of [51] that we presented in [52] and that can be used not only for gamut mapping but also in order to make an image more similar (or coherent) in terms of colour with respect to a given reference.

8.3.2 Gamut extension by applying LHE in AB channels of CIELAB

The GEA in RGB presented in the previous section, due to its inherent behaviour of expanding colours by increasing the contrast of the image, produces results with over-enhanced contrast, which in turn makes a few colours go towards black (loss of saturation), as it is visible in Fig. 8.6. It can be seen that the overall contrast of the reproduction is increased noticeably, making it depart from the original image. Also, the over-enhancement of contrast causes loss of colour details as it is shown in the area highlighted by a bounding box in Fig. 8.6.

To overcome these problems, we introduced in [53] a new GEA based on [51]; this method performs gamut extension using the same energy functional (Eq. (8.1)) and its corresponding evolution equation (Eq. (8.2)) as used by [51], but under the CIELAB colour space. This key modification eliminates not only the problems with saturation and contrast in the reproduced images, but also the need to perform any sort of preprocessing as it was the case with the GEA in RGB method of [51].

FIGURE 8.6 GE example

Gamut extension example. Left: input image. Middle: result of GEA in RGB from [51]. Right: result of GEA in Lab. Figure from [54].

To perform gamut extension, the method in [53] first converts the RGB input image to the CIELAB colour space, and then the evolution Eq. (8.2) is applied, independently, on channels a and b, while the lightness channel L is left untouched. First, we initialise our final image and set all its values to zero. Then, we run Eq. (8.2) for $\beta = 1$, $\alpha = 0$, and $\gamma = 0$ until we reach the steady state. For each pixel of the steady state solution we check if it accomplishes at the same time three different constraints, one in saturation, one in hue, and one in chroma:

- The saturation can't be less than in the original image.
- The hue angle can't be too far apart from the original.
- For pixels belonging to a critical region encompassing some memory colours and skin tones, the chroma value of the gamut extension result can't be too far apart from the original.

In case a pixel doesn't satisfy some or all of these constrains, we select its current value as the value of this pixel in the final image, and don't modify the pixel anymore. For the pixels that do satisfy the three constraints we keep the iteration process by increasing γ and setting $\alpha = \frac{\gamma}{20}$ until either the final image has been filled or until the gamut of the original image exceeds our destination gamut up to a threshold T, at which point the iteration process ends and the out-of-gamut pixels are brought into the target gamut by applying the GRA from [51].

To show how the evolution Eq. (8.2) extends the colour gamut, an example with several different gamuts (visible spectrum, source gamut, target gamut and reproduced gamut) on a chromaticity diagram is shown in Fig. 8.7. It is important to note that for each set of values for α, β, and γ, the evolution Eq. (8.2) has a steady state. For example, it is shown in Fig. 8.7 that when $\beta = 1$, $\alpha = 0$, and $\gamma = 0$ we obtain the original image as the steady state of the evolution equation. Moreover, it can be seen in the same figure that as we increase γ the steady state of Eq. (8.2) has a gamut which is gradually larger.

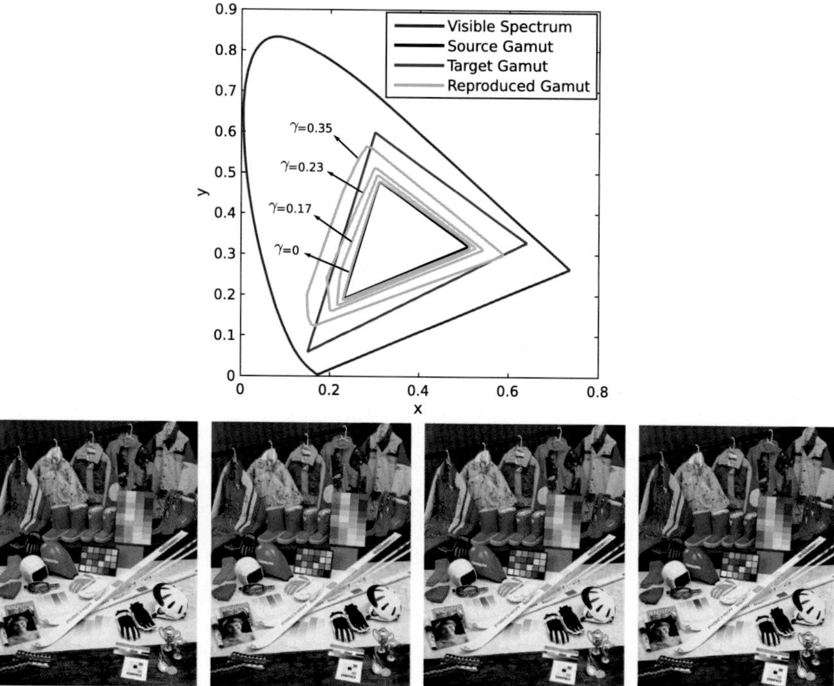

FIGURE 8.7 GE approach

Gamut extension approach. Top: gamuts con chromaticity diagram. Bottom, from left to right: input image ($\gamma = 0$), gamut extended image with $\gamma = 0.17$, $\gamma = 0.23$ and $\gamma = 0.35$. As the γ value increases the gamut becomes larger; notice the increment in saturation of helmets, socks, ski suits and shoes. Figure from [53].

8.3.2.1 Experiments and results

For the experiments, we have chosen the following approach: in cinematic conditions (large screen, dark room) we present observers with three versions of the same image, in the center of the screen will be a reference wide-gamut picture, and to each side of it there will be the results of two GEAs from which the observer has to pick the one that most closely resembles the reference. We generate a reduced-gamut version of each wide-gamut original, to which we apply our method and four competing GEAs: LCA, CE, HCM and SDS [44].

One could question the choice of evaluation criteria: why ask users to choose the most accurate result instead of the one they find most pleasant? The reason is that, for a GE technique to be adopted by the movie industry, it must yield gamut extended results that preserve as much as possible the artistic intent of the content's creator. Designing GEAs for pleasantness does not guarantee this, usually quite the opposite: apart from general inter-subject variability in terms of preference, there is also a strong correlation between colourfulness and perceived image quality, so user tests based on subject preference would rank higher those GEAs that increased colour saturation even if that implied a departure from a subdued colour palette chosen by the movie artist. User tests based on accuracy, on the other hand, are much less subjective (as users are simply asked to estimate differences with respect to the provided reference) and the preservation of the artistic intent is in a sense "built-in" in the procedure, since the ground truth acts as a stand-in for the content's creator intent.

To test the robustness of our approach with respect to different combinations of source and target gamuts, we defined two setups for our experiments:

1. *Setup 1: Mapping from small gamut to DCI-P3 gamut.* As quantum dot displays [55] and laser projectors [57] with their extended gamut capabilities are becoming popular, in the near future the common case will be to have large colour differences between the standard gamut and the gamut of the display. Therefore, this setup is created to investigate how different GEAs will perform when the difference between source and target gamuts is large. To this end, we map the source images from the small 'Toy' gamut (slightly smaller than the BT.709 gamut) to the large DCI-P3 gamut. On the chromaticity diagram, the difference in gamuts for this setup is nearly equal to the difference between BT.709 and BT.2020. This represents the future scenario where we need to show on a wide-gamut display some content that was mastered for TV.
2. *Setup 2: Mapping from BT.709 to DCI-P3 gamut.* In this setup we mimic the practical situation where the source material has BT.709 gamut and we map the source colours to the colours of the DCI-P3 gamut.

The room for the experiment has a low-light ambiance of 1 lux and the illumination measured at the screen was around 750 lux. The glare-free screen used in our experiments was 3 meters wide and 2 meters high. We had 15 observers (10 male and 5 female) all with correct colour vision and with ages between 27 and 44 years (average of 32 years). Observers were asked to sit approximately 5 meters away from the screen.

As already stated before, we used a forced-choice pairwise comparison. The observers were simultaneously shown 3 images: ground-truth (in the center) and the results of two gamut extension methods (located left and right of the ground-truth image). The selection instructions given to the observers were: a) if there are any sort of artifacts in one of the reproductions, choose the other, and b) if both of the reproductions have artifacts or are free from artifacts, choose the one which is perceptually closer to the ground truth.

Moreover, to further validate the robustness of the different GEAs we asked 9 experienced observers (who belong to the image processing community and participated in various psychophysical studies) to perform a second experiment. In this case, they were shown a pair of images side by side in the projection screen: the ground-truth image and the result for some particular method. Observers were asked to look for artifacts and hue shifts in the reproductions as compared with the original material.

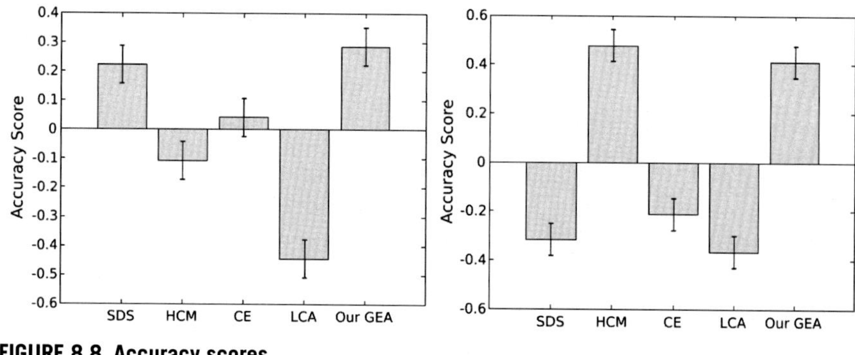

FIGURE 8.8 Accuracy scores

Accuracy scores using 15 observers and 30 images. Left: Setup 1. Right: Setup 2. Figure from [53].

8.3.2.2 Results

To compute accuracy scores from the raw psychophysical data we use the same approach as in [20] (Chapter 5), that is based on Thurstone's law of comparative judgement and that we briefly summarise in what follows.

In order to compare n methods with experiments involving N observers, we create a $n \times n$ matrix for each observer where the value of the element at position (i, j) is 1 if method i is chosen over method j. From the matrices for all observers we create a $n \times n$ frequency matrix where each of its elements shows how often in a pair one method is preferred over the other. From the frequency matrix we create a $n \times n$ z-score matrix, and an accuracy score A for each method is given by the average of the corresponding column in the z-score matrix. The 95% confidence interval is given by $A \pm 1.96\frac{\sigma}{\sqrt{N}}$, as A is based on a random sample of size N from a normal distribution with standard deviation σ. In practice $\sigma = \frac{1}{\sqrt{2}}$, because the z-score represents the difference between two stimuli on a scale where the unit is $\sigma * \sqrt{2}$ (in Thurstone's paper this set of assumptions is referred to as "Case V"); as the scale of A has units which equal $\sigma * \sqrt{2}$, then we get that $\sigma = \frac{1}{\sqrt{2}}$.

The higher the accuracy score is for a given method, the more it is preferred by observers over the competing methods in the experiment.

The analysis for the first and second setups is shown in Fig. 8.8. For the first setup we can see that, when there is a large difference among the source-target gamut pair, our GEA in LAB produces images that are perceptually more faithful to the original as compared with the other competing algorithms. Regarding the second setup the figure shows that when the difference between source and target gamut is smaller the ranking order of the GEAs changes dramatically. In this case the HCM algorithm is ranked as the most accurate method, with our GEA showing a comparable performance.

Results for the second experiment, on visibility of distortions (artifacts or hue shifts) are shown in Fig. 8.9. For the setup 1 subjects noticed artifacts in 25% of the

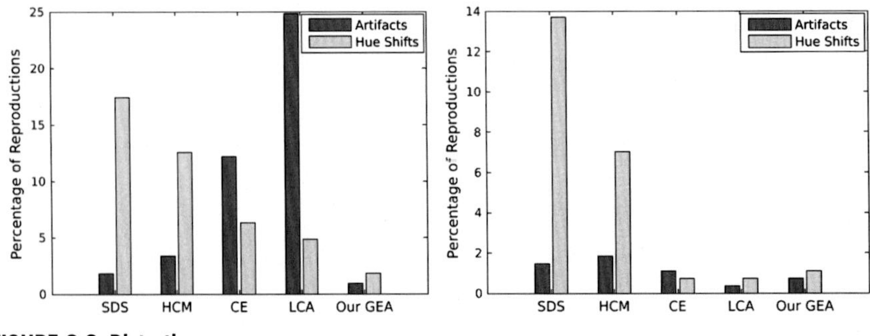

FIGURE 8.9 Distortions

Percentage of reproductions in which 9 experienced observers noticed visual distortions. Left: Setup 1. Right: Setup 2. Figure from [53].

FIGURE 8.10 Examples of artifacts

Examples of artifacts. The bottom row images show a zoomed-in view of the corresponding top image. From left to right: ground truth, output of CE algorithm [44], output of LCA algorithm [44], output of our GEA in LAB. Figure from [53].

reproductions obtained using the LCA algorithm and in 12% of the images in the case of the CE algorithm. The observers also confirmed that our GEA in LAB produces images with very low error rate, around 2%. In terms of hue shifts, both the SDS and HCM algorithms show strong hue shifts. Regarding the setup 2 we can see that the SDS and HCM algorithms produce gamut extended images with strong hue shifts for 13.6% and 7% of the input images, respectively. It can be seen in the same figure that none of the competing algorithms produces images with distinct visual artifacts for setup 2, in which there are small colour differences between source and target gamut.

Finally, Fig. 8.10 presents examples of artifacts found by the observers, while Fig. 8.11 presents examples of hue shifts. In both cases we can see that our method is free of the problems presented by the competing ones. Let us note that in these figures we are only displaying the colours that are inside the sRGB gamut, and masking those that are outside.

FIGURE 8.11 Examples of hue shifts

Examples of hue shifts. The bottom row images show a zoomed-in view of the corresponding top image. From left to right: ground truth, output of SDS algorithm [44], output of HCM algorithm [44], output of our GEA in LAB. Note that these are wide-gamut images where out-of-sRGB pixels are masked green. Figure from [53].

In order to examine the temporal consistency of the GEAs, we conducted a psychophysical study with 9 experienced observers and two colourful image sequences with different levels of motion that had been extended using the different GEAs.

In this experiment, each observer was asked to inspect the following attributes: temporal colour consistency (objects should retain the same hue, chroma and brightness), global flickering, local region flickering, and excessive noise. None of the observers noticed any temporal artifacts, which supports our choice to apply all competing GEAs on each frame independently.

8.4 Gamut mapping with modified LHE model

Finally we present a framework for gamut mapping, introduced in [56], that is based on vision science knowledge and the modified LHE model introduced in Chapter 7, and that allows us to perform both gamut reduction and gamut extension. It is computationally efficient and yields results that outperform state-of the-art methods, as validated using psychophysical tests. Another contribution of this framework is to highlight the limitations of existing image quality metrics when applied to the GM problem, as none of them, including two state-of-the-art deep learning metrics for image perception, trained over large and very large scale databases (20,000+ images in one case, 160,000+ in the other) is able to predict the preferences of the observers.

8.4.1 Some vision principles for gamut mapping

We will start by briefly recapping some principles and findings from the vision science literature that were introduced in earlier chapters and that form the basis of our GM framework, to be presented in the next section.

Light reaching the retina is transformed into electrical signals by photoreceptors, rods and cones. At photopic light levels, rods are saturated and the visual information comes from cones, of which there are three types, according to the wavelengths they are most sensitive to: L (long), M (medium), and S (short). The response of all photoreceptors is non-linear and, for a single cell without feedback, can be well approximated by the Naka-Rushton equation [58].

Photoreceptors do not operate individually though, they receive negative (inhibitory) feedback from horizontal cells, which receive excitatory input from cones and generate inhibitory input to cones. Cone output goes to bipolar cells, that also receive lateral inhibition from horizontal cells and from amacrine cells. Bipolars feed into retinal ganglion cells (RGCs), which also receive input from amacrine cells, and the axons of the ganglion cells form the optic nerve, sending visual signals to the lateral geniculate nucleus (LGN) in the thalamus, where the signals are re-organised into different layers each projecting to a specific layer in the cortex. There are numerous axons providing feedback from the cortex to the LGN, but their influence on colour vision is not known [59].

The lateral inhibition or center-surround processing, in which a cell's output corresponds to the difference between the activity of the cell's closest neighbours and the activity of the cells in the near (and possibly far) surround, allows to encode and enhance contrast therefore being key for efficient representation, and is present at every stage of visual processing from the retina to the cortex. The size of the receptive field (the visual region to which a neuron is sensitive to) tends to increase as we progress down the visual pathway. Lateral inhibition is often modelled as a linear operation, a convolution with a kernel shaped as a difference of Gaussians (DoG). In recent studies, the surround RF of RGCs is modelled as a sum of Gaussians [60]. RGCs produce an achromatic signal by combining information coming from the three cone types (L+M+S), and produce chromatic opponent signals by performing center-surround processing on signals coming from cones of different types: (L+M)–S roughly corresponds to "Yellow–Blue" opponency, and L–M to "Red–Green". Achromatic and colour-opponent signals are kept separate in the LGN and onto the cortex.

Aside from photoreceptors, most neurons transmitting feedforward excitation signals in the visual system are of type either ON or OFF. ON cells are excited by light increments but do not respond to decrements, while OFF cells respond only to light decrements; they are organised in parallel channels that separately transmit lightness and darkness, and that are maintained separate from the retina to the cortex throughout the whole visual pathway.

In the vision science literature the response of a cell is often (but not exclusively) modelled as a linear operation (weighted summation of the neighbours' activity, e.g. for lateral inhibition) followed by a non-linear operation (e.g. rectification, so as to consider only increments or decrements, but not both). For the linear part, DoG filters and oriented DoG (ODoG) filters are useful in predicting many perceptual phenomena [61], while common models for the non-linear part include rectification, divisive normalisation and power-laws. For instance, non-linear photoreceptor response followed by linear filtering produces bandpass contrast enhancement that correlates with the contrast sensitivity function of the human visual system [62].

The purity of a colour is represented by its saturation S that expresses the amount of white that the colour has: an achromatic colour has $S = 0$, and blood red colour has the same hue as pink, but higher saturation. The value of S can be computed from a combination of the achromatic and the chromatic signals. There is evidence in region V1 of the visual cortex, but not in the retina nor LGN, for cells tuned to S and for neural activity correlated with S [63], with a possible neural mechanism to this effect proposed in [64].

Finally, and very importantly for the GM problem, we shall recall the so-called Helmholtz–Kohlrausch effect, that implies that brightness perception depends on luminance and chrominance: colour patches of the same luminance but different hue appear to have different brightness, as well as colour patches of the same luminance and hue but different saturation (the higher the saturation, the higher the perceived brightness). As a consequence, if we were to modify the saturation of a colour while preserving all other attributes, its brightness would appear to change. The interaction between achromatic and chromatic signals that produces brightness perception has been shown to happen at V1 [65], not before; there are some models for this, e.g. [66,67], with a review in [68].

8.4.2 Proposed gamut mapping framework

8.4.2.1 Gamut extension

We have previously seen how the LHE contrast enhancement method produces gamut extension when applied independently to the R, G and B channels of an input image [51], and this is also the case when applied just on the colour opponent channels [53] or just on the saturation channel [69]. Based on this gamut extension ability that contrast enhancement has, and considering the visual principles enumerated in the previous section, we now propose the following basic gamut extension method: perform contrast enhancement on the saturation channel by center-surround filtering, followed by rectification so as to ensure that the saturation does not decrease, and finally modifying the brightness to account for the Helmholtz–Kohlrausch (H-K) effect.

Basic GE method:

1. The inputs are an image sequence, whose gamut we want to extend, and the specifications of the source and target gamuts.
2. Convert each input frame into HSV colour space. We will keep H constant.
3. Using the specifications of source and target gamuts, define a linear filter K_e similar to a DoG. This filter is then convolved with S, obtaining S_1 which is contrast-enhanced. Fig. 8.12 (left) shows an example K_e filter.
4. Add a constant value image C to S_1, obtaining S_2. This step attempts to preserve the mean of the original image.
5. Rectify $(S_2 - S)$ and produce $S_3 = S + rectified(S_2 - S)$, thus ensuring that $S_3(x) \geq S(x)$ for each and every pixel x.
6. Modify V to compensate for the Helmholtz–Kohlrausch effect, correcting V so that perceived brightness does not change for those colours whose saturation has been modified. This is done using a simplified version of the model by Pridmore [67] that yields $V_1 = V\left(\frac{S}{S_3}\right)^{\rho}$.
7. The final result is the image with channels (H, S_3, V_1).

See Fig. 8.13 comparing the original image (left) with the intermediate result replacing S with S_3 (middle) and the final result with both S_3 and V_1 (right).

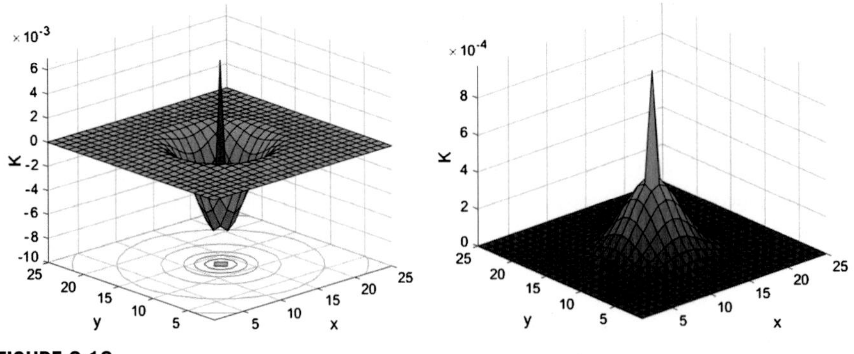

FIGURE 8.12

Left: gamut extension kernel. Right: gamut reduction kernel.

As an enhancement to the method we can add a logistic function τ after step (3) that linearly combines S_2 with the original S, giving more importance to S in the case of low-saturated values so as to preserve skin tones and other memory colours: $S_2{}' = (1 - \tau(S))S_2 + \tau(S)S$. The shape of function $\tau(S)$ is shown in Fig. 8.15 and the formula is:

$$\tau(S(x)) = 1 - \frac{1}{\left(1 + 0.55e^{-1.74S(x)}\right)^2} \tag{8.4}$$

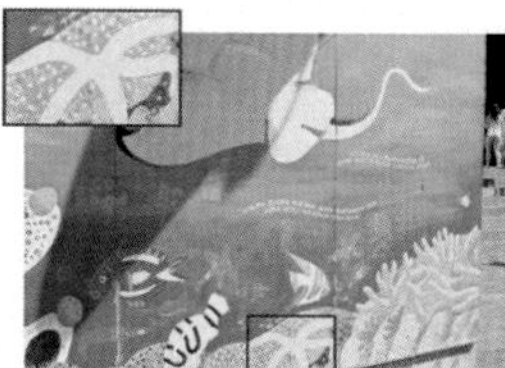
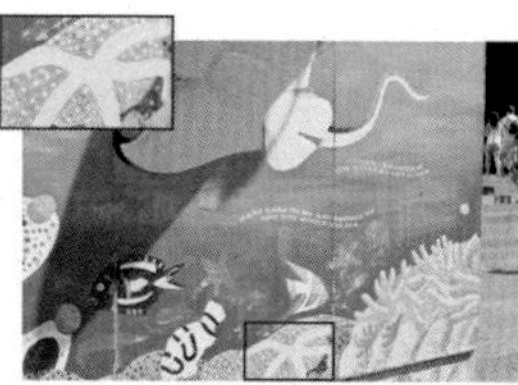
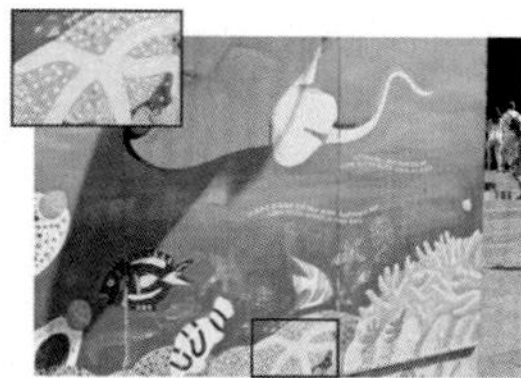

FIGURE 8.13 GE results and H-K effect

Comparison of gamut extension results: input image (left), extended-gamut image ignoring the H-K effect (middle), extended-gamut image considering the H-K effect (right).

FIGURE 8.14 GR results and H-K effect

Comparison of gamut reduction results: input image (left), reduced-gamut image ignoring the H-K effect (middle), reduced-gamut image considering the H-K effect (right).

The values used in Eq. (8.4) have been chosen based on tests we performed on several images with different colour characteristics. Let us note that these training images were not included our evaluation tests.

8.4.2.2 Gamut reduction

Essentially, gamut extension can be seen as the inverse of the gamut reduction problem [10]. Since GE can be achieved by contrast enhancement, GR can be obtained by decreasing contrast, as we proved in [51]. In [70] Yeonan-Kim et al. showed that convolution with K_e minimises a functional that has a term for contrast enhancement; if we change the sign of this term, then the minimisation of the functional performs contrast decrease and the solution is achieved by convolution with a new kernel K_r. Fig. 8.12 (right) shows an example K_r filter. Following the idea presented in section 8.4.2.1, where convolution of S with some kernel K_e yields GE, then GR could be performed by convolution with a kernel K_r that is the inverse (in Fourier space) of a kernel that would perform GE.

Basic GR method:

1. The inputs are an image sequence, whose gamut we want to reduce, and the specifications of the source and target gamuts.

2. Convert each input frame into HSV colour space. We will keep H constant.
3. Use a linear filter K_r, similar to a sum of Gaussians, to convolve with S, obtaining S_1 which is contrast-decreased.
4. Add a constant value image C to S_1, obtaining S_2. This step attempts to preserve the mean of the original image.
5. Rectify $(S - S_2)$ and produce $S_3 = S - rectified(S - S_2)$, thus ensuring that $S_3(x) \leq S(x)$ for each and every pixel x.
6. Modify V to compensate for the Helmholtz–Kohlrausch effect: $V_1 = V\left(\frac{S}{S_3}\right)^{\rho}$.
7. The final result is the image with channels (H, S_3, V_1).

See Fig. 8.14 comparing the original image (left) with the intermediate result replacing S with S_3 (middle) and the final result with both S_3 and V_1 (right).

While one pass of this basic method already performs GR, we have found that it gives better results to iterate steps (2) to (4) with a sequence of filters K_r of progressively larger spatial extent, keeping fixed after each iteration all pixels whose colours have become in-gamut. Fig. 8.16 shows the evolution of the filters, the image gamut and the image.

Both for GR and GE our method is applied independently to each frame of the video input. We have not needed to impose any sort of temporal consistency to our algorithm and this is due to the effectively large size of the kernels we use, which remove or strongly reduce the influence of sudden changes and make the results stable and the framework very robust.

8.4.2.3 Computation of the convolution kernel

The kernel K_e for GE is computed as

$$K_e = \mathcal{F}^{-1}\left(\frac{1}{1 - \gamma(\frac{19}{20} - \mathcal{F}(\omega))}\right), \tag{8.5}$$

where $\mathcal{F}$ denotes the Fourier transform, ω is a normalised 2D Gaussian kernel and γ is a positive constant. As mentioned above, the shape of K_e is similar to a difference of Gaussians, see an example in Fig. 8.12 (left).

The kernel K_r for GR is computed as

$$K_r = \mathcal{F}^{-1}\left(\frac{1}{1 - \gamma(\frac{21}{20} - \mathcal{F}(\omega))}\right), \tag{8.6}$$

where γ is in this case a negative constant. The shape of K_r is similar to a sum of Gaussians, see an example in Fig. 8.12 (right).

Both for K_e and K_r the only parameters are γ and the standard deviation of the Gaussian kernel ω. The sign of γ defines the type of GM and the shape of the kernel (see Fig. 8.12): in the case of gamut extension γ is a positive constant and in the case of gamut reduction it is negative.

The motivation for the form of these kernels is given in what follows. We saw in Chapter 7 how the original LHE model of [48] could be extended by modifying the energy functional, considering a local mean term instead of a global one and by replacing the absolute value function in the contrast term by a square function, as suggested in [70] and [71]:

$$E(I) = \frac{\alpha}{2}\sum_x (I(x)-\mu)^2 + \frac{\beta}{2}\sum_x (I(x)-I_0(x))^2 - \frac{\gamma}{2}\sum_x\sum_y w(x,y)(I(x)-I(y))^2. \tag{8.7}$$

With this modification the Euler–Lagrange equation can be solved directly:

$$\delta E(I) = \alpha(I(x)-\mu) + \beta(I(x)-I_0(x)) - \gamma\sum_x w(x,y)(I(x)-I(y)) = 0. \tag{8.8}$$

We can see that the last term is equal to the convolution of the image I and the kernel $w(x, y)$. Rearranging, and with an abuse of notation to avoid writing the pixel location x we reach to

$$(\alpha+\beta-\gamma)I + \gamma w * I = \alpha\mu + \beta I_0 \tag{8.9}$$

Applying the Fourier transform we obtain

$$(\alpha+\beta-\gamma)\mathcal{F}(I) + \gamma\mathcal{F}(w)\mathcal{F}(I) = \mathcal{F}(\alpha\mu+\beta I_0) \tag{8.10}$$

Rearranging and applying the inverse Fourier transform we obtain the solution

$$I = \mathcal{F}^{-1}\left(\frac{\mathcal{F}(\alpha\mu+\beta I_0)}{(\alpha+\beta-\gamma)+\gamma\mathcal{F}(w)}\right) \tag{8.11}$$

which can be rewritten in terms of convolving a kernel with the input image I_0 as

$$I = \beta I_0 * K + C \tag{8.12}$$

where C is a flat, single-value image with all pixels equal to the product of α, the mean of the input image channel, and the mean of the kernel K. The kernel K is computed as

$$K = \mathcal{F}^{-1}\left(\frac{1}{(\alpha+\beta-\gamma)+\gamma\mathcal{F}(w)}\right). \tag{8.13}$$

The sign of γ defines the type of GM and the shape of the kernel (see Fig. 8.12): in the case of gamut extension γ is a positive constant and in the case of gamut reduction it is negative.

From the above, convolving with K the three channels of an RGB image would yield a gamut-mapped result. Perhaps counter-intuitively, we also obtain a gamut-mapped result if we convolve K with the saturation channel S of an image in HSV colourspace, and then rectify the output. For instance, for GE, the constant γ is positive, which means that convolution with K will increase contrast and therefore locally maximal values of S will go up, while some S values will go down, but then rectification ensures that the S values that decreased are brought back to their original value; the net result is that after convolution and rectification, S has increased, and therefore the gamut has been extended. The same reasoning can of course be applied to the gamut reduction case.

In our application, for performing GM we fix in Eq. (8.13) the parameter values $\beta = 1$ and $\alpha = \frac{|\gamma|}{20}$, which leads to the kernels presented in Eqs. (8.5) and (8.6).

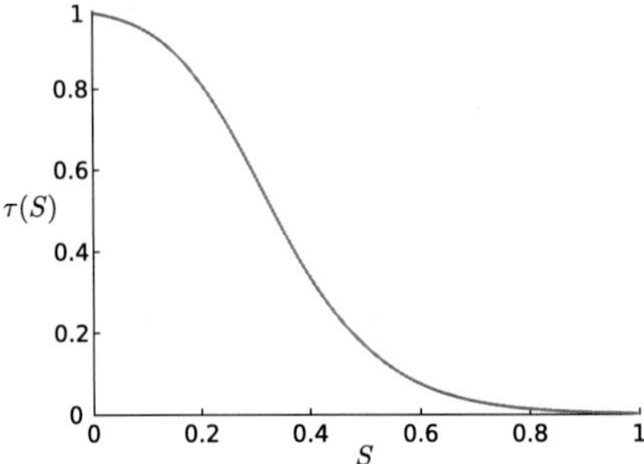

FIGURE 8.15

Logistic function used to give weights to each pixel of the input image.

8.4.3 Psychophysical evaluation

We performed pair-comparison psychophysical tests in cinematic conditions (large screen, cinema projector, dark room), in the same way as described in Section 8.3.2. Fifteen observers took part in each of the experiments we performed in our lab, as this is the number of observers for pair comparison tests that is suggested by several technical recommendation documents (e.g. [72,73]). Some of the test images were taken from publicly available datasets [74,75], while others were from [53] and from mainstream feature films.

8.4.3.1 Experiment 1: Evaluation of GEAs

For the evaluation of GEAs we have used the same two setups defined in Section 8.3.2.1.

For each set-up, we compare the proposed method with the top-ranked GEAs in [53]:

- **Same Drive Signal (SDS).** This method linearly maps the RGB primaries of the source gamut to the RGB primaries of the destination device gamut, therefore making full use of the gamut of the target display.

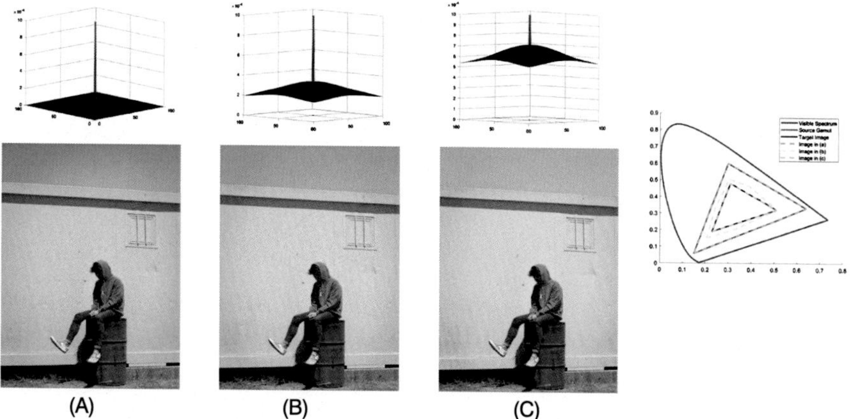

(A) (B) (C)

FIGURE 8.16 Effect of kernel size on gamut

Effect of increasing kernel size on image gamut. Row 1: example of kernels with progressively larger spatial extent. Row 2: reduced-gamut images corresponding to each kernel. Last column: evolution of image gamut that progressively decreases with an increase in the spatial extent of the kernel.

Table 8.1 Primaries of gamuts.

Gamuts	Red primaries		Green primaries		Blue primaries	
	x	y	x	y	x	y
BT.2020	0.708	0.292	0.170	0.797	0.131	0.046
BT.709/sRGB	0.640	0.330	0.300	0.600	0.150	0.060
DCI-P3	0.680	0.320	0.265	0.690	0.150	0.060
Toy	0.570	0.320	0.300	0.530	0.190	0.130

- **Hybrid Colour Mapping (HCM).** This GEA is a combination of the SDS algorithm and the true-colour algorithm, that represents the input image in the target gamut without applying any extension. The method of HCM aims at preserving natural colours by leaving unchanged the low-saturated colours such as flesh tones, while mapping the high-saturated colours using the SDS method. The HCM algorithm [44] analyses the saturation of the input image and then linearly combines the output of the true-colour method and the SDS method:

$$\begin{bmatrix} R \\ G \\ B \end{bmatrix}_{HCM} = (1-\kappa)\begin{bmatrix} R \\ G \\ B \end{bmatrix}_{true\text{-}colour} + \kappa\begin{bmatrix} R \\ G \\ B \end{bmatrix}_{SDS}, \quad (8.14)$$

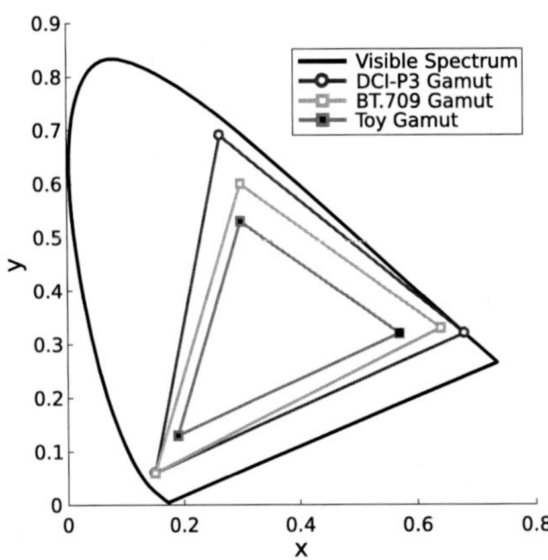

FIGURE 8.17

Gamuts on chromaticity diagram.

where κ is a mixing factor that works as a function of saturation:

$$\kappa(S) = \begin{cases} 0, & \text{if } S \leq S_L \\ \frac{S-S_L}{S_H-S_L}, & \text{if } S_L < S < S_H \\ 1, & \text{if } S \geq S_H \end{cases} \tag{8.15}$$

- **GEA of Zamir et al. [53]** is the one detailed in Section 8.3.2.

Fig. 8.18 (left) presents the accuracy scores computed by analysing the psychophysical data of the setup 1 where it can be seen that, when the difference between the source gamut and the destination gamut is large, the proposed GEA yields images that are perceptually more faithful to the ground truth images than the other competing algorithms. The observers declared SDS [44] as the least accurate method, whereas the algorithm of [53] ranked second.

In Fig. 8.18 (right) we present results for the setup 2 where it can be seen that, when the colour difference between the source-destination gamut pair is small, our algorithm ranks first, followed by the HCM algorithm [44] and the method of [53].

8.4.3.2 Experiment 2: Evaluation of GRAs

All the competing GRAs receive as input the wide-gamut DCI-P3 images and generate reproductions for the following two different experimental setups.

1. *Setup 1: Mapping from DCI-P3 gamut to a small gamut.* We created this particular setup with a large difference between source and target gamuts, nearly as large as it is between BT.2020 and BT.709 gamuts. An experimental setup with such large difference in gamuts allows us to not only evaluate the performance of competing GRAs reliably but also provides us with an indication of how these GRAs might perform when BT.2020 content becomes commonly available and needs to be

mapped to BT.709 displays or DCI-P3 cinema projectors. To compute the results using the competing GRAs, we map the colours of 15 DCI-P3 test images to the challenging smaller 'Toy' gamut. (See Table 8.1 and Fig. 8.17.)

2. *Setup 2: Mapping from DCI-P3 to BT.709 gamut.* Colourists perform this gamut reduction procedure by using 3D LUTs (as we mentioned in more detail in the introduction.) Therefore, we engaged a professional colourist from a post-production company to use their own in-house 3D LUTs and apply them on our DCI-P3 test images in order to create reduced-gamut BT.709 images. We also perform GR using the following competing GRAs.

We compared the proposed GRA with the following methods:

- **LCLIP [11]** clips the chroma of the out-of-gamut colours to the destination gamut boundary along lines of constant hue and lightness.
- **Hue Preserving Minimum ΔE (HPMINDE) [12]** involves clipping of the out-of-gamut colour to the closest colour, in terms of ΔE error, on the boundary of the destination gamut along lines of constant hue.
- **Alsam and Farup [30]** proposed an iterative GRA that at iteration level zero behaves as a gamut clipping algorithm, and as the number of iterations increases the solution approaches spatial gamut mapping.
- **Schweiger et al. [24]** make use of a compression function that squeezes colours near the destination gamut boundary in order to accommodate the out-of-gamut colours. This is a method proposed and used by the British Broadcasting Corporation (BBC).

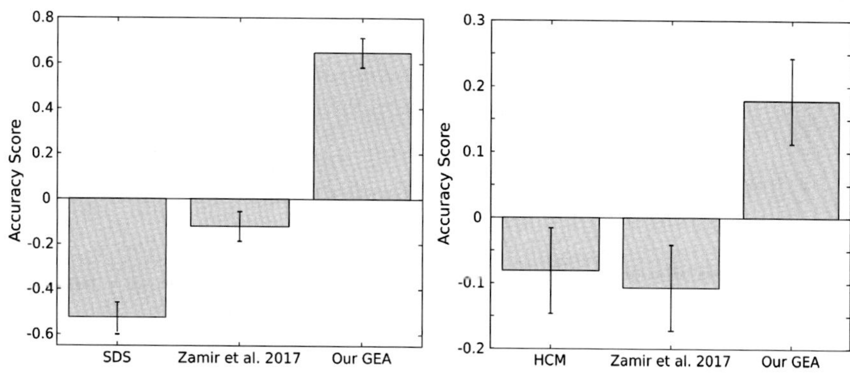

FIGURE 8.18 Accuracy scores for GEAs

Accuracy scores of competing GEAs: 15 observers took part in each experiment and 30 images were used. Left: setup 1. Right: setup 2.

The analysis of psychophysical data for setup 1 is presented in Fig. 8.19 (left) showing that the proposed GRA produces images that are perceptually more faithful to the original images than any other competing method.

For the experimental setup 2 we also ran the psychophysical tests with 15 observers, of which 9 had experience in image processing and the other 6 were *skilled* technicians (colourists and editors) from a post-production company. In order to reduce the number of pair comparisons, in this particular setup we opted to use the reproduced images of the top three ranked methods from setup 1 and the reduced-gamut images created by using the custom LUT of the same post-production company. Fig. 8.19 shows the results for setup 2. It can be seen that observers preferred the in-house LUT results over the other methods, with our GRA being ranked second.

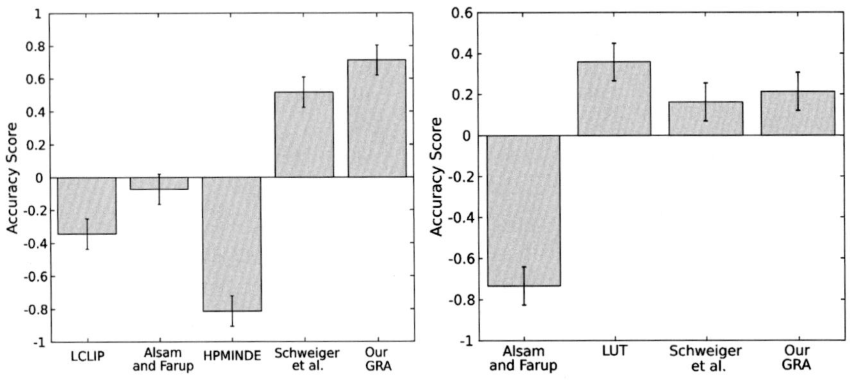

FIGURE 8.19 Accuracy scores for GRAs

Accuracy scores of competing GRAs: 15 observers took part in each experiment and 15 images were used. Left: setup 1. Right: setup 2.

In order to test the temporal coherence we apply the proposed gamut reduction and gamut extension methods to all frames of videos independently. We confirm that the results produced by our algorithms are free from artifacts.

8.4.4 Does any error metric approximate our psychophysical results?

In this section we evaluate if there exists any image metric able to predict the result of our psychophysical test, following the same strategy we used for the GE case in [53]. To this end we consider a total of 10 metrics: a perceptually-based colour image difference (CID) metric [76] particularly tailored for the gamut mapping problem, its more recent extension iCID [38], CIEΔE00 [77], the metrics presented in [78] such as Laplacian mean square error (LMSE), structural content (SC), normalised absolute error (NAE), peak signal-to-noise ratio (PSNR), and absolute difference (AD), and finally two very recent, state-of-the-art deep learning metrics for perceived appearance, PieAPP [79] and LPIPS [80], learned from human judgements on large image databases. Let us note that all these metrics are full-reference, and therefore they have access to the ground-truth, as do the observers in our experiments.

	GE_S1			GE_S2			GR_S1					GR_S2			
	SDS	Zamir et al. [56]	Ours	Zamir et al. [56]	HCM	Ours	HPMINDE [12]	LCLIP [11]	Alsam et al. [29]	Schweiger et al. [24]	Ours	Alsam et al. [29]	Schweiger et al. [24]	Ours	3D LUT
CID	-0.209	-0.372	0.582	-0.465	0.279	0.186	0.102	-0.686	-0.279	0.61	0.254	-0.627	0.593	0.356	-0.322
iCID	-0.279	-0.279	0.559	-0.419	0.14	0.279	-0.178	-0.61	-0.254	0.584	0.457	-0.559	0.322	0.457	-0.22
Delta E00	-0.349	-0.279	0.628	-0.326	0.163	0.163	0.559	-0.279	-0.635	-0.152	0.508	-0.491	0.254	0.728	-0.491
LMSE	-0.652	0.303	0.349	0.465	-0.396	-0.07	0.127	-0.711	-0.152	0.686	0.051	-0.423	0.627	-0.119	-0.085
SC	-0.349	0.628	-0.279	0.582	-0.047	-0.535	-0.356	0.686	-0.711	0.102	0.279	-0.288	0.254	0.728	-0.694
NAE	-0.372	-0.116	0.489	-0.116	0.07	0.047	0.406	-0.61	-0.356	0.406	0.152	-0.593	0.356	0.525	-0.288
PSNR	0.396	0.023	-0.419	0.07	-0.093	0.023	-0.279	0.737	0.025	-0.559	0.076	0.593	-0.525	-0.186	0.119
AD	-0.559	0.675	-0.116	0.628	-0.326	-0.303	-0.356	0.711	-0.584	-0.025	0.254	-0.085	0.051	0.728	-0.694
PieAPP	0.14	-0.303	0.163	-0.396	0.326	0.07	0.051	0.229	-0.711	0.051	0.381	-0.728	0.288	0.627	-0.186
LPIPS	-0.209	-0.233	0.442	-0.116	0.163	-0.047	-0.051	-0.61	-0.152	0.584	0.229	-0.322	0.457	0.152	-0.288
Observers' data	-0.525	-0.121	0.647	-0.107	-0.081	0.178	-0.812	-0.343	-0.073	0.516	0.712	-0.733	0.162	0.213	0.358

FIGURE 8.20 Image metrics and psychophysical evaluation

Comparison between the results of different image metrics and the results from psychophysical evaluation. Metrics were considered as observers in a pair comparison experiment. Each experiment is colour coded individually. Colour codes are green (gray in print version) for the best result and red (dark gray in print version) for the worst one.

In order to perform a fair comparison to our experiment, we consider the metrics as if they were observers in our pair comparison test. This means that, for each metric, we will run all the possible comparisons, and in each comparison we will give a 1 to the image with better metric value and a 0 to the image with worse metric value. Later, we will apply the Thurstone Case V analysis [81] to each of the image metrics to end up with the preference values for each of the methods. These preference values will therefore be comparable to the ones shown for the psychophysical analysis in our previous section.

Fig. 8.20 shows the result of the aforementioned analysis. Each of the experimental setups is individually coloured with a colour code where the hue goes from pure red (dark gray in print version) for the lowest value to pure green (gray in print version) for the highest one. Therefore, for any metric to be able to predict the psychophysical results, its colour code should match that of the results of the observers, shown in the last row. We can see that there is only one specific case where we could argue that this is happening, the NAE metric in the first setup of gamut extension. However, the same metric is not able to predict the observers' response in any of the other three cases.

It is interesting to mention that the CID and iCID metrics, which were specifically developed for gamut mapping, do not match the observer data; but let us note that these metrics were designed for input images in the BT.709 gamut, while in this paper the input images are in DCI-P3, which may explain the limitations of CID and iCID in the context of our problem.

Another significant result is that two state-of-the-art deep learning metrics, PieAPP [79] and LPIPS [80], designed to predict perceptual image error like human observers and based on large scale datasets (20K images in one case, 160K images in the other) labelled with pair-comparison preferences and, in the case of [80], using close to 500K human judgements, are not able to predict the observers' preference in any of the experimental set-ups that we have tested. This result is important, as it suggests that current deep learning approaches are not accurate enough for validating (and therefore developing) GM methods for cinema applications, although it's not a surprising result in the sense that several very recent works have also shown how large or very large scale image databases (250,000+ images) can be used to train deep neural networks to predict user preference but whose performance decays remarkably when used on images that belong to some other dataset different from the one used for training [82,80].

8.5 Conclusion

We have introduced a GM framework based on some basic vision science principles and models. The algorithms for GE and GR are simple, of low computational complexity, produce results without artifacts and outperform state-of-the- art methods according to psychophysical experiments.

We also tested a number of established and state-of-the-art objective metrics on the results produced by the GM methods we have compared, and the conclusion is that there does not seem to be an adequate metric that is able to predict the preference of observers for GM results. This has two important consequences.

First, that in order to evaluate gamut mapping methods, we still need to rely only on psychophysical studies. However, conducting subjective studies is operationally difficult, hard to replicate from experiment to experiment, economically costly and may require special equipment (cinema projector, large screen, etc.).

Second, that we cannot develop GM methods or optimise GM results simply by maximising an image appearance metric or minimising an error metric (as it is done in other contexts, e.g. [83]), which of course would be extremely practical.

Therefore, the results present in this section point out the importance of working towards defining better metrics for the gamut mapping problem that are able to predict the observers' preference, which we strongly believe would be of great importance for the colour imaging community.

References

[1] SMPTE. D-cinema quality – reference projector and environment. Report 431-2:2011; 2011.

[2] ITU-R. Recommendation BT.709-5: parameter values for the HDTV standards for production and international programme exchange; 2002.

[3] ITU-R Recommendation. BT.2020: parameter values for ultra high definition television systems for production and international programme exchange; 2012.

[4] Pointer MR. The gamut of real surface colours. Color Research & Application 1980;5(3):145–55.

[5] Bertalmío M. Image processing for cinema, vol. 4. CRC Press, Taylor & Francis; 2014.

[6] Bankston D. The color-space conundrum, part one. American Cinematographer 2005:6.

[7] Kennel G. Color and mastering for digital cinema: digital cinema industry handbook series. Taylor & Francis US; 2007.

[8] Kusakabe Y, Iwasaki Y, Nishida Y. Wide-color-gamut super hi–vision projector. In: Proc. ITE annual convention; 2013 (in Japanese).

[9] Silverstein BD, Kurtz AF, Bietry JR, Nothhard GE. A laser-based digital cinema projector. SID Symposium Digest of Technical Papers 2011;42(1):326–9.

[10] Morovič J. Color gamut mapping, vol. 10. Wiley; 2008.

[11] Sara JJ. The automated reproduction of pictures with nonreproducible colors. Ph.D. thesis, Massachusetts Institute of Technology (MIT); 1984.

[12] Murch GM, Taylor JM. Color in computer graphics: manipulating and matching color. Eurographics Seminar: Advances in Computer Graphics V 1989:41–7.

[13] Masaoka K, Kusakabe Y, Yamashita T, Nishida Y, Ikeda T, Sugawara M. Algorithm design for gamut mapping from UHDTV to HDTV. Journal of Display Technology 2016;12(7):760–9.

[14] Katoh N, Ito M. Gamut mapping for computer generated images (ii). In: Proc. of 4th IS&t/SID color imaging conference; 1996. p. 126–9.

[15] Marcu G, Abe S. Gamut mapping for color simulation on CRT devices. In: Proc. of color imaging: device-independent color, color hard copy, and graphic arts; 1996.

[16] Montag ED, Fairchild MD. Gamut mapping: evaluation of chroma clipping techniques for three destination gamuts. In: Color and imaging conference; 1998. p. 57–61.

[17] Gentile RS, Walowitt E, Allebach JP. A comparison of techniques for color gamut mismatch compensation. Journal of Imaging Technology 1990;16:176–81.

[18] Herzog PG, Müller M. Gamut mapping using an analytical color gamut representation. In: Proc. of color imaging: device-independent color, color hard copy, and graphic arts; 1997. p. 117–28.

[19] Johnson AJ. Perceptual requirements of digital picture processing. Paper presented at IARAIGAI symposium, and printed in part in Printing World; 1979.

[20] Morovič J. To develop a universal gamut mapping algorithm. Ph.D. thesis, UK: University of Derby; 1998.

[21] UGRA. UGRA GAMCOM Version 1.1: Program for the Color Gamut Compression and for the comparison of calculated and measured values. Tech. rep., St Gallen: UGRA; 17 July 1995.

[22] Yuan J, Hardeberg JY, Chen G. Development and evaluation of a hybrid point-wise gamut mapping framework. In: Colour and visual computing symposium; 2015. p. 1–4.

[23] Pytlarz J, Thurston K, Brooks D, Boon P, Atkins R. Real time cross-mapping of high dynamic range images. IET Conference Proceedings 2016.

[24] Schweiger F, Borer T, Pindoria M. Luminance-preserving colour conversion. In: SMPTE annual technical conference and exhibition; 2016. p. 1–9.

[25] Bala R, Dequeiroz R, Eschbach R, Wu W. Gamut mapping to preserve spatial luminance variations. Journal of Imaging Science and Technology 2001;45:122–8.

[26] Meyer J, Barth B. Color gamut matching for hard copy. In: Proc. of SID digest; 1989. p. 86–9.

[27] Zhu M, Hardeberg JY, Wang N, Sun B. Spatial gamut mapping based on guided filter. In: Proc. of IS&t/SPIE electronic imaging; 2016. p. 1–4.

[28] Zolliker P, Simon K. Retaining local image information in gamut mapping algorithms. IEEE Transactions on Image Processing 2007;16(3):664–72.

[29] McCann JJ. A spatial colour gamut calculation to optimize colour appearance. In: Colour image science: exploiting digital media; 2002. p. 213–33.

[30] Alsam A, Farup I. Spatial colour gamut mapping by orthogonal projection of gradients onto constant hue lines. In: Proc. of 8th international symposium on visual computing; 2012. p. 556–65.

[31] Farup I, Gatta C, Rizzi A. A multiscale framework for spatial gamut mapping. IEEE Transactions on Image Processing 2007;16(10):2423–35.

[32] Gatta C, Farup I. Gamut mapping in RGB colour spaces with the iterative ratios diffusion algorithm. In: Proc. of IS&t/SPIE electronic imaging; 2017. p. 12–20.

[33] Kimmel R, Shaked D, Elad M, Sobel I. Space-dependent color gamut mapping: a variational approach. IEEE Transactions on Image Processing 2005;14:796–803.

[34] Lau C, Heidrich W, Mantiuk R. Cluster-based color space optimizations. In: Proc. of IEEE international conference on computer vision; 2011. p. 1172–9.

[35] Li Y, Song G, Li H. A multilevel gamut extension method for wide gamut displays. In: Proc. of international conference on electric information and control engineering (ICEICE); 2011. p. 1035–8.

[36] Nakauchi S, Hatanaka S, Usui S. Color gamut mapping based on a perceptual image difference measure. Color Research & Application 1999;24(4):280–91.

[37] Preiss J, Urban P. Image-difference measure optimized gamut mapping. In: Proc. of IS&t/SID 20th color imaging conference; 2012. p. 230–5.

[38] Preiss J, Fernandes F, Urban P. Color-image quality assessment: from prediction to optimization. IEEE Transactions on Image Processing 2014;23(3):1366–78.
[39] Kang BH, Morovič J, Luo MR, Cho MS. Gamut compression and extension algorithms based on observer experimental data. ETRI Journal 2003;25(3):156–70.
[40] Kim MC, Shin YC, Song YR, Lee SJ, Kim ID. Wide gamut multi-primary display for HDTV. In: Proc. of 2nd European conference on color graphics, imaging and vision; 2004. p. 248–53.
[41] Anderson H, Garcia E, Gupta M. Gamut expansion for video and image sets. In: Proc. of the 14th international conference of image analysis and processing – workshops; 2007. p. 188–91.
[42] Pan H, Daly S. A gamut-mapping algorithm with separate skin and non-skin color preference controls for wide-color-gamut TV. SID Symposium Digest of Technical Papers 2008;39(1):1363–6.
[43] Casella SE, Heckaman RL, Fairchild MD. Mapping standard image content to wide-gamut displays. In: Color and imaging conference; 2008. p. 106–11.
[44] Laird J, Muijs R, Kuang J. Development and evaluation of gamut extension algorithms. Color Research & Application 2009;34(6):443–51.
[45] Heckaman RL, Sullivan J. Rendering digital cinema and broadcast TV content to wide gamut display media. SID Symposium Digest of Technical Papers 2011;42(1):225–8. https://doi.org/10.1889/1.3621279.
[46] Meng X, Song G, Li H. A human skin-color-preserving extension algorithm for wide gamut displays. In: Proc. of international conference on information technology and software engineering. Lecture notes in electrical engineering; 2013. p. 705–13.
[47] Song G, Cao H, Huang H. Hue preserving multi-level expansion method based on saturation for wide gamut displays. Journal of Information & Computational Science 2014;11(2):461–72.
[48] Bertalmío M, Caselles V, Provenzi E, Rizzi A. Perceptual color correction through variational techniques. IEEE Transactions on Image Processing 2007;16(4):1058–72.
[49] Bertalmío M, Caselles V, Provenzi E. Issues about Retinex theory and contrast enhancement. International Journal of Computer Vision 2009;83(1):101–19.
[50] Kodak. http://r0k.us/graphics/kodak/, 1993.
[51] Zamir SW, Vazquez-Corral J, Bertalmío M. Gamut mapping in cinematography through perceptually-based contrast modification. IEEE Journal of Selected Topics in Signal Processing 2014;8(3):490–503.
[52] Vazquez-Corral J, Bertalmío M. Spatial gamut mapping among non-inclusive gamuts. Journal of Visual Communication and Image Representation 2018;54:204–12.
[53] Zamir SW, Vazquez-Corral J, Bertalmío M. Gamut extension for cinema. IEEE Transactions on Image Processing 2017;26(4):1595–606.
[54] Zamir SW, Vazquez-Corral J, Bertalmío M. Gamut extension for cinema: psychophysical evaluation of the state of the art, and a new algorithm. In: Proc. of IS&t/SPIE electronic imaging; 2015.
[55] Jian C, Hardev V, Yurek J. Quantum dot displays: giving LCDs a competitive edge through color. Nanotechnology Law & Business 2014;11:4–13.
[56] Zamir SW, Vazquez-Corral J, Bertalmío M. Vision models for wide color gamut imaging in cINEMA. IEEE Transactions on Pattern Analysis and Machine Intelligence 2019.
[57] Beck B. Lasers: coming to a theater near you. http://spectrum.ieee.org/consumer-electronics/audiovideo/lasers-coming-to-a-theater-near-you, 2014.
[58] Shapley R, Enroth-Cugell C. Visual adaptation and retinal gain controls. Progress in Retinal Research 1984;3:263–346.
[59] Gouras P. Colour vision. eLS 2009.
[60] Yeonan-Kim J, Bertalmíot M. Retinal lateral inhibition provides the biological basis of long-range spatial induction. PLoS ONE 2016;11(12):e0168963.
[61] Blakeslee B, McCourt ME. A unified theory of brightness contrast and assimilation incorporating oriented multiscale spatial filtering and contrast normalization. Vision Research 2004;44(21):2483–503.
[62] Watson AB, Solomon JA. Model of visual contrast gain control and pattern masking. JOSA A 1997;14(9):2379–91.
[63] Hanazawa A, Komatsu H, Murakami I. Neural selectivity for hue and saturation of colour in the primary visual cortex of the monkey. European Journal of Neuroscience 2000;12(5):1753–63.

[64] Billock VA, Tsou BH. Sensory recoding via neural synchronization: integrating hue and luminance into chromatic brightness and saturation. JOSA A 2005;22(10):2289–98.

[65] Xing D, Ouni A, Chen S, Sahmoud H, Gordon J, Shapley R. Brightness–color interactions in human early visual cortex. Journal of Neuroscience 2015;35(5):2226–32.

[66] Ayama M, Ikeda M. Brightness-to-luminance ratio of colored light in the entire chromaticity diagram. Color Research & Application 1998;23(5):274–87.

[67] Pridmore RW. Chroma, chromatic luminance, and luminous reflectance. Part ii: Related models of chroma, colorfulness, and brightness. Color Research & Application 2009;34(1):55–67.

[68] Withouck M, Smet KA, Ryckaert WR, Pointer M, Deconinck G, Koenderink J, et al. Brightness perception of unrelated self-luminous colors. JOSA A 2013;30(6):1248–55.

[69] Zamir SW, Vazquez-Corral J, Bertalmío M. Perceptually-based gamut extension algorithm for emerging wide color gamut display and projection technologies. In: SMPTE annual technical conference & exhibition; 2016. p. 1–11.

[70] Kim J, Batard T, Bertalmío M. Retinal processing optimizes contrast coding. Journal of Vision 2016;16(12):1151.

[71] Bertalmío M, Batard T, Kim J. Correcting for induction phenomena on displays of different size. Journal of Vision 2016;16(12).

[72] ITU-T RECOMMENDATION P. Subjective video quality assessment methods for multimedia applications. ITU; 2008.

[73] ITU-R RECOMMENDATION B. Methodology for the subjective assessment of the quality of television pictures. ITU; 2012.

[74] Froehlich J, Grandinetti S, Eberhardt B, Walter S, Schilling A, Brendel H. Creating cinematic wide gamut HDR-video for the evaluation of tone mapping operators and HDR-displays. In: Proc. of IS&t/SPIE electronic imaging; 2014.

[75] Andriani S, Brendel H, Seybold T, Goldstone J. Beyond the kodak image set: a new reference set of color image sequences. In: IEEE international conference on image processing; 2013. p. 2289–93.

[76] Lissner I, Preiss J, Urban P, Lichtenauer MS, Zolliker P. Image-difference prediction: from grayscale to color. IEEE Transactions on Image Processing 2013;22(2):435–46.

[77] Luo MR, Cui G, Rigg B. The development of the CIE 2000 colour-difference formula: CIEDE2000. Color Research & Application 2001;26(5):340–50.

[78] Eskicioglu AM, Fisher PS. Image quality measures and their performance. IEEE Transactions on Communications 1995;43(12):2959–65.

[79] Prashnani E, Cai H, Mostofi Y, Sen P. Pieapp: Perceptual image-error assessment through pairwise preference. In: The IEEE conference on computer vision and pattern recognition (CVPR); 2018.

[80] Zhang R, Isola P, Efros AA, Shechtman E, Wang O. The unreasonable effectiveness of deep features as a perceptual metric. In: CVPR; 2018.

[81] Thurstone LL. A law of comparative judgment. Psychological Review 1927;34(4):273–86.

[82] Talebi H, Milanfar P. Nima: Neural image assessment. IEEE Transactions on Image Processing 2018;27(8):3998–4011.

[83] Cyriac P, Batard T, Bertalmío M. A nonlocal variational formulation for the improvement of tone mapped images. SIAM Journal on Imaging Sciences 2014;7(4):2340–63.

CHAPTER

9 Vision models for tone mapping in cinema

Many challenges that deal with processing of HDR material remain very much open for the film industry, whose extremely demanding quality standards are not met by existing automatic methods. Therefore, when dealing with HDR content, substantial work by very skilled technicians has to be carried out at every step of the movie production chain.

Based on recent findings and models from vision science, we present in this chapter effective tone mapping and inverse tone mapping algorithms for production, post-production and exhibition. These methods are automatic and real-time, and they have been both fine-tuned and validated by cinema professionals, with psychophysical tests demonstrating that the proposed algorithms outperform both the academic and industrial state-of-the-art.

We believe these methods bring the field closer to having fully automated solutions for important challenges for the cinema industry that are currently solved manually or sub-optimally.

Another contribution of our research is to highlight the limitations of existing image quality metrics when applied to the tone mapping problem, as none of them, including two state-of-the-art deep learning metrics for image perception, are able to predict the preferences of the observers.

9.1 Introduction

We mentioned at the beginning of this book that all the possibilities that the HDR format could offer, in terms of revenue and market growth for companies and in terms of improved user experience for the viewer, depend upon the existence of a fully functional HDR ecosystem. But we are currently a long way from such an ecosystem, as stated in recent reports from several international organisations and standardisation bodies [1–3] as well as very recent publications from the motion picture and TV community [4–9].

This is due to open challenges happening at all stages of the production chain, from capture to display. While some of these challenges are technical (e.g. developing HDR cinema projectors [4]), most of them stem from the fact that the interactions of HDR images with the human visual system are quite complex and not yet fully understood, not even in the vision science community, for a number of reasons (that may

Vision Models for High Dynamic Range and Wide Colour Gamut Imaging. https://doi.org/10.1016/B978-0-12-813894-6.00014-4

include the preference of vision studies to use synthetic instead of natural stimuli, and modelling vision as a cascade of linear and nonlinear filters, which might be too restricting: see [10] for a comprehensive review on the subject). Furthermore, cinema artists complain that the software tools they work with are based on very basic vision science; for instance, noted cinematographer Steve Yedlin says that: *"Color grading tools in their current state are simply too clunky [...]. Though color grading may seem complex given the vast number of buttons, knobs and switches on the control surface, that is only the user interface: the underlying math that the software uses to transform the image is too primitive and simple to achieve the type of rich transformations [...] that define the core of the complex perceptual look"* [11].

Among the consequences of this lack of accurate vision models we can cite the following:

- There are no guidelines on how to shoot and post-produce HDR footage in a way that maximises the expressive capabilities of the HDR medium while considering the visual perception of HDR images.
- There is a need for colour management and grading tools so that the mastering process has a simple workflow that can deliver consistent images across the whole spectrum of possible display devices and environments.
- There is a need for high quality real-time tools for conversion from HDR to SDR formats (so-called tone mapping (TM) operators) and viceversa (inverse tone mapping (ITM) algorithms), and among standard and wide colour gamuts, which are essential for HDR capture, monitoring, post-production, distribution and display, for cinema and TV, given that SDR and HDR displays will co-exist for the foreseeable future.

Regarding this latter point, starting from the mid 1990s the academic literature has seen a *very* significant number of works on the subject of TM (and, to a lesser degree, ITM), following diverse goals (e.g. emulating perception or improving appearance), based on diverse ideas (e.g. emulating photoreceptor response or emulating techniques by photographers), intended for still images or video sequences, and normally evaluated through user tests with regular observers or with objective metrics specific for tone mapping. The literature on the topic is so vast that we have decided to concentrate just on the seminal works on vision-based tone mapping that introduced a number of ideas that became widespread in the field, of which we'll give an overview in Section 9.3.1.5. Later, in section 9.4, we also describe the various states of the art TM and ITM methods that we compare our approach with. For comprehensive surveys that include methods not based on vision models we refer the interested reader to [12] and [13].

But despite the fact that there are a number of very effective TM and ITM algorithms, that the field is mature enough, and that the development of new TM methods seems to have peaked a few years back, the cinema industry does not rely on automatic methods and resorts instead to intensive manual process by very skilled technicians and colour artists. For accurate on-set monitoring of HDR footage, 3D LUTs are created beforehand by cinematographers and colourists. The first pass of the colour grade, called technical grade, where the image appearance is homogenised among shots and made to look as natural as possible [14], is performed manually by the colourist. The mastering and versioning processes are assisted by propietary automated transforms, the most popular ones coming from Baselight, Resolve, Transkoder/Colourfront and Dolby Vision (which are the ones we will be comparing our framework with), but there is a subsequent and quite substantial input from the colourist, who's required to perform trim passes.

These practices can be explained by the fact that the cinema industry has almost impossibly high image quality standards, beyond the reach of most automated methods, but another, very important point that can be inferred is the following: cinema professionals have the ability to modify images so that their appearance on screen matches what the real-world scene would look like to an observer in it [14]. Remarkably, artists and technicians with this ability are able to achieve what neither state-of-the-art automated methods nor up-to-date vision models can. Put in other words, the manual techniques of cinema professionals seem to have a "built-in" vision model.

We present in this chapter effective tone mapping and inverse tone mapping algorithms for production, post-production and exhibition. These methods have been developed in collaboration with Praveen Cyriac and David Kane, with contributions form Trevor Canham. The advantages of the proposed algorithms are the following:

- They outperform, in terms of visual appearance, the state-of-the-art algorithms from academia as well as the standard industry methods, according to psychophysical validation tests performed by cinema professionals.
- They have very low computational complexity and can be executed in real-time.

Other contributions in this chapter are:

- We show the limitations of existing objective image quality metrics for TM and ITM, since they do not correlate well with the preferences of observers.
- Our TM and ITM algorithms are based on vision models whose parameters are fine-tuned by cinema professionals. This idea, clearly useful for our cinema applications, might also prove helpful in the opposite direction, for vision science: colourists may assist the improvement or development of novel, more accurate vision models.

In the next section we will start by briefly reviewing a few principles and findings from the vision science literature that we introduced in Chapters 4 and 5, that are relevant to the HDR imaging problem and that form the basis of our framework.

9.2 Some vision principles relevant for HDR imaging

One fundamental challenge that our visual system must tackle is imposed by the limited dynamic range of spiking neurons, of around two orders of magnitude, while in the course of a day the ambient light level may vary over 9 orders of magnitude [15]. In order to deal with this the visual system uses a number of strategies.

One of them is adapting sensitivity to the average light intensity. This process, termed light adaptation, is fully accomplished by the retina. It starts already in the photoreceptors, whose sensitivity declines inversely with the ambient light level over a wide range of intensities [16]: the net result is that retinal output can become relatively independent from illumination and encode instead the reflectance of objects, which provides a key survival ability [15].

Another one is encoding local contrast instead of absolute light level. Local contrast is defined as percent change in intensity with respect to the average or background level. In a typical scene large changes in the absolute level of illumination do not have an impact on the range of contrast values, that remain fairly constant and below two orders of magnitude [17], and therefore can be *fit* into the limited dynamic range of neurons.

A third strategy is splitting contrast signals into ON and OFF parallel channels. Visual signals are split at the retina into no less than 12 parallel pathways (also called visual streams or channels) that provide a condensed representation of the scene that can pass through the anatomical bottleneck that is the optic nerve [18]. Some of these channels go all the way from the retina to the primary visual cortex, like the ON pathway, that encodes positive contrast, and the OFF pathway, that encodes negative contrast [19]. That is, ON neurons are mostly responsive to increases in light intensity with respect to the average level, while OFF neurons mainly respond to light intensities that go below the average level. The existence of these two channels has been reported in many animal species, from flies to primates, that use vision to move about in their surroundings [20]. In [21] it's hypothesised that the reason for the existence of the ON and OFF pathways is efficiency: neuron spikes consume energy, so rather than having neurons firing at some significant level for the average light intensity so that the firing rate can decrease for low intensities and increase for high intensities, it appears more convenient to have the neurons be silent or firing at low rates for the average light intensity, with ON neurons devoting their whole dynamic range for increments over the average and OFF neurons for decrements.

These vision properties just described are only three of the many manifestations of efficient representation or efficient coding principles in the visual system, a very popular school of thought within vision science [22] that postulates that all resources in the visual system are optimised for the type of images we encounter in the natural world. Furthermore, this optimisation is not static, the visual system must modify its computations and coding strategy as the input stimuli changes, adapting itself to the local characteristics of the stimulus statistics [23].

Adaptation is an essential feature of the neural systems of all species, a change in the input–output relation of the system that is driven by the stimuli [24]. Through adaptation the sensitivity of the visual system is constantly adjusted taking into account multiple aspects of the input stimulus, matching the gain to the local image statistics through processes that aren't fully understood and contribute to make human vision so hard to emulate with devices [25].

There are very different timescales for adaptation [26]: retinal processes that adapt to local mean and variance take place in less than 100 msec, the fixation time between consecutive rapid eye movements (microsaccades) [25], while global adaptation processes are usually in the order of a few seconds [15].

In the early visual system, adaptation is concerned mainly with changes in two statistical properties of the light intensity: its mean and variance [27]. Adaptation to the mean is what's referred to as light adaptation. Contrast adaptation tailors the performance of the visual system to the range of fluctuations around the mean, i.e. the contrast, and for instance when the contrast increases (under constant mean) the retinal ganglion cells become less sensitive [26].

By adapting to the statistical distribution of the stimulus, the visual system can encode signals that are less redundant and this in turn produces metabolic savings by having weaker responsiveness after adaptation, since action potentials are metabolically expensive [28]. Atick [29] makes the point that there are two different types of redundancy or inefficiency in an information system like the visual system:

1. **If some neural response levels are used more frequently than others.** For this type of redundancy, the optimal code is the one that performs *histogram equalisation*. There is evidence that the retina is carrying out this tye of operation [30]: Laughlin showed in 1981 how the photoreceptors of a fly had a response curve that closely matched the cumulative histogram of the average luminance distribution of the fly's environment.
2. **If neural responses at different locations are not independent from one another.** For this type of redundancy the optimal code is the one that performs decorrelation. There is evidence in the retina, the LGN and the visual cortex that receptive fields act as optimal "whitening filters", decorrelating the signal. It should also be pointed out that a more recent work [31] contends that decorrelation is already performed by the rapid eye movements that happen during fixations, and therefore the signal arrives already decorrelated at the retina: the subsequent spatial filtering performed at the retina and downstream must have other purposes, like enhancing image boundaries.

Neural adaptation performs a (constrained) signal equalisation by matching the system response to the stimulus mean and variance [25], thus ensuring visual fidelity under a very wide range of lighting conditions.

Fig. 9.1 (left) shows that when the mean light level is high, the nonlinear curve that models retinal response to light intensity is a sigmoid function with less steep slope than when the mean light level is low. Fig. 9.1 (right) shows that at a given ambient level, the slope of the sigmoid is lower when the contrast is higher. The same behaviour has been observed for the nonlinear functions that model lightness perception in HDR images [32].

In both cases, the data is consistent with the nonlinearity of the neural response to light performing histogram equalisation, since the nonlinearity behaves as the cumulative histogram (which is the classical tool used in image processing to equalise

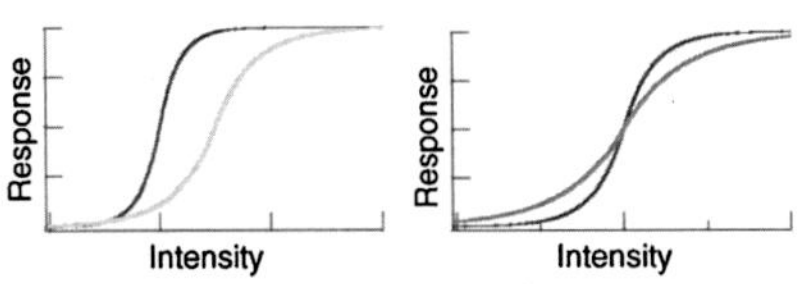

FIGURE 9.1 Neural adaptation to mean and variance

Neural adaptation to mean and variance. Left: neural response to higher (in green (dark gray in print version)) and lower (in blue (black in print version)) mean luminance. Right: neural response to higher (in red (mid gray in print version)) and lower (in blue (black in print version)) luminance variance. Adapted from [25].

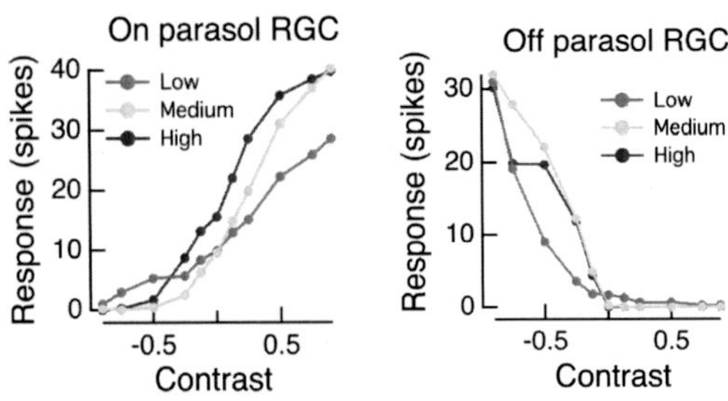

FIGURE 9.2

ON and OFF cells have different nonlinearities. Figure from [36].

a histogram) does: darker images and images with lower contrast typically have less variance and therefore their cumulative histograms are steeper.

We have performed psychophysical experiments where the observers manipulate a display nonlinearity in order to optimise image appearance [33], and our results corroborate that the visual system performs histogram equalisation by showing how observers prefer display nonlinearities that allow the displayed image to be perceived as having a brightness distribution as close to uniform (i.e. with an equalised histogram) as possible.

Recent works from neurophysiology prove that OFF cells change their gain more than ON cells during adaptation [34], and, very importantly for our applications, that the nonlinear responses of retinal ON and OFF cells are different [35–37], see Fig. 9.2. This data on neural activity is consistent with psychophysical data [38] that demonstrates that our sensitivity to brightness is enhanced at values near the average or background level, so the brightness perception nonlinearity has a high slope at the average level, different arcs at each side of it, and changes shape as the background level changes, see Fig. 9.3 (left).

Fig. 9.3 (right) shows that, in linear coordinates, the brightness perception curve can't be adequately modelled by a simple power-law function as it's usually done in classic models (e.g. [39]), and rather it should be represented by two power-laws, one for values below the average level, another for values above it. We can also see in Fig. 9.3 (right) that when using a logarithmic axis, the brightness perception curve

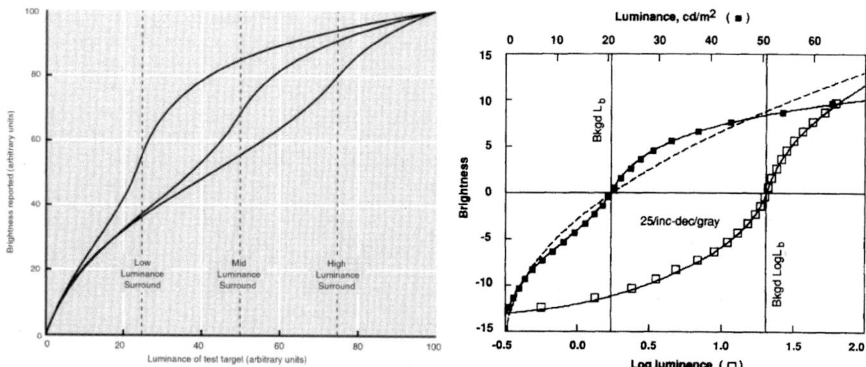

FIGURE 9.3 Brightness perception curve changes shape at background intensity level

Left: the shape of the brightness perception nonlinearity is different for values below and above the background level (from [40]). Right: the brightness perception curve is more adequately modelled with two power-laws (linear coordinates) or an asymmetric sigmoid (log coordinates) (from [38]). This psychophysical data is consistent with neural data showing ON and OFF channels having different nonlinearities. Nundy S., Purves D., A probabilistic explanation of brightness scaling, Proceedings of the National Academy of Sciences 2002; 99(22):14482–14487. ©National Academy of Sciences, USA.

is not a regular, symmetric sigmoid function like the Naka-Rushton equation used to model photoreceptor responses [16].

The consequence of the phenomena just described and depicted in Figs. 9.2 to 9.3 is that power-law models of the nonlinear retinal response to the input stimulus should have different exponents for the values below the average light level than for those above the average level.

Once more we have an element that corroborates the efficient coding theory, which is the following. The average luminance histogram of natural images is, in log-log coordinates, piecewise linear [41], with two lines of different slope at each side of the average level, see Fig. 9.4. It's easy to see that the cumulative histogram will also be piecewise linear in log coordinates. Since a linear function in log coordinates corresponds to a power-law in linear coordinates, the retinal nonlinearity modelled as two power-laws can act as the cumulative histogram and therefore it can perform histogram equalisation. This histogram equalisation process performed by the retinal nonlinearity would be *complete* if the image had a histogram that perfectly matched the piecewise form of the *average* histogram of natural images, but as this is not the case in general, this nonlinearity performs just an approximation to the full histogram equalisation.

The slope or first derivative of the nonlinear retinal response is the gain, and the slope of the brightness perception curve is the sensitivity. They both follow Weber–Fechner's law and decay with increasing stimulus, highlighting again the remarkable similarity between retinal response and perceptual data. Fig. 9.5 (left) shows that neu-

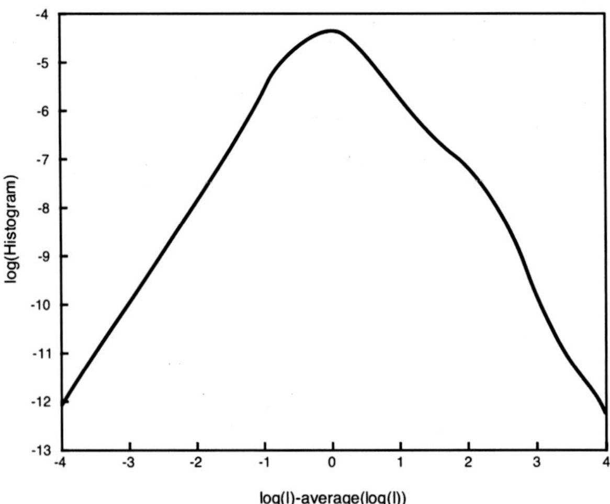

FIGURE 9.4 Average histogram of natural scenes

Average histogram of natural scenes, in log-log coordinates, adapted from [42].

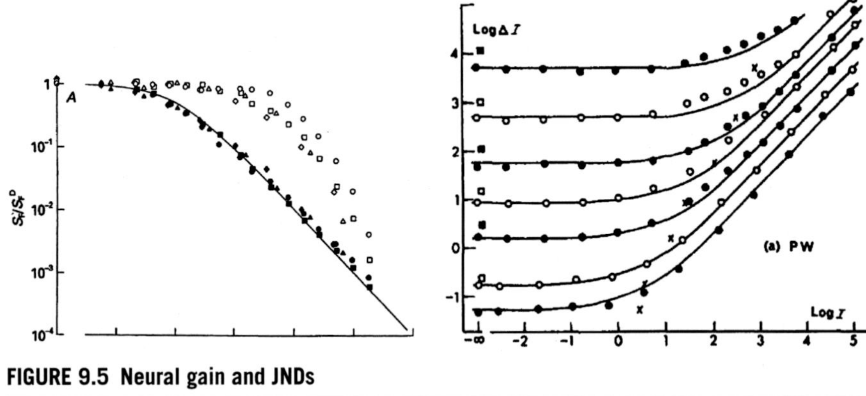

FIGURE 9.5 Neural gain and JNDs

Left: neural gain vs light intensity (from [43]). Right: JND vs light intensity (from [44]).

ral gain decreases with light intensity, while Fig. 9.5 (right) shows that the detection threshold (the inverse of sensitivity) increases with intensity.

Very importantly for our application, sensitivity plateaus for low intensities, suggesting that for the darkest levels there's a limit imposed on the slope of the brightness perception nonlinearity.

At a given ambient level, contrast enhancement is different for achromatic and chromatic signals. A common way of representing vision models is as cascades of linear and nonlinear operations [45], where the linear part is expressed as a convolution with a filter: at retinal level these filters have the form of difference of Gaussians (DoG), so the response to the cell's receptive field surround is subtracted from the response to the receptive field center and this process, termed lateral inhibition or center-surround modulation, produces contrast enhancement. Smaller surrounds enhance more the contrast. Type H2 horizontal cells are the only known retinal cells that sum responses from L, M and S cones [46], their signal is therefore achromatic, and they provide feedback and lateral inhibition to cones. Their surround size is smaller than that of the horizontal feedback to the chromatic or colour-opponent channels (that encode the difference between cone signals) [47] and as a consequence there is more contrast enhancement for the achromatic channel than for the colour-opponent channels.

Coupling among horizontal cells allows them to have, apart from a narrow surround, a much wider surround than was assumed until recently [48], and it's accepted that these cells provide cones with an estimate of average ambient luminance [49] that is not localised but quite global.

Natural image statistics do change with the dynamic range of the scene, but basic retinal transforms (like photoreceptor response and center-surround inhibition at the level of retinal ganglion cells) have been shown to make the signal statistics almost fully independent of the dynamic range [41]. Indeed, most of the vision literature on dynamic range focused on how objects appear invariant over changing lighting conditions, thus omitting irradiance information [50]. But tracking changes in irradiance is advantageous in terms of survival, and there is a type of retinal neuron called ipRGC (for "intrinsical photosensitive retinal ganglion cell") whose outputs are proportional to scene irradiance; individual ipRGCs activate within a range of irradiance values so that collectively they can span and encode light intensity levels from moonlight to full daylight. ipRGCs directly control the pupillary reflex and the circadian rhythm, that modulates the release of dopamine in the retina, which in turn alters the size of the receptive field of horizontal cells performing local contrast enhancement on the cone signals [51]. The key implication is that contrast enhancement will increase as the ambient level is reduced.

As mentioned above, light adaptation takes place fully in the retina, and a substantial part of the contrast adaptation as well. Some of these were phenomena that used to be attributed to cortical processing, but there's now a growing consensus that the retina performs many more tasks and more complicated processes than it was previously assumed [18,15,52].

9.3 Proposed framework

Our framework consists of three methodologies for dealing with HDR material in three different scenarios that cover the cinema production pipeline, from shooting to exhibition. We will start by introducing our approach for TM of ungraded footage (coming directly from the camera) that follows the vision principles enunciated in the previous section. This method can be used for on-set monitoring during the shoot, and for creating the technical grade during postproduction. Next we will detail what modifications must be made to this TM algorithm in order to perform TM of graded content. The intended application of this second method is for generating SDR versions from an HDR master. Finally we will introduce our method for ITM, intended to allow HDR cinema projectors and displays to enhance on-the-fly the dynamic range of their SDR input.

9.3.1 Tone mapping of ungraded footage

The foundations for the TM problem were laid in the seminal works of [53] and [54], that stated how the tone-mapped material presented on a screen should produce *"a subjective experience that corresponds with viewing the real scene"*, and therefore the tone-mapping method must behave as *"the concatenation of the real-world observer, the inverse of the display observer, and the inverse of the display model"*.

Let H denote the HDR image representing the real-world scene, V_1 a function that represents the real-world observer (in a sense that will be clarified shortly), V_2 a function that represents the display observer, and D a function that represents the nonlinear display response. With this notation, matching the subjective experiences of seeing the real-world scene and the tone-mapped image on the screen is expressed as

$$V_1(H) = V_2(D(TM(H))), \tag{9.1}$$

from which we get that

$$TM(H) = D^{-1} \circ V_2^{-1} \circ V_1(H) \tag{9.2}$$

and the basic principle recalled above about how a TM method should behave can be stated as:

$$TM \equiv D^{-1} \circ V_2^{-1} \circ V_1 \tag{9.3}$$

In our case, we choose to consider V_i, $i \in \{1, 2\}$, as a function that encapsulates the retinal processes of global nonlinear response followed by contrast enhancement that we saw in Section 9.2 account for light adaptation and contrast adaptation in the retina. Therefore, V_i does not represent the full image percept (it's ignoring higher-level processes that involve for instance edge orientation, memory colours, etc.) and just aims to emulate retinal output. Nonetheless, if two scenarios produce the same retinal output then by necessity they will produce the same subjective experience, so for our application we are not losing generality by defining V_i in this way.

The other choice we make is that our goal is for TM to approximate V_1, and have cinema experts optimise its parameters in a reference environment so that Eq. (9.1) holds: in other words, we propose as tone mapping operator a vision model that is fine-tuned by cinema professionals. From Eq. (9.3), if our goal is met then this implies that $D = V_2^{-1}$, which corresponds to the simplified model used in practice where the display nonlinearity is very similar to the inverse of the brightness perception nonlinearity of the visual system [55].

This similarity between V_2 and D^{-1} is a good approximation that has been established through experiments using simple synthetic stimuli, with no context adaptation and in laboratory conditions. For natural images as it is our case, we conjecture that the adaptation effects, context influence and all other aspects that are required for a perceptual match (Eq. (9.1)) are handled by the $TM \equiv V_1$ function, *thanks to it having been optimised by cinema professionals.* As we mentioned in the introduction, cinema professionals have the ability to modify images to produce a perceptual match between the real-world scene and the display, and remarkably they are able to achieve what neither state-of-the-art automated methods nor up-to-date vision models can. Put in other words, the manual techniques of cinema professionals seem to have a "built-in" vision model.

In summary, we will propose a TM algorithm that emulates retinal response to a real-world scene. For that we will proceed in three consecutive steps, first emulating light adaptation on a single frame, next emulating contrast adaptation, and finally ensuring temporal coherence on the video sequence.

9.3.1.1 Light adaptation

The first stage of our TM algorithm takes as input the HDR source and emulates light adaptation by passing the image values through a global response curve. This is a dynamic nonlinearity that adapts to the image statistics, aiming to perform histogram equalisation as per the theory of efficient representation, but with some constraints that comply with neuroscience data cited in Section 9.2:

- The slope is limited for the darkest levels.
- For intensity values above the average, the response curve is approximated by a power-law, of some exponent γ^+.
- For intensity values below the average, the response curve is approximated by a power-law, of some exponent γ^-, which is generally different from γ^+.

We estimate the average μ of the light intensity based on the median of the luminance channel Y of the HDR source; the median provides a better fit than the mean to the piecewise linear shape, in log-log coordinates, of the average histogram of natural scenes [41]. The values for γ^+, γ^- are obtained as the slopes of the piecewise linear fit to the cumulative histogram of the luminance. In this way they can produce a nonlinearity that, if it were applied to Y, would tend to equalise its histogram, making the distribution of Y closer to being uniform. We define the nonlinearity $\hat{NL}(\cdot)$ to be applied to a normalised signal $C \in [0, 1]$ as a power-law $p(\cdot)$:

$$\hat{NL}(C) = C^{p(C)}, \tag{9.4}$$

where $p(\cdot)$ smoothly goes from γ^- to γ^+ with a transition at μ of slope n:

$$p(C) = \gamma^+ + (\gamma^- - \gamma^+)\frac{\mu^n}{C^n + \mu^n}, \tag{9.5}$$

and both n and the dependence of μ on the median have been determined with the help of cinema professionals, as will be discussed in Section 9.3.4.

For the darkest levels, the curve should not be approximated by a power-law because their derivative tends to infinity as the intensity goes to zero, in the same manner that the gamma correction curve for the Rec.709 standard, that approximates brightness perception, is a power-law for most of the intensity range except for the lower values, where it's a linear function [55]. For this reason we modify $\hat{NL}$ simply by limiting its slope for very low values of C. This is key for our application, because otherwise we would be amplifying the noise in dark scenes, a common issue with video TM algorithms [12].

Although the definining parameters μ, γ^-, γ^+ of the resulting nonlinearity $NL(\cdot)$ have been automatically computed from the distribution of Y, the curve NL is not applied on Y but to each colour channel $C \in \{R, G, B\}$, thus emulating the achromatic horizontal feedback to cones mentioned in Section 9.2:

$$C' = NL(C). \tag{9.6}$$

Fig. 9.6 shows, on the top left, an HDR image linearly mapped i.e. the minimum image value is mapped to 0 and the maximum is mapped to 1, with all in-between values scaled linearly between 0 and 1. Next to it is its luminance histogram, heavily slanted towards the left because most values are very small compared with the few high values of the highlights in the scene. Finally on the top right of Fig. 9.6 we can see the cumulative histogram and, superimposed, the tone curve NL. Fig. 9.6, bottom left, shows the result of performing tone mapping just with tone curve NL. The other plots on the same row show how the application of NL has had the effect of flattening the histogram and making the cumulative histogram closer to the identity function, hence increasing the efficiency of the representation.

9.3.1.2 Contrast adaptation

The second stage of our TM algorithm emulates contrast adaptation by performing contrast enhancement on the intermediate result coming from the light adaptation stage of our method. The motivation is as follows. The real-world scene, source of the HDR material, has an ambient light level that is typically much higher than that of the environment where the screen is located and where the tone-mapped version will be viewed. We saw in Section 9.2 that when the mean light level is high, the nonlinear curve that models retinal response to light intensity is a sigmoid function with less steep slope than when the mean light level is low. We also mentioned that, at a given ambient level, the slope of the sigmoid is lower when the contrast is higher. From these two elements we conclude that, if we want to emulate at the low ambient level of the screen environment the response to light intensity that is produced at the high ambient level of the real-world scene, we need to increase the contrast of the tone-mapped image, so that when we look at it the retinal response curve becomes shallower, thus emulating the response to the real-world scene. In short, we have to perform contrast enhancement. Furthermore, following the efficient coding principles and the *equalisation* role of neural adaptation, the contrast has to be modified so that the resulting image has a variance value closer to that of an image with a uniform distribution.

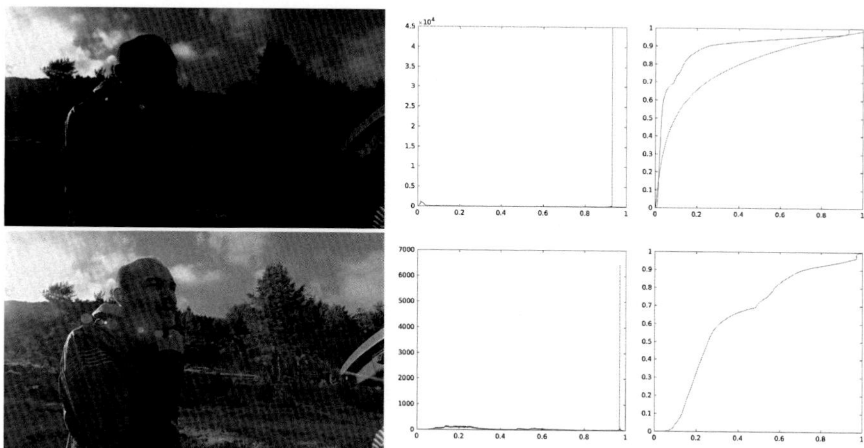

FIGURE 9.6 Luminance histogram, cumulative histogram and tone mapping

Top row, from left to right: HDR image linearly mapped, luminance histogram and cumulative histogram showing also NL. Bottom row, from left to right: HDR image mapped with NL, luminance histogram and cumulative histogram. Original image from [56].

As the visual system performs contrast enhancement differently on the achromatic signal than on the chromatic channels, we express in a colour-opponent space the light-adapted image coming as the output of the first stage of our method. We use for this the *IPT* colourspace [57], where *I* corresponds to the achromatic channel and *P* and *T* are the colour-opponent channels. Then we perform contrast enhancement by a process that while in essence emulates convolution with a DoG kernel where the extent of the surround is larger for the chromatic channels *P, T* than for channel *I*, thus achieving more contrast enhancement on *I* than on *P* and *T*, in practice is simplified by performing contrast normalisation on the *I* channel and leaving untouched the *P* and *T* channels (equivalent to using a DoG with a very wide surround):

$$\{I', P', T'\} = RGB2IPT(\{R', G', B'\}) \tag{9.7}$$

$$I'' = SI' + \frac{m}{\sigma}(I' - SI') \tag{9.8}$$

$$P'' = P' \tag{9.9}$$

$$T'' = T', \tag{9.10}$$

where $RGB2IPT$ is the transform from RGB to IPT colourspace, SI' is a smoothed version of I' obtained by convolution with a regularising kernel (the sum of two Gaussians, which is the dual narrow-wide form of the lateral inhibition produced by retinal interneurons, as explained in [48] and mentioned previously in Section 9.2; in our case the Gaussians are of standard deviation 5 and 25 pixels), σ is the standard deviation of I', and m is a constant.

9.3.1.3 Temporal consistency on sequences

In order to tone-map a frame, first we compute temporary values for μ, γ^+, γ^- following the procedure described above; these are averaged with the parameter values from the previous frames, yielding the final parameter values to be applied on the current frame. This is a standard way, in the TM literature, to extend to video processing a TMO devised for still images [58].

In practice the temporal averaging equation we use is the following, but of course many other options are possible:

$$P_f(i) = (30 * P_f(i-1) + P(i))/31, \tag{9.11}$$

where $P_f(i)$ is the final value of the parameter P (be it μ, γ^+ or γ^-) for frame i after the temporal average of the single-frame estimate $P(i)$ and the previous frame value $P_f(i-1)$.

9.3.1.4 Computational complexity

The proposed TM method requires computing basic statistics from the luminance histogram in order to define the global nonlinear transform, and performing a convolution with a fixed kernel for the contrast adaptation process. Therefore, the computational complexity of the method is very low. For 2K images and with our current, non-fully optimised GPU implementation, the method is taking 90 ms per frame or about 11fps. We are confident that by parallelising the histogram computation we can bring the implementation to achieve real-time execution at 24fps.

9.3.1.5 Related TM methods

A very basic version of our algorithm, with a simpler vision model and no input from cinema professionals, was introduced in [59]. In that work the slope of the nonlinearity is not limited for low intensity values, which produces abundant noise in common scenarios, while the contrast enhancement process is performed in the same way on the *R*, *G* and *B* channels, which creates problems in the case of highly saturated colours; see Fig. 9.7 for some examples.

There are some significant novelties in our approach, to be discussed in Section 9.6, that make our method more effective than the state of the art, as we will show in Section 9.4.1. Nonetheless, our proposed method is grounded on ideas that are recurrent in the TM literature and in most cases date back to the earliest works of the mid 1990s: vision emulation, retinal models, histogram re-distribution, power-laws, slope limitation, global operations followed by local enhancement, etc.

The following are very brief overviews of the seminal works on vision-based tone mapping that introduced the above-mentioned ideas, all of which have become very extended in the field, e.g. see [13] and references therein:

- We mentioned above how Tumblin and Rushmeier [53] laid the foundations of the TM problem, they posed it as the quest for an operator that causes "a close match between real-world and display brightness sensations". They based their TMO on approximating brightness perception through a power-law, using the model

FIGURE 9.7 Artifacts produced by preliminary version of our TMO

Top row: example results of a preliminary version of our algorithm [59], with a simple vision model and no input from cinema professionals. Notice the noise and colour problems. Bottom row: results with the method proposed in Section 9.3.1. Source images from ARRI and Mark Fairchild's HDR Survey.

and perceptual data of Stevens, and modelling the display response with another power-law nonlinearity.

- The work of Ward Larson et al. [54] was also foundational, they defined what they considered to be the two most important criteria for TM and these criteria have been adopted by the community and explicitly or implicitly assumed by all subsequent works:

  1. Reproduction of visibility: an object is seen on the screen if and only if it's seen in the real scene.
  2. The subjective experience, the overall impression of brightness, contrast and colour, is the same when viewing the image on the screen as when viewing the real scene.

  They propose a TM method that performs histogram equalisation, but instead of directly using the cumulative histogram as tone curve (which is what regular histogram equalisation does) they approximate brightness perception with a logarithmic function, compute the cumulative histogram curve of the brightness image, and then impose constraints on this curve based on models of contrast perception. In this way they aim to comply with the reproduction of visibility requirement, i.e. they don't want the tone curve to introduce details and contrast that are not visible in the real scene.

- Pattanaik et al. [60] used for tone mapping a global nonlinearity that models photoreceptor responses, the Naka-Rushton equation, with parameters that change in time mimicking the visual response during adaptation.
- The TMO by Ashikhmin [61] estimates a local adaptation luminance for each pixel, then local tone mapping is applied based on perceptual discrimination thresholds, and finally there is a contrast enhancement stage.
- The iCAM06 colour appearance model of Kuang et al. [62] performs tone mapping with the Naka-Rushton equation in RGB, and converts the result to IPT before applying difference enhancements to the chroma and colour channels.

9.3.2 Tone mapping of graded content

Graded HDR footage has had its contrast and colours modified according to the creator's intentions, and its appearance has been optimised for an HDR display of given peak luminance in a dim or dark surround. The tone-mapping of this source material should preserve the average picture level, not making the images brighter nor darker but simply reducing the intensity of the highlights; the colour intent must also be maintained as much as possible.

A TM algorithm must take the above into account, so for TM of graded content we introduce a number of modifications to the method presented in Section 9.3.1. Let Y_1 be the normalised (i.e. in the range $[0, 1]$) luminance of the HDR source, μ_1 the median of Y_1, D_1 the peak luminance of the display for which the HDR source material was graded, and D_2 the peak luminance of the reduced dynamic range display where the tone-mapped results will be presented. Therefore, the median luminance (in light intensity units, i.e. without normalisation) of the source is $\mu_1 D_1$, and we want the tone-mapped result to have the same median luminance. When normalising by the corresponding peak luminance, and using a power-law transform of exponent γ, this translates into the following requirement:

$$\mu_1^{\gamma} = \mu_1 \frac{D_1}{D_2},$$

from which γ can be solved as:

$$\gamma = \frac{log(\mu_1) + log(\frac{D_1}{D_2})}{log(\mu_1)}. \tag{9.12}$$

(In practice, we limit γ to be larger or equal than 0.45: $\gamma = max(0.45, \gamma)$).

So γ is the power-law exponent that must be used for the average light intensity, but as we saw before the exponents are, in general, different for intensities below the average than for intensities above it. We now define γ^+, γ^- in the following way:

$$\gamma^+ = (1-k)\gamma; \quad \gamma^- = (1+k)\gamma, \tag{9.13}$$

with k a small parameter.

The tone curve characterised by μ_1, γ^+, γ^- is applied to Y_1 yielding a tone-mapped luminance image Y_{TM}. Each colour channel $C \in \{R, G, B\}$ is tone-mapped resulting in a final output channel C', in this manner:

$$C' = \left(\frac{C}{Y_1}\right)^s * Y_{TM}, \tag{9.14}$$

where s is a parameter that adjusts the saturation of the output, a common practice in the TM literature.

The contrast adaptation described in Section 9.3.1.2 for TM of ungraded content is omitted now. The temporal consistency is enforced in the same way as in Section 9.3.1.3.

9.3.3 Inverse tone mapping of graded content

Our method for inverse tone mapping of graded content is based on the following ideas:

1. The global curve for inverse tone mapping of graded content should be the inverse of the curve for tone mapping of graded content.
2. The average picture level should be preserved, just increasing the intensity of the highlights.
3. The colour intent must also be maintained as much as possible.

From the first item above we see that, since the global curve for TM of graded content is expressed as two power-laws with different exponents for values below and above the average, the global curve for ITM can also be expressed as two power-laws with different exponents for values below and above the average.

From item number 2 above we can find the exponent γ that preserves the median luminance in a way analogous to the one used in Section 9.3.2, so γ can now be solved as:

$$\gamma = \frac{log(\mu_2) + log(\frac{D_2}{D_1})}{log(\mu_2)}, \tag{9.15}$$

where μ_2 is the median of the normalised luminance Y_2 of the source SDR graded content, D_2 is the peak luminance of the display for which the source was graded, and D_1 is the peak luminance of the HDR display intended for the output of the ITM algorithm.

We again define γ^+, γ^- as:

$$\gamma^+ = (1 + k)\gamma; \quad \gamma^- = (1 - k)\gamma, \tag{9.16}$$

with k a small parameter.

The tone curve characterised by μ_2, γ^+, γ^- is applied to Y_2 yielding an inverse-tone-mapped luminance image Y_{ITM}. Each colour channel $C \in \{R, G, B\}$ is inverse-tone-mapped resulting in a final output channel C', in this manner:

$$C' = \left(\frac{C}{Y_2}\right)^s * Y_{ITM}, \tag{9.17}$$

where s is a parameter that adjusts the saturation of the output.

The temporal consistency is enforced in the same way as in Section 9.3.1.3.

While the literature on inverse tone mapping is not as extensive as that for tone mapping, there are several ITM operators based on inverting sigmoidal compression functions or that take ideas from human vision, e.g. see [63–65].

9.3.4 Model tuning by cinema professionals

For the TMO for ungraded content proposed in Section 9.3.1, input from cinematographers and colourists allowed us to fine-tune the model and to determine an adequate form for the slope n in the nonlinear function NL for the transition between exponent γ^- and γ^+, such that

$$n = -4.5/\mu \tag{9.18}$$

$$\mu = log(median(Y)) - 2 \tag{9.19}$$

For the equations in the contrast adaptation section we only need to perform contrast enhancement on the achromatic channel, and an adequate value for m is 0.33.

For the TM and ITM methods for graded content presented in Sections 9.3.2 and 9.3.3, we conducted psychophysical experiments in a major post-production facility. Three colourists took part in the tests, where they were asked to match the general appearance of a series of images produced by our methods to the appearance of images produced manually using the current state-of-the-art workflow for TM and ITM of graded content. Images were matched via the adjustment of two algorithm parameters, corresponding to the global contrast and mid-tone luminance value. All the experiments and grading were conducted on calibrated reference displays in a standard reference viewing environment. For the ITM method there was a significant correlation ($R = 0.86$) between the value for γ obtained in Eq. (9.15) and the average value chosen by the colourists, which corroborates the efficacy of the model.

A value of $\gamma^- = 1.4$ provides a good correlation with the observers' settings, from which we can derive k and γ^+ from Eq. (9.16) on an image-by-image basis. Correspondingly, for the TM method for graded content, we set $\gamma^- = 1/1.4$.

The saturation parameter s was set to $s = 0.9$ in Eq. (9.14) and $s = 1.4$ in Eq. (9.17).

Finally, we want to stress that the parameter tuning was performed on images and video sequences that were different from the ones used for validation and comparisons with other methods, to be discussed in the following section.

9.4 Validation: psychophysical tests

9.4.1 TM of ungraded footage

We will compare our TMO introduced in Section 9.3.1 with the three methods that performed best in the very recent survey by [12]:

- Mantiuk et al. [66] proposed a piecewise linear curve to perform dynamic range compression, where the tone curve parameters are chosen so as to minimise the difference between the estimated response of the human visual system model for the output and the original image. The method can adapt to the particularity of the display and the viewing condition.
- Eilertsen et al. [67] extended the above approach by performing a decomposition into base and detail layers while considering the noise level in the image. The detail layer, after discounting for the noise visibility, is added to the transformed base layer to obtain the final tone mapped image.
- Boitard et al. [58] proposed a two-stage post-processing technique for adapting a global TMO for video. In the first stage, the tone mapped video frames are analyzed to find an anchor frame, then as a second stage the frames are divided into brightness zones and each zone is adjusted separately to preserve the brightness coherence with the anchor frame zone.

We will also compare our TMO with the most popular TM methods used in the cinema industry during production and post-production: camera manufacturer look-up tables (LUTs) (Canon provided LUT, Arri Alexa V3 to Rec709 LUT, and REDRAW default processing in Resolve), Resolve Luminance Mapping, Baselight TM (Truelight CAM DRT).

Test content. A range of 11 test video clips, of 10 seconds each, have been chosen which possess the following characteristics: natural/on-set lighting, bi-modal and normal luminance distributions, range of average picture level (APL), varying digital cinema capture platforms (i.e. differing capture dynamic ranges), skin tones, memory colours, portraits and landscapes, colour/b+w, minimal visual artifacts, realistic scenes for cinema context. Sample frames from the chosen sequences are shown in Figs. 9.8 and 9.11. The original test clips were transformed through each of the tone mapping methods listed above.

Observers. A total of 17 observers participated in the experiment. This was a diverse group of cinema professionals involved with motion picture production and experienced with image evaluation (digital imaging technicians, directors of photography, colourists, editors and visual effects specialists)

Pairing. In order to make an appropriate ranking, every combination of academic methods was paired. Industrial methods were paired directly against our proposed TMO and not with each other. The order of pair presentation was random as to make for a "blind" test.

Observer task and experimental procedure. The observer task was pairwise preferential comparison between the outputs of the chosen methods on the grounds of abstract realism (i.e. the "naturality" of the results). In the experiment, a motion

FIGURE 9.8 Sample frames from videos used in our validation tests

Sample frames from 7 of the 11 videos used in our validation tests for TM of ungraded content. Samples from the other 4 videos are shown in Fig. 9.11. Original video sources from [68,69,56,70].

picture professional was ushered into the lab, seated in front of the two monitors, and briefed with the instructions. These explained the cadence and task of the experiment – that observers should use the keyboard controls to select the clip which appears most natural or realistic to them. The instructions also included the stipulation that observers could take as much time as they'd like to view the clips, but that they should view each new clip in its entirety at least once in order to consider all content details in their decision.

To compute accuracy scores from the raw psychophysical data we use the same approach as in [71] (Chapter 5), that is based on Thurstone's law of comparative judgement. We briefly summarised this method in Chapter 8, but we shall recall the description here.

In order to compare n methods with experiments involving N observers, we create a $n \times n$ matrix for each observer where the value of the element at position (i, j) is 1 if method i is chosen over method j. From the matrices for all observers we create a $n \times n$ frequency matrix where each of its elements shows how often in a pair one method is preferred over the other. From the frequency matrix we create a $n \times n$ z-score matrix, and an accuracy score A for each method is given by the average of the corresponding column in the z-score matrix. The 95% confidence interval is given by $A \pm 1.96\frac{\sigma}{\sqrt{N}}$, as A is based on a random sample of size N from a normal distribution with standard deviation σ. In practice $\sigma = \frac{1}{\sqrt{2}}$, because the z-score represents the difference between two stimuli on a scale where the unit is $\sigma * \sqrt{2}$ (in Thurstone's paper this set of assumptions is referred to as "Case V"); as the scale of A has units which equal $\sigma * \sqrt{2}$, then we get that $\sigma = \frac{1}{\sqrt{2}}$.

The higher the accuracy score is for a given method, the more it is preferred by observers over the competing methods in the experiment. Fig. 9.9 shows the accuracy scores of the methods from the academic literature, where we can see that our proposed TMO performs best.

For these methods from the academic literature we also performed another experiment where five cinema professionals were asked to rate the outputs on a scale of

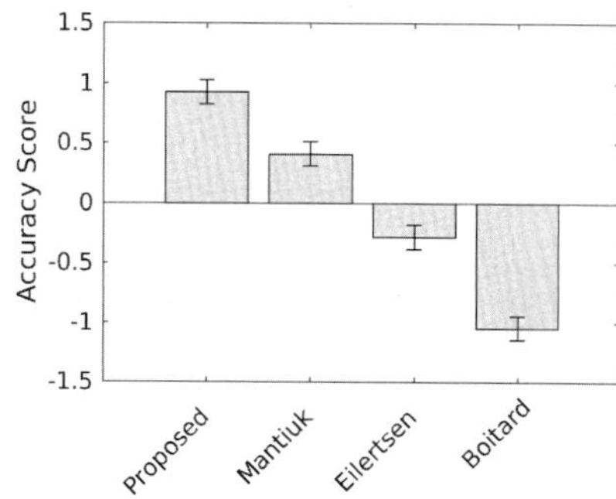

FIGURE 9.9 Observers' preferences of TM methods applied to ungraded content

Observers' preferences of TM methods applied to ungraded content. From left to right: proposed method, Mantiuk et al. [66], Eilertsen et al. [67], Boitard et al. [58].

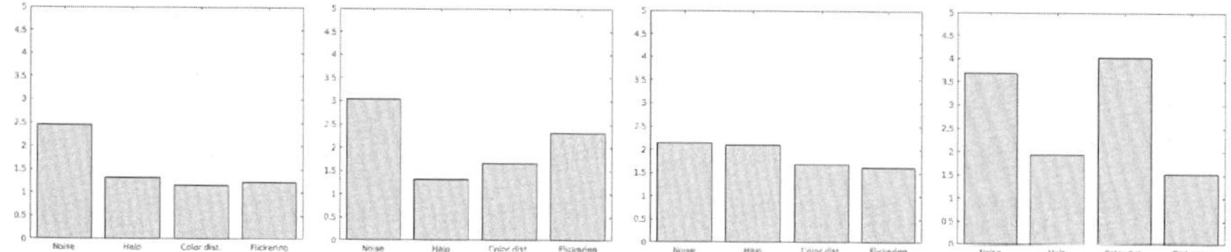

FIGURE 9.10 Observers' rating of artifact amount produced by TM methods applied to ungraded content

Observers' rating of artifact amount produced by TM methods applied to ungraded content. From left to right: proposed method, Mantiuk et al. [66], Eilertsen et al. [67], Boitard et al. [58].

one to ten in terms of the presence of the following artifacts: noise, haloing, colour distortions/clipping and, very importantly for cinema, temporal flickering. A higher rating implies a more noticeable artifact. Observers were allowed to randomly select between a set of video clips produced with different tone mapping methods applied to them, thus eliminating the need for an anchoring system as observers can re-adjust ratings after seeing all the available options. A manually tone mapped version of the clip was also included among the tested methods for observers as a control point. Results are shown in Fig. 9.10, where we can see that our proposed TMO consistently produces the least artifacts of all the four types, including noise, in which our approach compares well with the method by Eilertsen et al. [67] that is specifically designed to reduce noise in the tone-mapped output.

Regarding the industrial methods, our TMO is preferred over Baselight 66% of the time (with a statistical p-value of $p = 5.4 \cdot 10^{-7}$), over Resolve 73% of the time ($p = 1.4 \cdot 10^{-9}$), and over the camera LUT 58% of the time ($p = 0.0095$).

Fig. 9.11 shows sample results from the seven TM methods compared on 4 of the 11 test sequences. Notice the improved colour and contrast of our results, appearing in the first row.

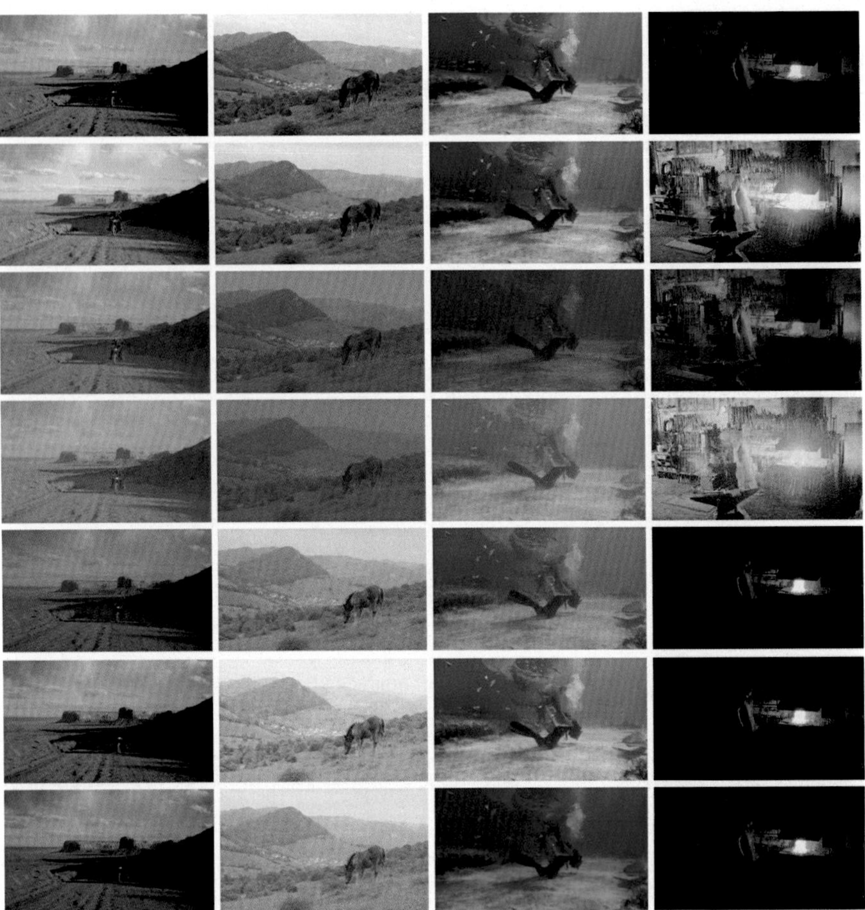

FIGURE 9.11 Sample TM results

Sample TM results. From top to bottom: proposed TMO, Mantiuk et al. [66], Eilertsen et al. [67], Boitard et al. [58], Baselight, Resolve, camera LUT. Original video sources from [68,69,56].

9.4.2 TM of graded footage

We validated the performance of our TM for graded content (TMG) method, introduced in Section 9.3.2, via pairwise comparison with the methods of [66,67], our TMO for ungraded content introduced in Section 9.3.1, and the industrial method of Dolby Vision. Video clips with the various methods applied to them were displayed to observers on a professional SDR reference monitor, who were instructed to choose the one which best matches a manually tone-mapped SDR reference (created by a colourist), in a two-alternative-forced-choice (2AFC) paradigm. This experiment was

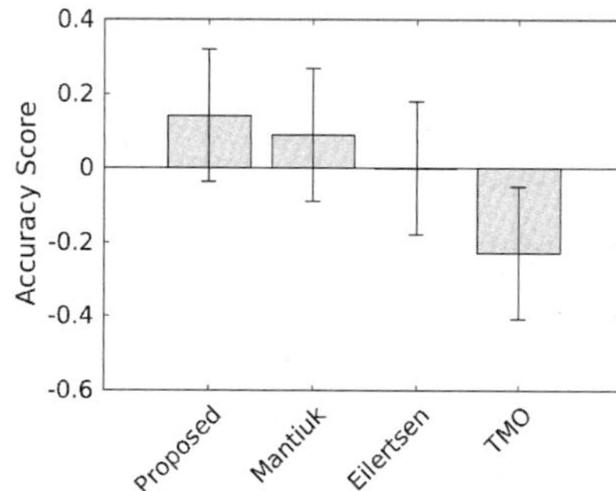

FIGURE 9.12 Observers' preferences of TM methods applied to graded content

Observers' preferences of TM methods applied to graded content. From left to right: proposed TM method for graded content, Mantiuk et al. [66], Eilertsen et al. [67], proposed TMO for ungraded content introduced in Section 9.3.1.

conducted at Deluxe, the largest post-production house in Spain, where a total of 6 observers, all professional cinema colourists, participated.

Fig. 9.12 shows the accuracy scores of the methods from the academic literature, where we can see that our proposed TMG algorithm performs best (although the inter-subject variability is larger in this experiment than in the one in Section 9.4.1). It's also interesting to note that our TMO for ungraded content performs worst, highlighting the importance of dealing differently with graded and ungraded content when developing TM methods.

Regarding the industrial method, our TMO is preferred over Dolby Vision 55% of the time; the p-value in this case is $p = 0.25$, not statistically significant, consistent with the very similar performance of both methods.

Fig. 9.13 shows sample results from all the methods compared. Notice how our results, first row, look quite similar to the manual grading results, second row.

9.4.3 ITM of graded footage

The performance of our method for inverse tone mapping of graded content (ITMG) was validated via pairwise comparison with the Transkoder/Colourfront industry software and three very recent state-of-the-art ITM methods from the academic literature:

- Masia et al. [72] proposed a simple gamma curve as the inverse tone mapping operator, where the gamma value is calculated by a linear combination of image key value, the geometric mean of the luminance and the number of overexposed pixels.
- In order to preserve the lighting style aesthetic, Bist et al. [73] also used a gamma curve, which is calculated from the median of the L* component of the input image in the CIE Lab colour space.

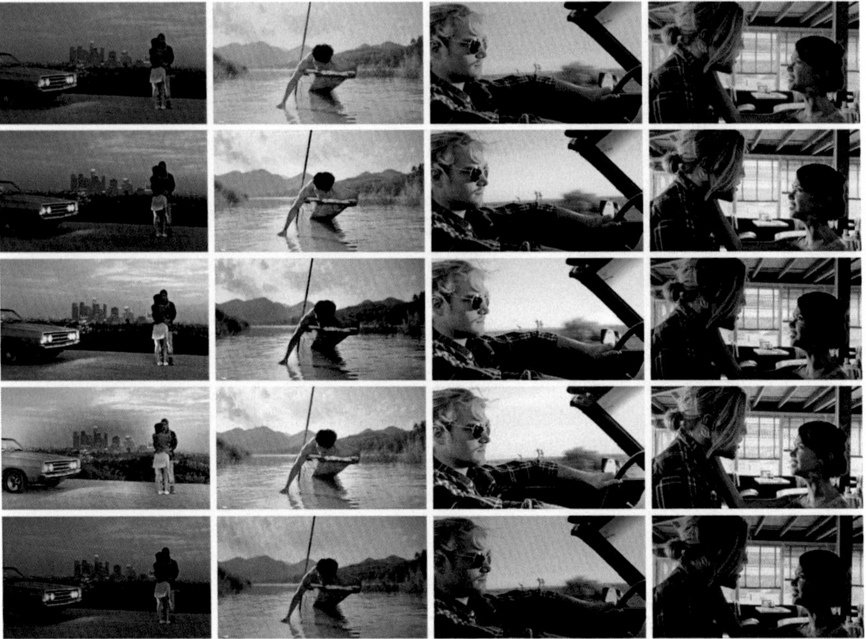

FIGURE 9.13 Result comparison for TM of graded content

Result comparison for TM of graded content. From top to bottom: proposed method, manual grading, Mantiuk et al. [66], Eilertsen et al. [67], Dolby. Original video sources from [70].

- Luzardo et al. [74] proposed a simple non-linear function that maps the middle gray value (0.214 for sRGB linear LDR images) of the input image to that of the output HDR image, where the mid-gray of the HDR image is computed by a linear combination of the geometric mean, the contrast and the percentage of overexposed pixels of the luminance channel of the input image.

Let us note that there are also some recent works on ITM with deep learning [75,76]. They produce results that are not display-referred, meaning that they are not encoded to be reproduced on any particular display and the tone mapping is not designed with any consideration for image appearance in this sense (e.g. the maximum value of the result is not associated to a given peak luminance of a display). Therefore these methods would require an additional tone mapping step in order to present their results on a display and perform a comparison with our method, whose intention as we have made clear is entirely to preserve display-referred appearance between SDR and HDR. Furthermore, one of the main goals of the methods in [75,76] is to recover image regions where values have been clipped due to over-exposure, but this was not the intention of the film professionals when they created the HDR reference images

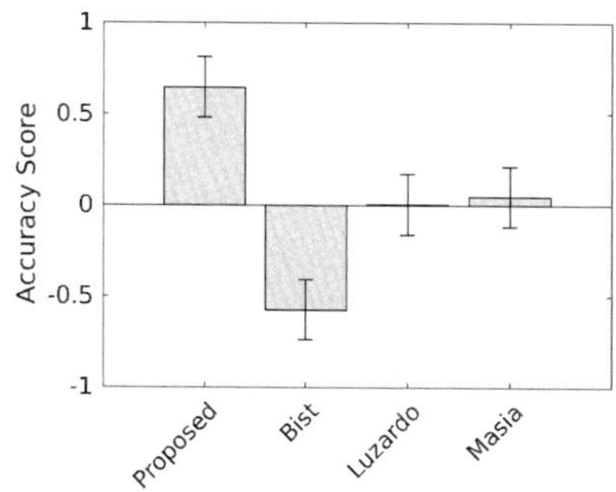

FIGURE 9.14 Observers' preferences of ITM methods applied to graded content

Observers' preferences of ITM methods applied to graded content. From left to right: proposed ITM, Bist et al. [73], Luzardo et al. [74], Masia et al. [72].

for our experiments by manual grading of professional, correctly-exposed footage. For these reasons, we decided not to include these two methods in our comparisons.

The experiment was conducted at Deluxe with the participation of 7 observers, all of them cinema professionals, and the details are as follows. The reference HDR footage consisted of 10 SDR clips of 10 seconds each, which were selected and manually graded for a 1000 cd/m^2 HDR display. Motion picture results from the methods discussed above were displayed to observers on a professional HDR reference monitor, in sequence. Observers were instructed to choose the result which best matches the manually graded HDR reference, shown on a second HDR reference monitor (2AFC).

Fig. 9.14 shows the accuracy scores of the methods from the academic literature, where we can see that our proposed ITM algorithm performs best.

Regarding the industrial method, our TMO is preferred over Transkoder 54% of the time; the p-value in this case is $p = 0.28$, again not statistically significant and consistent with the very similar performance of both methods.

9.5 Limitations of objective metrics

Evaluating the user preference of TM and ITM results is a challenging problem. Subjective studies like the ones described in the previous sections involve very time-consuming preparations, require special hardware, have to be performed under controlled viewing conditions in an adequate room, etc. Therefore, instead of having to perform user studies, it would be ideal to be able to use image quality metrics to automatically evaluate and rank TM methods, and for this reason a number of objective metrics have been proposed over the years that deal with HDR and tone-mapped images.

In this section we evaluate if any of the most widely-used image metrics for HDR applications is able to predict the results of our psychophysical tests. The metrics we

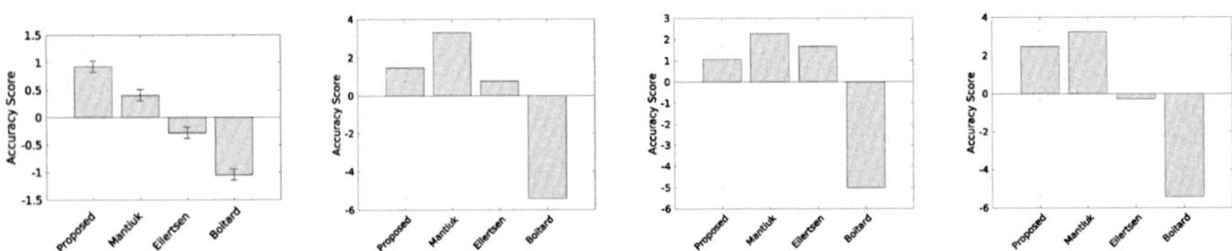

FIGURE 9.15 Validation of metrics for the case of TM of ungraded content

Validation of metrics for the case of TM of ungraded content. From left to right: observers' preference, prediction by FSITM, prediction by TMQI, prediction by FSITM-TMQI.

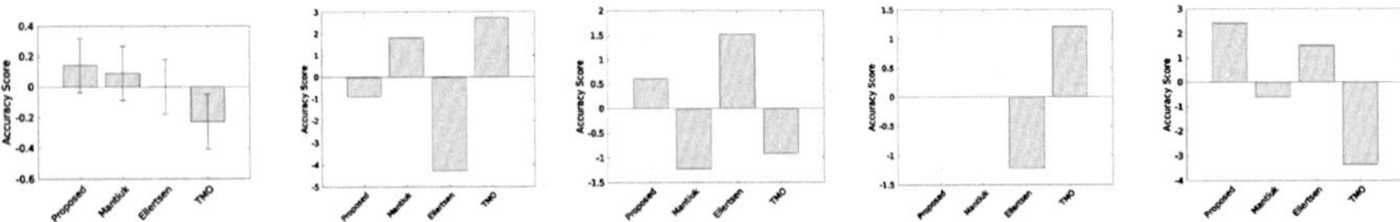

FIGURE 9.16 Validation of metrics for the case of TM of graded content

Validation of metrics for the case of TM of graded content. From left to right: observers' preference, prediction by HDR-VQM, prediction by HDR-VDP2, prediction by PieAPPs, prediction by LPIPS.

have chosen are TMQI [77], FSITM and FSITM-TMQI [78], HDR-VDP2 [79] and HDR-VQM [80]. FSITM, TMQI and FSITM-TMQI compare the TM results with the original HDR source, while HDR-VDP2 and HDR-VQM compare the manual TM or ITM results with the output of the automatic methods. HDR-VQM was specially developed for video assessment so we have applied it directly on the full test sequences; the other metrics are computed for 1 every 5 frames in each sequence, and intra-frame scores are averaged using TMQI's memory effect.

Furthermore we will also use two very recent, state-of-the-art deep learning metrics for perceived appearance, PieAPP [81] and LPIPS [82]. These two metrics are designed to predict perceptual image error like human observers and are based on large scale datasets (20K images in one case, 160K images in the other) labelled with pair-comparison preferences and, in the case of LPIPS, using close to 500K human judgements. As these metrics were trained on SDR image datasets, we'll use them just for the case of TM of graded content.

In order to evaluate the performance of all the above metrics in the context of our experimental setting we will consider the metrics as if they were observers in our pair comparison experiments. This means that, for each metric, we will run all the possible comparisons, and in each comparison we will give a score of 1 to the image with better metric value and a score of 0 to the image with worse metric value. Then, we apply the Thurstone Case V analysis to obtain the accuracy scores that each metric predicts for the TM and ITM methods tested.

Figs. 9.15, 9.16 and 9.17 compare the accuracy scores predicted by the metrics with the actual accuracy scores yielded by the observers. These figures show that the

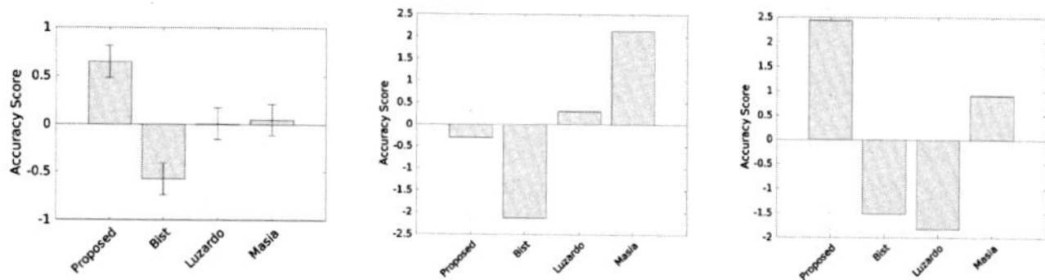

FIGURE 9.17 Validation of metrics for the case of ITM

Validation of metrics for the case of ITM. From left to right: observers' preference, prediction by HDR-VQM, prediction by HDR-VDP2.

FIGURE 9.18 The deep-learning metric LPIPS [82] gives scores that contradict observers' preferences

The lower the LPIPS score is, the closest the image is supposed to be to the reference in terms of perception. From left to right, top to bottom: reference (manual grading); output of [58], LPIPS=0.324; result from Dolby, LPIPS=0.376; output of [66], LPIPS=0.39; result from [67], LPIPS=0.435. Notice how LPIPS considers the second image to be the most similar to the reference, while observers tend to consider the second image as the worst result among these four methods. Original image from [70].

metrics were not able to accurately predict the choices made by observers for any application (TM of ungraded or graded content, ITM).

Only in the case of LPIPS (Fig. 9.16, rightmost plot) it could be argued that the metric is not inconsistent with observers' choices. But Fig. 9.18 shows examples where LPIPS clearly contradicts observers' preferences.

We are aware that no strong conclusions should be derived from this reduced set of tests, but the limitations of objective metrics that we have observed are in line with the results recently reported in [83].

9.6 Discussion

We would like to share some thoughts on possible reasons for the effectiveness of our approach, as well as its potential for vision science research.

While our proposed TMO for ungraded content is based on a simple retinal model that is consistent with neurophysiological data, it can also be seen as a combination of ideas that have been recurrent in the TM literature, as mentioned in Section 9.3.1.5. The advantage of our method comes, we believe, from incorporating into our algorithm a number of properties that have been observed in neural response data and that visibly impact the results. Possibly the most important of these properties are:

- The very wide receptive field of horizontal cells supports the use of a global nonlinearity, which is guaranteed to produce no halo artifacts.
- The different response of ON and OFF channels supports modelling the global nonlinearity as an asymmetric sigmoid (in log coordinates) or as two power-laws (in linear coordinates), which allows to better perform histogram equalisation.
- The constant neural gain for the lowest light levels supports limiting the slope of the nonlinearity for the darkest values, which allows to reduce the noise.
- The different extent of the horizontal cell feedback for chromatic and achromatic pathways supports performing contrast enhancement differently for the luminance and the chroma channels, which allows for a better reproduction of both colour and contrast.

But the core of our method isn't just the accurate modelling of retinal processes: the contribution of cinema professionals like cinematographers and colourists has been essential in order to properly adjust and fine-tune our framework. Colourists are capable of adjusting the appearance of images so that, when we look at them on a screen, our subjective impression matches what we would perceive with the naked eye at the real-world scene. As such, these cinema professionals behave as "human TMOs", complying with the TM requirements stated in [53,54]. We have shown in the current work (and we're definitely not the first ones in doing this) how it's appropriate to develop a TMO after a model of retinal processes, and therefore by optimising our TMO one could argue that the cinema artists have been optimising a vision model, simple as it might be. It follows that the work of cinema professionals could potentially be useful as well for vision science research, normally limited by the use of synthetic stimuli and restrictive models in a way that has raised important concerns in the vision science community [10,84]: the input of colourists and cinematographers might help to better find parameter values for the vision model, as they have done in our case for our algorithms, but going forward their contribution could be even more significant, helping to develop and validate models that lie beyond the restrictive setting of L+NL cascades [84,85].

9.7 Conclusion

We have proposed a framework for TM and ITM that is useful for the cinema industry over all stages of the production chain, from shooting to exhibition.

Our framework is very simple, its effectiveness stems from adapting data from the vision neuroscience literature to propose a model that is then fine-tuned by cinema professionals that aim for the best perceptual match between on-screen and real-world images.

The algorithms have very low computational complexity and they could be implemented for real-time execution. The visual quality of the results is very high, surpassing the state-of-the-art in the academic literature and outperforming as well (although sometimes by a narrow margin) the most widely used industrial methods. Given that industrial methods are the result of considerable research into very specific tasks using specific technologies and are thus practical solutions to practical problems, that our approach can work as well as these methods across a broad range of problems and technologies indicates that the theoretical underpinning of our methodology has general applicability and operates to a high standard.

The proposed methods have been validated by expert observers, and we have shown that these user preferences can't be predicted by existing objective metrics, highlighting the need to develop more accurate metrics for HDR imaging.

Going forward, we are currently working on adding more features to our basic TM method so that more perceptual phenomena can be simulated.

We are also working on developing a new type of user interface for colour-grading software suites that will allow colourists to modify the appearance of images in a way that echoes how the visual system processes information, and in that manner operating the software system becomes (hopefully) more intuitive.

References

[1] SMPTE. SMPTE HDR study group report. https://www.smpte.org/standards/reports, 2015.

[2] ITU-R. Report itu-r bt. 2390-0, 2016.

[3] UHDForum. http://ultrahdforum.org/wp-content/uploads/2016/04/Ultra-HD-Forum-Deployment-Guidelines-V1.1-Summer-2016.pdf, 2016.

[4] Boitard R, Smith M, Zink M, Damberg G, Ballestad A. Using high dynamic range home master statistics to predict dynamic range requirement for cinema. In: SMPTE 2018; 2018. p. 1–28.

[5] Gordon A. Beyond better pixels: how hdr perceptually and emotionally effects storytelling. In: SMPTE 2018; 2018. p. 1–12.

[6] Reinhard E, Stauder J, Kerdranvat M. An assessment of reference levels in hdr content. In: SMPTE 2018; 2018. p. 1–10.

[7] Routhier PH. What are the "killer apps" for hdr? Analysis of sdr assets to predict the potential of hdr. In: SMPTE 2018; 2018. p. 1–17.

[8] Ploumis S, Boitard R, Jacquemin J, Damberg G, Ballestad A, Nasiopoulos P. Quantitative evaluation and attribute of overall brightness in a high dynamic range world. In: SMPTE 2018; 2018. p. 1–16.

[9] Vandenberg J, Andriani S. A survey on 3d-lut performance in 10-bit and 12-bit hdr bt. 2100 pq. In: SMPTE 2018; 2018. p. 1–19.

[10] Carandini M, Demb JB, Mante V, Tolhurst DJ, Dan Y, Olshausen BA, et al. Do we know what the early visual system does? Journal of Neuroscience 2005;25(46):10577–97.

[11] Yedlin S. On color science. http://www.yedlin.net/OnColorScience/, 2016.

[12] Eilertsen G, Mantiuk RK, Unger J. A comparative review of tone-mapping algorithms for high dynamic range video. Computer graphics forum, vol. 36. Wiley Online Library; 2017. p. 565–92.

[13] Reinhard E, Heidrich W, Debevec P, Pattanaik S, Ward G, Myszkowski K. High dynamic range imaging: acquisition, display, and image-based lighting. Morgan Kaufmann; 2010.

[14] Van Hurkman A. Color correction handbook: professional techniques for video and cinema. Pearson Education; 2013.
[15] Gollisch T, Meister M. Eye smarter than scientists believed: neural computations in circuits of the retina. Neuron 2010;65(2):150–64.
[16] Shapley R, Enroth-Cugell C. Visual adaptation and retinal gain controls. Progress in Retinal Research 1984;3:263–346.
[17] Wandell BA. Foundations of vision, vol. 8. Sunderland, MA: Sinauer Associates; 1995.
[18] Nassi JJ, Callaway EM. Parallel processing strategies of the primate visual system. Nature Reviews Neuroscience 2009;10(5):360.
[19] Wässle H. Parallel processing in the mammalian retina. Nature Reviews Neuroscience 2004;5(10):747.
[20] Jansen M, Jin J, Li X, Lashgari R, Kremkow J, Bereshpolova Y, et al. Cortical balance between on and off visual responses is modulated by the spatial properties of the visual stimulus. Cerebral Cortex 2018;29(1):336–55.
[21] Hubel DH. Eye, brain, and vision. Scientific American Library/Scientific American Books; 1995.
[22] Olshausen BA, Field DJ. Natural image statistics and efficient coding. Network: Computation in Neural Systems 1996;7(2):333–9.
[23] Wark B, Lundstrom BN, Fairhall A. Sensory adaptation. Current Opinion in Neurobiology 2007;17(4):423–9.
[24] Wark B, Fairhall A, Rieke F. Timescales of inference in visual adaptation. Neuron 2009;61(5):750–61. https://doi.org/10.1016/j.neuron.2009.01.019.
[25] Dunn FA, Rieke F. The impact of photoreceptor noise on retinal gain controls. Current Opinion in Neurobiology 2006;16(4):363–70.
[26] Baccus SA, Meister M. Fast and slow contrast adaptation in retinal circuitry. Neuron 2002;36(5):909–19.
[27] Demb JB. Functional circuitry of visual adaptation in the retina. The Journal of Physiology 2008;586(18):4377–84.
[28] Kohn A. Visual adaptation: physiology, mechanisms, and functional benefits. Journal of Neurophysiology 2007;97(5):3155–64.
[29] Atick JJ. Could information theory provide an ecological theory of sensory processing? Network Computation in Neural Systems 1992;3(2):213–51.
[30] Olshausen BA, Field DJ. Vision and the coding of natural images: the human brain may hold the secrets to the best image-compression algorithms. American Scientist 2000;88(3):238–45.
[31] Rucci M, Victor JD. The unsteady eye: an information-processing stage, not a bug. Trends in Neurosciences 2015;38(4):195–206.
[32] Radonjić A, Allred SR, Gilchrist AL, Brainard DH. The dynamic range of human lightness perception. Current Biology 2011;21(22):1931–6.
[33] Kane D, Bertalmío M. System gamma as a function of image-and monitor-dynamic range. Journal of Vision 2016;16(6).
[34] Ozuysal Y, Baccus SA. Linking the computational structure of variance adaptation to biophysical mechanisms. Neuron 2012;73(5):1002–15.
[35] Kremkow J, Jin J, Komban SJ, Wang Y, Lashgari R, Li X, et al. Neuronal nonlinearity explains greater visual spatial resolution for darks than lights. Proceedings of the National Academy of Sciences 2014:201310442.
[36] Turner MH, Rieke F. Synaptic rectification controls nonlinear spatial integration of natural visual inputs. Neuron 2016;90(6):1257–71.
[37] Turner MH, Schwartz GW, Rieke F. Receptive field center-surround interactions mediate context-dependent spatial contrast encoding in the retina. bioRxiv 2018:252148.
[38] Whittle P. Brightness, discriminability and the "crispening effect". Vision Research 1992;32(8):1493–507.
[39] Stevens J, Stevens SS. Brightness function: effects of adaptation. JOSA 1963;53(3):375–85.
[40] Nundy S, Purves D. A probabilistic explanation of brightness scaling. Proceedings of the National Academy of Sciences 2002;99(22):14482–7.

[41] Grimaldi A, Kane D, Bertalmío M. Statistics of natural images as a function of dynamic range. Journal of Vision 2019;19(2).

[42] Huang J, Mumford D. Statistics of natural images and models. In: Computer vision and pattern recognition. IEEE computer society conference on., vol. 1. IEEE; 1999.

[43] Matthews H, Fain G, Murphy R, Lamb T. Light adaptation in cone photoreceptors of the salamander: a role for cytoplasmic calcium. The Journal of Physiology 1990;420(1):447–69.

[44] Whittle P, Challands P. The effect of background luminance on the brightness of flashes. Vision Research 1969;9(9):1095–110.

[45] Martinez-Garcia M, Cyriac P, Batard T, Bertalmío M, Malo J. Derivatives and inverse of cascaded linear+ nonlinear neural models. PLoS ONE 2018;13(10):e0201326.

[46] Lee BB, Martin PR, Grünert U. Retinal connectivity and primate vision. Progress in Retinal and Eye Research 2010;29(6):622–39.

[47] Schmidt BP, Neitz M, Neitz J. Neurobiological hypothesis of color appearance and hue perception. JOSA A 2014;31(4):A195–207.

[48] Yeonan-Kim J, Bertalmío M. Retinal lateral inhibition provides the biological basis of long-range spatial induction. PLoS ONE 2016;11(12):e0168963.

[49] Bloomfield SA, Völgyi B. The diverse functional roles and regulation of neuronal gap junctions in the retina. Nature Reviews Neuroscience 2009;10(7):495.

[50] Milner ES, Do MTH. A population representation of absolute light intensity in the mammalian retina. Cell 2017;171(4):865–76.

[51] Chapot CA, Euler T, Schubert T. How do horizontal cells 'talk' to cone photoreceptors? Different levels of complexity at the cone–horizontal cell synapse. The Journal of Physiology 2017;595(16):5495–506.

[52] Masland RH. The neuronal organization of the retina. Neuron 2012;76(2):266–80.

[53] Tumblin J, Rushmeier H. Tone reproduction for realistic images. IEEE Computer Graphics and Applications 1993;13(6):42–8.

[54] Ward G, Rushmeier H, Piatko C. A visibility matching tone reproduction operator for high dynamic range scenes. IEEE Transactions on Visualization & Computer Graphics 1997;(4):291–306.

[55] Poynton C. Digital video and HD: algorithms and interfaces. Elsevier; 2012.

[56] Pascual A. Unrealeased footage. http://albertpascualcinema.blogspot.com/, 2018.

[57] Ebner F, Fairchild MD. Development and testing of a color space (ipt) with improved hue uniformity. In: Color and imaging conference, vol. 1998. Society for Imaging Science and Technology; 1998. p. 8–13.

[58] Boitard R, Cozot R, Thoreau D, Bouatouch K. Zonal brightness coherency for video tone mapping. Signal Processing. Image Communication 2014;29(2):229–46.

[59] Cyriac P, Kane D, Bertalmío M. Optimized tone curve for in-camera image processing. Electronic Imaging 2016;2016(13):1–7.

[60] Pattanaik SN, Tumblin J, Yee H, Greenberg DP. Time-dependent visual adaptation for fast realistic image display. In: Proceedings of the 27th annual conference on computer graphics and interactive techniques. ACM Press/Addison-Wesley Publishing Co.; 2000. p. 47–54.

[61] Ashikhmin M. A tone mapping algorithm for high contrast images. In: Proceedings of the 13th eurographics workshop on rendering, eurographics association; 2002. p. 145–56.

[62] Kuang J, Johnson GM, Fairchild MD. icam06: A refined image appearance model for hdr image rendering. Journal of Visual Communication and Image Representation 2007;18(5):406–14.

[63] Banterle F, Ledda P, Debattista K, Chalmers A. Inverse tone mapping. In: Proceedings of the 4th international conference on computer graphics and interactive techniques in Australasia and Southeast Asia. ACM; 2006. p. 349–56.

[64] Rempel AG, Trentacoste M, Seetzen H, Young HD, Heidrich W, Whitehead L, et al. Ldr2hdr: on-the-fly reverse tone mapping of legacy video and photographs. ACM transactions on graphics (TOG), vol. 26. ACM; 2007. p. 39.

[65] Huo Y, Yang F, Dong L, Brost V. Physiological inverse tone mapping based on retina response. The Visual Computer 2014;30(5):507–17.

[66] Mantiuk R, Daly S, Kerofsky L. Display adaptive tone mapping. ACM transactions on graphics (TOG), vol. 27, vol. 27. ACM; 2008. p. 68.
[67] Eilertsen G, Mantiuk RK, Unger J. Real-time noise-aware tone mapping. ACM Transactions on Graphics (TOG) 2015;34(6):198.
[68] Froehlich J, Grandinetti S, Eberhardt B, Walter S, Schilling A, Brendel H. Creating cinematic wide gamut hdr-video for the evaluation of tone mapping operators and hdr-displays. In: Digital photography X, vol. 9023. International Society for Optics and Photonics; 2014. p. 90230X.
[69] Cinema RD. Sample r3d files. https://www.red.com/sample-r3d-files, 2019.
[70] ARRI. Enhanced capture material for hdr4eu project d2.2. https://www.upf.edu/web/hdr4eu/publications, 2018.
[71] Morovic J. To develop a universal gamut mapping algorithm. Ph.D. thesis, UK: University of Derby; 1998.
[72] Masia B, Serrano A, Gutierrez D. Dynamic range expansion based on image statistics. Multimedia Tools and Applications 2017;76(1):631–48.
[73] Bist C, Cozot R, Madec G, Ducloux X. Tone expansion using lighting style aesthetics. Computers & Graphics 2017;62:77–86.
[74] Luzardo G, Aelterman J, Luong H, Philips W, Ochoa D, Rousseaux S. Fully-automatic inverse tone mapping preserving the content creator's artistic intentions. In: 2018 picture coding symposium (PCS). IEEE; 2018. p. 199–203.
[75] Endo Y, Kanamori Y, Mitani J. Deep reverse tone mapping. ACM Transactions on Graphics 2017;36(6).
[76] Eilertsen G, Kronander J, Denes G, Mantiuk RK, Unger J. Hdr image reconstruction from a single exposure using deep cnns. ACM Transactions on Graphics (TOG) 2017;36(6):178.
[77] Yeganeh H, Wang Z. Objective quality assessment of tone-mapped images. IEEE Transactions on Image Processing 2013;22(2):657–67.
[78] Nafchi HZ, Shahkolaei A, Moghaddam RF, Cheriet M. Fsitm: a feature similarity index for tone-mapped images. IEEE Signal Processing Letters 2015;22(8):1026–9.
[79] Mantiuk R, Kim KJ, Rempel AG, Heidrich W. Hdr-vdp-2: a calibrated visual metric for visibility and quality predictions in all luminance conditions. In: ACM transactions on graphics (TOG), vol. 30. ACM; 2011. p. 40.
[80] Narwaria M, Da Silva MP Le Callet P. Hdr-vqm: an objective quality measure for high dynamic range video. Signal Processing. Image Communication 2015;35:46–60.
[81] Prashnani E, Cai H, Mostofi Y, Sen P. Pieapp: Perceptual image-error assessment through pairwise preference. In: The IEEE conference on computer vision and pattern recognition (CVPR); 2018.
[82] Zhang R, Isola P, Efros AA, Shechtman E, Wang O. The unreasonable effectiveness of deep features as a perceptual metric. In: CVPR; 2018.
[83] Krasula L, Narwaria M, Fliegel K, Le Callet P. Preference of experience in image tone-mapping: dataset and framework for objective measures comparison. IEEE Journal of Selected Topics in Signal Processing 2017;11(1):64–74.
[84] Olshausen BA, Field DJ. How close are we to understanding v1? Neural Computation 2005;17(8):1665–99.
[85] Olshausen BA. 20 years of learning about vision: questions answered, questions unanswered, and questions not yet asked. In: 20 years of computational neuroscience. Springer; 2013. p. 243–70.

CHAPTER

Extensions and applications 10

We start this chapter by discussing how to combine gamut mapping and tone mapping so as to be able to transform images where input and output can both be of any type of dynamic range and colour gamut.

Next we recount how to extend our tone mapping methodology to take into account surround luminance and display characteristics.

This is followed by a number of applications where our tone mapping framework has proven useful: improved encoding of HDR video, photorealistic style transfer, image dehazing and enhancement of backlit images.

Finally, we present two common types of colour problems that might be encountered with HDR image creation and editing, where the encoding curve plays a role, and we propose methods to solve these issues.

10.1 Combining gamut mapping and tone mapping

In the chapter on gamut mapping the GM input is assumed to be of standard dynamic range, which is the usual approach in the GM literature. This would correspond, going back to the vision principles mentioned in Section 8.4.1, to having as input image for our GM framework a picture that emulates the (SDR) photoreceptor output, generated for instance with a sigmoid curve like the Naka-Rushton equation, since we saw in Chapter 4 that modelling photoreceptor responses with this equation optimises the performance efficiency of cones and rods, adapting the HDR input intensities to the SDR representation capabilities of neurons [1].

Therefore, if the input video to be gamut-mapped is in HDR, our framework requires that it is tone-mapped first with a method that preserves the colour gamut, like the TMOs we propose in Sections 9.3.1 and 9.3.2; then we can process the result with our GM algorithm. This is consistent with the workflow for GM of HDR content proposed in [2].

The output of our GM method will also be in SDR form. If it were required for it to be in HDR, then we should apply to the GM output an inverse tone mapping method that preserves the colour gamut, like the one introduced in Section 9.3.3.

In summary, from the gamut mapping and tone mapping methods presented in Chapters 8 and 9 we can build a general framework for mapping between images, that allows for input and output of any type of dynamic range and colour gamut. If

Vision Models for High Dynamic Range and Wide Colour Gamut Imaging. https://doi.org/10.1016/B978-0-12-813894-6.00015-6

the input is in HDR we must start by performing tone mapping. Next we apply, if needed, GR or GE. Finally, if the output must be in HDR then we perform inverse tone mapping.

10.2 Tone mapping and viewing condition adaptation

We saw in Chapter 5 how viewing conditions and the display's capabilities in terms of contrast and peak luminance can significantly affect the perceived image quality. Ideally, the colour grading stage in postproduction should take into account all these factors, although a conventional simplified solution is to apply a power-law adjustment (system gamma) of exponent γ_{adj} that depends on the intended viewing condition.

Based on this line of research, in [3] we set out to identify the power-law value γ_{adj} needed for the results of the tone mapping algorithm we had proposed in [4] to look optimal for different surround conditions and displays. For this we performed the following steps:

1. Conduct psychophysical experiments where subjects modify the power-law value γ_{adj} to be applied to the tone-mapped image O until the appearance of O^{γ}_{adj} is optimal (the γ_{adj} exponent is applied independently to each of the RGB colour channels of O).
2. Record several physical measurements including minimum and maximum luminance of the display, surround luminance and ambient illuminance.
3. Propose a model for γ_{adj} that fits the data.

The stimuli for the experiment are 20 tone mapped versions of the HDR images from the HDR survey by Mark Fairchild [5]. The images were chosen to cover a variety of scenarios: night images, indoor scenes, bright outdoor scenes, landscapes, etc. The HDR images were tone mapped by the method of [4]. A schematic of the experimental setup is shown in Fig. 10.1. Subjects were asked to adjust the gamma non-linearity via a scroll bar such that the image achieves an optimal appearance. The viewing conditions that we considered were:

- *Two surround environments.*
  Office room: ambient illuminant of 47 cd/m^2 and average near surround luminance of 65 cd/m^2.
  Dark room: ambient illuminant of 0.3 cd/m^2 and average near surround luminance $\simeq 0$ cd/m^2.
- *Three display types.*
  LCD: ASUS VS197D LCD monitor set to sRGB mode.
  OLED: Sony Trimaster PVM.
  HDR: SIM2 HDR47ES4MB monitor set to HDR mode.

Fig. 10.2 shows the average subject choice for γ_{adj} for two observers, 20 images, 3 display types and two surround environments.

FIGURE 10.1 Psychophysical experiment setup

Psychophysical experiment setup. Source image from [5].

Display	Dark room	Office room
LCD	1.11	1
OLED	1.09	0.98
HDR	-	1.5

FIGURE 10.2

Subject choice for γ_{adj}.

For each combination of display type and surround condition we measured both the sequential and ANSI contrast. Both contrasts are given by the ratio of the average white luminance to the average black luminance. In the case of the sequential contrast, the luminance measurement is recorded when the whole display is black or white. On the other hand, in the case of the ANSI contrast the luminance measurement is recorded while displaying the pattern shown in Fig. 10.3.

FIGURE 10.3 Checker pattern to measure ANSI contrast

Checker pattern to measure ANSI contrast.

Display	Dark room			Office room		
	Min luminance	Max luminance	Sequential contrast	Min luminance	Max luminance	Sequential contrast
LCD	0.35 nits	170 nits	486	2.3 nits	170 nits	74
OLED	0.001 nits	97 nits	97000	1.2 nits	97 nits	80
HDR	-	-	-	1.7 nits	2700 nits	1588

FIGURE 10.4

Sequential contrast measurement.

Display	Dark room			Office room		
	Min luminance	Max luminance	ANSI contrast	Min luminance	Max luminance	ANSI contrast
LCD	0.7 nits	170 nits	242	2.6 nits	170 nits	65
OLED	0.3 nits	97 nits	323	1.3 nits	97 nits	74
HDR	-	-	-	2 nits	2700 nits	1350

FIGURE 10.5

ANSI contrast measurement.

A Konica Minolta LS 100 photometer was used to measure the luminance and the reading was taken at a distance at which an observer views the display (approximately 3 times the display height). Results are shown in Figs. 10.4 and 10.5.

In general, the ANSI contrast is substantially smaller than the sequential contrast in the dark surround condition. This is mainly because some of the light emitted by the display is reflected back onto the screen by objects in the surround, resulting in raising the effective minimum value of the display. For example, the minimum luminance measured on the OLED is 0.001 cd/m^2 in a dark surround when the full screen is black, whereas the measured luminance is 0.3 cd/m^2 when the pattern shown in Fig. 10.3 is displayed. This result is consistent with the result of Schuck and Lude [6] and Tydtgat et al. [7], where they measured both sequential contrast and the effective contrast by varying the image white content on cinema projectors. We see again in these measurements that the effective contrast produced by a display depends not only on its maximum contrast ratio but also on the surround and the reproduced image content.

Analysing the results of the psychophysical experiment and the contrast measurement, we found that the subject-preferred non-linear adjustment value γ_{adj} can be predicted from the ANSI contrast and the maximum peak luminance of the display by the following formula:

$$\gamma_{adj} = (1 + 0.2|C|)^{sign(C)}, \tag{10.1}$$

where

$$C = log_{10}\left(\frac{L_{vc}^{peak}}{L_{grad}^{peak}}\right) + log_{10}(ANSI_{vc}) - log_{10}(ANSI_{grad}). \tag{10.2}$$

In Eq. (10.2), L_{vc}^{peak} and $ANSI_{vc}$ are the peak luminance and the ANSI contrast of the intended display in which the image is to be viewed. L_{grad}^{peak} and $ANSI_{grad}$ are the peak luminance and the ANSI contrast of the grading display: an LCD display in an office environment in our experiments, since the parameter values for the tone mapping method in [4] were selected to optimise image appearance in that scenario. Fig. 10.6 shows that the model predicts well the result of the psychophysical experiment.

Display	Dark room		Office room	
	Subject choice	Model prediction	Subject choice	Model prediction
LCD	1.11	1.12	1	1
OLED	1.09	1.1	0.98	0.98
HDR	-	-	1.5	1.51

FIGURE 10.6 γ_{adj} **model accuracy**

Comparison between user chosen and model predicted γ_{adj}.

In Fig. 10.7 we show the result produced by our method for three different viewing conditions. As the effective contrast of the display (ANSI contrast) increases, our method tries to compensate it by decreasing the contrast of the lower mid-intensities of the input image. Note that these images will look optimal only under the intended viewing conditions and display type.

LCD office (65) OLED dark room (323) SIM2 office (1350)

FIGURE 10.7 Results of γ_{adj} **model**

Results of method in [3] applied to HDR video frame from the ARRI dataset [8] for three different viewing conditions. ANSI contrast given in parentheses. Figure from [3].

10.2.1 Updates to the γ_{adj} model

After improving the tone mapping method to the one presented in Chapter 9, we decided to do new experiments similar to the one just described, so as to verify or re-define the model of Eq. (10.1). This work was performed by Trevor Canham and Praveen Cyriac.

10.2.1.1 Influence of display luminance and contrast

A set of ten static test images was chosen to cover a variety of real-world shooting scenarios. Images originated from a range of capture platforms and covered a wide range of scene content. Outdoor and indoor shooting locations, landscapes, closeups, and skin tones/human subjects were all included in the test. Images were rendered via the following signal processing chain. Initial images were linearly encoded in Rec.2020 colour space and stored in an EXR file container. The tone mapping operator was applied in this state, and the results were transformed to produce an sRGB output. In its native behaviour the tone mapping algorithm adds a degree of non-linearity to the linear image tonescale. However, this non-linearity is optimised for a particular display in an office viewing environment. Thus in the experiment an additional corrective gamma function is added to the image and adjusted by observers to optimise for the tested displays in a dark surround viewing environment.

The laboratory was equipped with three different displays for testing, a Sony trimaster PVM OLED reference monitor, a mid-2018 MacBook pro labtop, and a Samsung EU65KS9800 TV. The MacBook was tested at three different settings in order to get multiple data points from different peak luminance levels with no additional image rendering variations introduced by the display. All displays were viewed in dark surround ambient conditions. Preliminary measurements were taken on the displays in order to characterise their native gamma decoding behaviour as well as their ANSI contrast. Both of these measurements were taken using a Konica Minolta LS 100 photometer.

The OLED was viewed at a distance of approximately 50 cm so that 34 pixels subtended 1 degree of visual angle and the full display subtended 57 by 34 degrees. MacBook was viewed at a distance of approximately 30 cm so that 50 pixels subtended 1 degree of visual angle and the full display subtended 51 by 33 degrees. Samsung TV was viewed at a distance of approximately 120 cm so that 64 pixels subtended 1 degree of visual angle and the full display subtended 60 by 37 degrees. The presented image covers roughly 85% of the display area. Observers were instructed to adjust the gamma non-linearity of the images using a scroll bar in the user interface so that the image achieves an optimal appearance. This task was repeated for 10 different images at two different initial presentation points. After all image trials are complete the task is repeated for the various displays and display settings. A total of seven subjects were tested.

Shown in Fig. 10.8 are the display peak luminance, ANSI contrast measurements, and the resulting average value chosen by observers. Minimum and maximum luminance were based on the measurements from local regions of the display whereas ANSI contrast is based on the average over different regions. It can be seen that the average responses were primarily influenced by the peak luminance level of the display and secondarily influenced by ANSI contrast.

Display	Min. Luminance	Max. Luminance	ANSI Contrast	Average γ_{adj}
Sony OLED	0.37	188	415	0.985
Samsung TV	1.18	580	395	1.067
Mac Low	0.55	132	240	0.954
Mac Mid	1.45	350	240	1.026
Mac High	1.61	449	280	1.033

FIGURE 10.8 Luminance and contrast measurements, observers' γ_{adj} responses

Minimum and maximum luminance readings (cd/m^2), resulting ANSI contrast, and observers responses averaged between all trials.

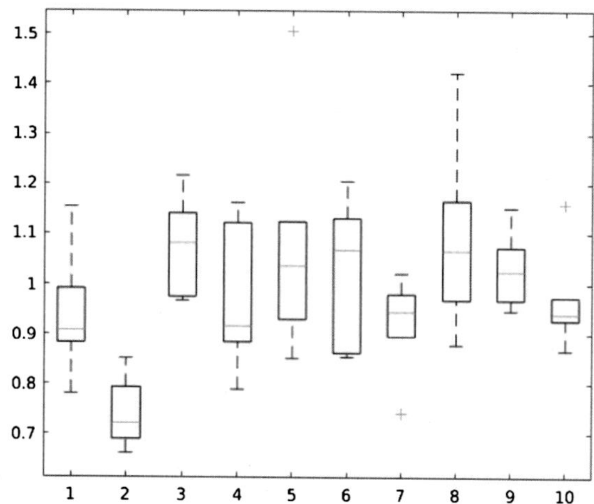

FIGURE 10.9 Example observer response plot for OLED display

The horizontal axis represents the different images tested and the vertical axis represents the responses of observers.

The plot shown in Fig. 10.9 displays the responses of each observer for each individual image. This plot highlights the high degree of variance in responses, as is common for this type of experiment task, as well as the impact of image dependence.

In relation to the previous experiment, the average responses can be predicted by a slight modification of the original model:

$$\gamma_{adj} = (1 + 0.1|C|)^{sign(C)} \tag{10.3}$$

This modification can be explained by the updated performance of the tone mapping algorithm from the previous experiment as well as changes to the experiment. The previous experiment tested less observers with different images on different displays, and the tone mapping algorithm applied to experimental stimulus produced a differing output compared to the current iteration. Thus, this adjustment of the formula could potentially be caused by one or all of these alterations.

As several observers reported and as the plot in Fig. 10.9 shows, responses were heavily influenced by the content of the image being evaluated. This effect was also present in the results of Bartleson [9]. It can be seen that a lower γ_{adj} level was consistently selected for image two. A likely reason for this is that observers desired an increase in the midtones of the image so that indoor detail could be better seen. Observers also reported that in the case of relatively homogeneous luminance scenes, the adjustment of gamma behaved similarly to an exposure adjustment and their final decision was reflective of the exposure of the scene that they most preferred. On the other hand, for scenes with a bimodal luminance distribution the selection of gamma value relied on balancing the relative luminance of bright and dark image areas. Finally, observers reported their selections being influenced by different image artifacts, e.g. some reported adjusting their final decision to avoid noise in the shadows. While image dependence is a known phenomenon in psychophysics, the specifics of the observer reports will be helpful in compiling image sets for future experiments.

In addition to image content, the variability in image rendering between displays used in the experiment also had an effect on observer responses. In some cases image noise was reported to be more or less visible when displayed on different devices, and also other artifacts were introduced by the displays which could have influenced observer results. Additionally, the displays had varying colour performance, screen sizes, and luminance stability across the panel area. While the decoding tonescale behaviour for all displays was characterised and accounted for in image presentation, these tangential display properties could have an effect on observers final responses.

It can be seen in Fig. 10.9 that results of the experiment had a large degree of inter-observer variability. This variability was expected due to the nature of the observer preference task and is explained by the differing tastes and experiences of observers. In future experiments, a set of more specific parameters with which to judge images or some sort of pairwise comparison experiment could be developed to limit this variance.

10.2.1.2 Influence of surround luminance: extension of TMG method

We've discussed how varying ambient luminance in the viewing environment produces changes in the perceived brightness and contrast (local and global) due to luminance adaptation as well as obscured shadow detail and raised black levels due to screen flare, and how these effects also vary in impact according to the capabilities of the display (its peak and minimum luminance levels.)

In response to this, we decided to extend the method of tone mapping of graded content (TMG) introduced in Chapter 9, Section 9.3.2, so that it takes into account surround luminance. We recall that in the TMG method, we denote as Y_1 the normalised (i.e. in the range $[0, 1]$) luminance of the HDR source, μ_1 is the median of Y_1, D_1 is the peak luminance of the display for which the HDR source material was graded, and D_2 is the peak luminance of the reduced dynamic range display where the tone-mapped results will be presented. Therefore, the median luminance (in light intensity units, i.e. without normalisation) of the source is $\mu_1 D_1$, and we want the tone-mapped result to have the same median luminance. When normalising by the corresponding peak luminance, and using a power-law transform of exponent γ, this translates into the following requirement:

$$\mu_1^{\gamma} = \mu_1 \frac{D_1}{D_2}, \tag{10.4}$$

from which γ can be solved as:

$$\gamma = \frac{log(\mu_1) + log(\frac{D_1}{D_2})}{log(\mu_1)}. \tag{10.5}$$

Now we want to take into account the viewing conditions, by raising the tone-mapped result to an exponent γ_{adj}. This is equivalent to multiplying γ in Eqs. (10.4) and (10.5) by γ_{adj}, which gives a new equation to compute γ:

$$\gamma = \frac{log(\mu_1) + log(\frac{D_1}{D_2})}{log(\mu_1)\gamma_{adj}} = \frac{log(\mu_1) + log(\frac{D_1}{D_2})}{log(\mu_1)(1 + 0.1|C|)^{sign(C)}} \tag{10.6}$$

The second modification to the TMG method is the inclusion of the contrast enhancement step that is part of the algorithm for tone mapping of ungraded content introduced in Section 9.3.1 but was omitted for the TMG algorithm. This is intended to account for the shift in contrast sensitivity which can occur due to luminance adaptation.

In summary, we have an updated TMG method, call it TMG', where γ is now computed as in Eq. (10.6) and the algorithm has a second stage of contrast enhancement.

An optimisation experiment was performed for this TMG' method, in which observers adjusted model parameters to match the appearance of an input image, shown under reference (dark) viewing conditions, to the appearance of the image processed through TMG' under two different viewing conditions, with increased surround luminance. Instead of having observers perform this task via memory matching or some other time-consuming method involving adaptation periods, a set of reference images were manually graded to relate the appearance of the images under the reference viewing condition to the two test conditions. While this method hinges on a single technician's ability to relate appearance by memory between the two viewing conditions (detailed in Fig. 10.10), we believe this to be a better alternative as the tools of a full colour grading suite would be made available to them as opposed to a limited number of algorithmic parameters, and the technician could take significantly more time to prove the matches that were made and make adjustments if necessary.

The grading process was conducted as follows. The grading suite was set up for the lighting conditions that would be tested in the experiment (shown in Fig. 10.11). Next a DaVinci Resolve project timeline was set up with two repetitions of each of

Factor	Type	Levels
Ambient Surround	Variable	40 nits, 150 nits average
Display Flare	Variable	1 nit, 3 nits average
Stimuli	Variable	20 Still Images
Experiment Task	Constant	Match to reference
Display	Constant	Sony PVM-250 OLED, 100 nits
Viewing Distance	Constant	2 Picture heights

FIGURE 10.10

Viewing condition optimisation factors.

the images in the test set, and an 18% gray patch between each image. The UI monitor was then turned off and moved to the side of the desk so that the remainder of the grading process could be conducted purely through the Tangent Element control panel. The lights in the suite were turned off and the technician was given a one-minute adaptation period to adapt to the dark surround while viewing the 18% gray patch on the monitor. Then the patch was switched to the first image in the test set and the first memorisation period was initiated. The technician was careful to select image content references at a range of luminance levels for memorisation and to observe the noise level in the image as well as the appearance of high frequency detail. Then the image was switched to the next gray patch, the lighting in the suite was turned on to the first ambient luminance level for another one-minute adaptation period. After this, the image was switched to the second instance of the first test image, and lift, gamma, gain, and image sharpness were adjusted to match the reference points from the memorisation period. Once a satisfactory first pass had been made, the lighting was turned off once again and another one-minute adaptation period followed. Then, the image was switched back to the first instance of the first image to prove the match. If the match was determined to be satisfactory the next image was presented, and the process was repeated. After all images were graded for a given lighting condition, a second proofing pass was conducted to verify the accuracy of the matches. This process was then repeated for the second lighting condition.

These reference images were then used in an experiment in which five professional observers adjusted several algorithmic parameters of the TMG' method while viewing the output under the lighting condition that the reference image was graded under. The parameters included the γ and k values from the method, as well as the intensity of contrast enhancement. The experiment was conducted on the single OLED display, and observers were given a control keyboard which allowed them to freely switch between reference and test images, as well as to adjust parameters in any order they preferred. They were, however, encouraged to first adjust the γ and k parameters to make a mid-tone level and global contrast match before moving on to the local contrast parameter. Observers reported being able to make close matches between the test and reference conditions, with the exception of matching the highlight and black levels which were not directly modified by the model parameters as mentioned before. To match with the observer data, we modified the C function defined above in Eq. (10.2) by adding an extra term that depends on the surround luminance. The

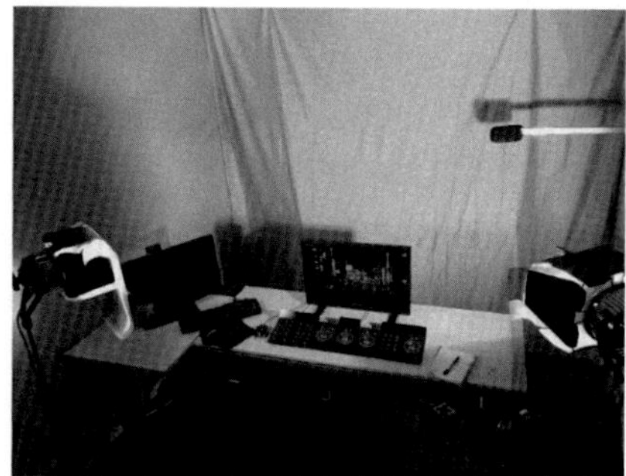

FIGURE 10.11 Grading suite setup

Two tungsten light sources (800w with daylight gels, 1kw with diffusion and daylight gels) are placed at 45 degrees to the room normal. A light gray sheet was hung behind the monitor to give an achromatic reference and to provide a similar reflectance factor to the desk. Surround luminance readings were taken close to each side of the display and averaged. Flare readings are taken with 5% ANSI contrast grating on screen and averaged between four different points.

modified C is given as:

$$C = log_{10}\left(\frac{L_{vc}^{peak}}{L_{grad}^{peak}}\right) + log_{10}(ANSI_{vc}) - log_{10}(ANSI_{grad}) - 0.1 log_{10}\left(\frac{surr_{vc}}{surr_{grad}}\right). \tag{10.7}$$

In this equation L_{vc}^{peak} and $ANSI_{vc}$ are the peak luminance and the ANSI contrast of the intended display in which the image is to be viewed, L_{grad}^{peak} and $ANSI_{grad}$ are the peak luminance and the ANSI contrast of the grading display, in this case an OLED display in a dark room, and $surr_{vc}$ and $surr_{grad}$ are the surround luminance values for the target and source displays respectively.

While the parameter optimisation of this experiment may provide accurate results for viewing on standard dynamic range displays, it will be important to repeat this experiment on an HDR monitor for the case of inverse tone mapping. However, since the model is invertible it is our expectation that the optimised parameters will hold for the HDR case.

10.3 Tone mapping for improved HDR video coding

We have seen in Chapter 5 that in HDR video processing it is crucial to select an appropriate transfer function (TF) that performs perceptual linearisation, transforming scene linear luminance into a perceptual non-linear signal value. A good TF successfully manages both darkness and brightness in luminance while maintaining colour vibrancy in chroma. HDR-television systems employ two different kinds of standardised TF, which are PQ and HLG. In this section we review the method of Sugito et al. presented in [10], where we introduced a new TF based on our tone mapping formulation of [4].

We termed this new transform NISTF for "Natural Image Statistics Transfer Function", as the TM method it's based on was developed by taking into account the average histogram of natural images. While the TMO of [4] has two stages, the first a global tone curve and the second a contrast enhancement step, both of which are image-dependent, we have chosen to define NISTF as a fixed, image-independent global tone curve, as in this way our function doesn't require metadata for parameters.

The definition of the NISTF transform is:

$$I' = I^{\gamma}(I), \tag{10.8}$$

with

$$\gamma(I) = \gamma_H + (\gamma_L - \gamma_H)\left(1 - \frac{I^n}{I^n + M^n}\right), \tag{10.9}$$

where I is a normalised linear image with RGB components, $\gamma_L = 0.45$, $\gamma_H = 0.268$, $M = 4.47 \times 10^{-3}$ and $n = 0.45$. The normalisation of I is done as in HLG. Fig. 10.12 compares the curves for PQ, HLG and NISTF.

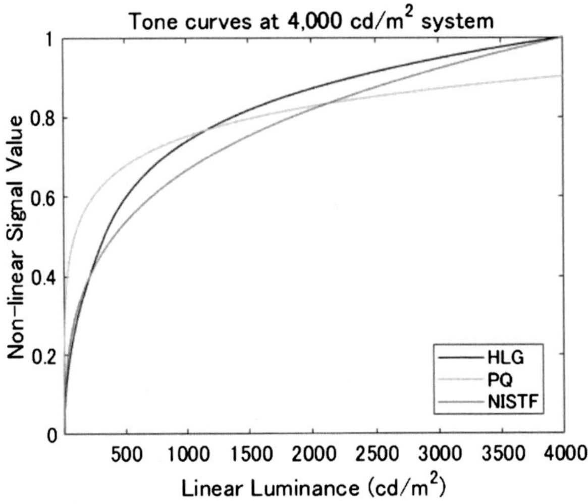

FIGURE 10.12 HDR TFs and the new non-linear function

HDR TFs and the new non-linear function. Figure from [10].

These parameters were selected as the resulting NISTF curve works well on various video coding sequences (different from the ones later used for validation). We

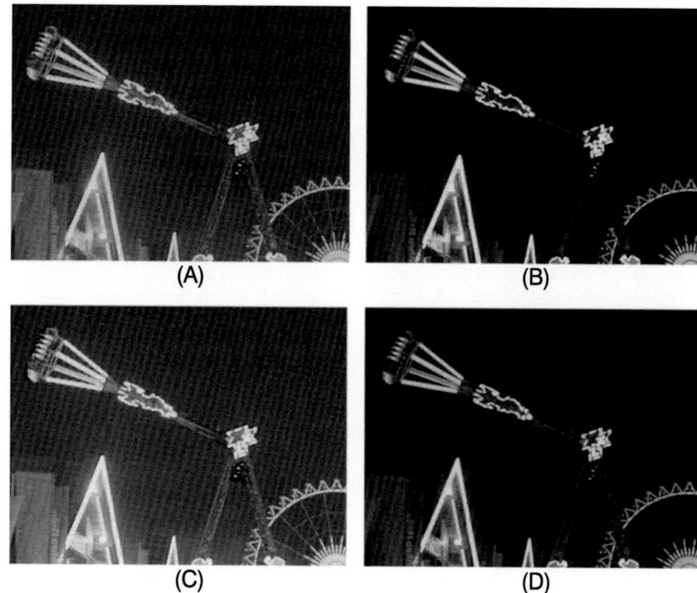

FIGURE 10.13 Examples of the luma image

Examples of the luma image. Top left: result of PQ. Top right: HLG. Bottom left: result of tone curve of [3]. Bottom right: result with NISTF. Figure from [10].

noticed how the tone curve of [4] produces, on dark sequences, results that share with the results of PQ some negative characteristics, namely that they are brighter and have more noise than those obtained with HLG. This can be observed in Fig. 10.13. Having darker, less noisy results is better for coding efficiency, so we wanted NISTF to try to replicate this behaviour of HLG.

Fig. 10.14 shows the tone curves for PQ, HLG, the TMO of [4] and NISTF, in log-log coordinates. We can see how the TMO curve and PQ allot a much larger value of non-linear luminance than HLG for low light intensities. This is why we have set γ_L to be 0.45, which is the conventional value for gamma correction. As a result, NISTF is almost parallel to HLG in dark areas, since the slope of HLG is 0.5.

In HDR video coding there is a pre-processing stage consisting of conversion from RGB to Y'CbCr, linear scaling of range from [0,1] to [0,1023] and rounding to integer (so as to have the signal in 10 bit precision), and finally a subsampling of the chroma components Cr and Cb to a 4:2:0 format. This means that the horizontal and vertical dimensions of the chroma channels are half of the original, which when using PQ causes noticeable visual artifacts that must be compensated with some extra processing called luma adjustment [11]. In Fig. 10.15 we can see an example: the PQ result shows artifacts in the upper left of the image, artifacts that are not present in

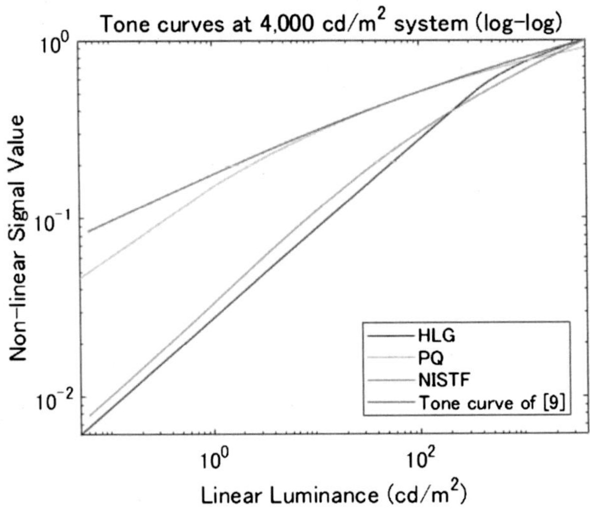

FIGURE 10.14 HDR TFs in log-log axes

HDR TFs in log-log axes. Figure from [10].

the original image nor in the HLG or NISTF results. This suggests that our NISTF method doesn't require the luma adjustment processing required by PQ.

10.3.1 HDR video coding experiments

We compared NISTF with HLG and PQ on a number of standard sequences, following the common test condition of MPEG and using the BD-rate on the HDR-VQM metric for quantitative evaluation of the results.

10.3.1.1 Test sequences

As shown in Fig. 10.16, we selected four sequences from the colour graded version of the HDR database of [8]. The sequences were graded for Rec.2020 primaries and from 0.005 to 4,000 cd/m^2. Each sequence is a 10-second clip at 25 fps and image dimensions of 1920 × 1080 pixels. For all sequences, each frame is provided as a 16-bit tiff file with PQ encoding.

Fig. 10.17 shows the encoder input video generation process. For NISTF and HLG we undo the PQ encoding of the input.

Fig. 10.18 shows the dynamic range and the median of linear luminance for each frame of the test sequences. The characteristics of the four sequences are different. Carousel Longshot and Moving Carousel are night scenes, and self-illuminated objects and dark surroundings exist at the same time. In Carousel Longshot, the dynamic range and the median of luminance are relatively constant, while both fluctuate in Moving Carousel. Cars Longshot and Fishing Longshot are sunlit scenes, and their luminance median values are close for the latter part of the sequences. The dynamic

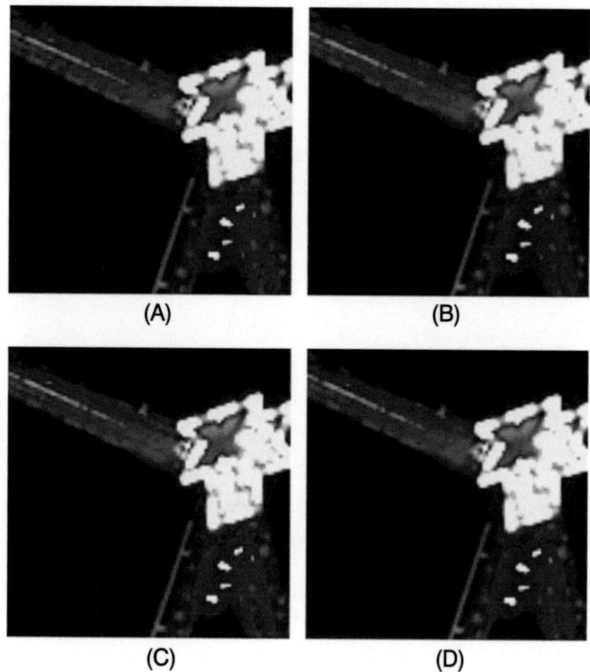

FIGURE 10.15 Artifacts due to chroma subsampling

Artifacts due to chroma subsampling. Top left: result of PQ. Top right: HLG. Bottom left: original (no subsampling). Bottom right: result with NISTF. Figure from [10].

range of Cars Longshot changes over the sequence, while it's nearly constant for Fishing Longshot.

10.3.1.2 Metrics

As a quantitative objective metric of the results we decided to use HDR-VQM [12]. Then, we calculated the Bjontegaard delta rate (BD-rate) [13] from the normalised HDR-VQM value and the bit-rate of the encoding results. BD-rate is a metric used to compute the average bit-rate saving ratio.

Fig. 10.19 shows an example of video coding results. In the graph, the horizontal and the vertical axes indicate bit-rate in kilobit per second (kbps) and normalised HDR-VQM, respectively. Comparing the two results at HDR-VQM value x, the bit-rate of the new method is two-thirds that of the reference. In such a case, the bit-rate saving ratio of the new method based on the reference is −33.3% ((2b/3-b)/b*100%).

To compute the BD-rate, we conducted video coding experiments with four different quantisation parameters, which control the bit-rate of the encoded video, and considered PQ as the reference. Fig. 10.20 shows the BD-rate results, where we can

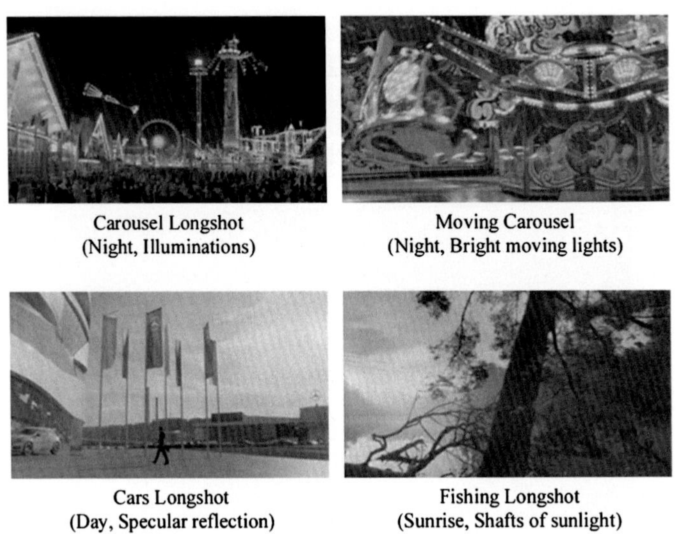

FIGURE 10.16 HDR test sequences used for video coding experiments

HDR test sequences used for video coding experiments. Original footage from [8].

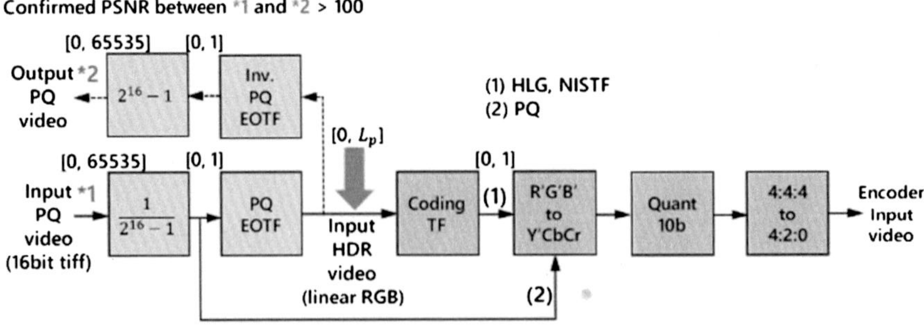

FIGURE 10.17 Diagram of encoder input video generation process

Diagram of encoder input video generation process. Figure from [10].

see that NISTF improves upon PQ by a larger margin (higher bit-rate savings) than HLG does.

Fig. 10.21 shows for each sequence the graphs of the normalised HDR-VQM vs bit-rate for the three TF compared. NISTF results are better than HLG especially in night scenes. And, as is the case for HLG, we did not observe with NISTF any sort of visual artifact due to the chroma subsampling process.

Nonetheless, we must point out that this study predates the work by Sugito and Bertalmío [14] where we showed the limitations of HDR metrics such as HDR-VQM to predict observers' responses. We are currently working on validating the effective-

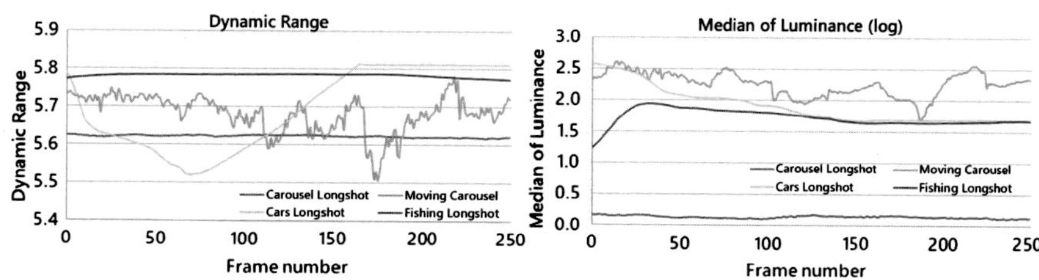

FIGURE 10.18 Sequence characteristics

Sequence characteristics. Left: dynamic range for each frame. Right: log_{10} of linear luminance for each frame. Figure from [10].

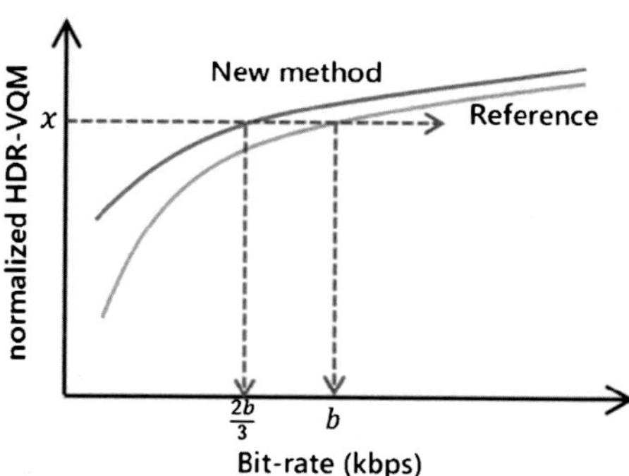

FIGURE 10.19 Examples of video coding results

Examples of video coding results. Figure from [10].

	HLG	NISTF
Carousel Longshot	-8.2%	**-15.6%**
Moving Carousel	-11.6%	**-24.3%**
Cars Longshot	-29.2%	**-32.0%**
Fishing Longshot	-20.7%	**-25.9%**

FIGURE 10.20 BD-rate results of experiments

BD-rate results of experiments. Figure from [10].

ness of alternative transfer functions such as NISTF using observer tests and more effective metrics.

Considering practical applications of a camera or other devices, both coding NISTF and its inverse would be calculated using a look-up-table, with no need for metadata. Thus, the complexity of NISTF is comparable with that of existing methods.

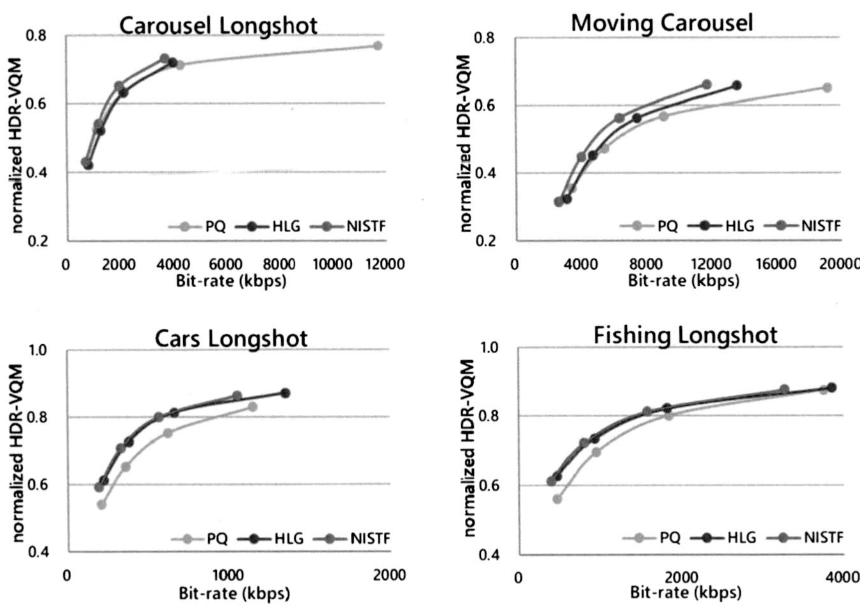

FIGURE 10.21 HDR-VQM value vs bit-rate for each sequence

HDR-VQM value vs bit-rate for each sequence. Figure from [10].

10.4 Tone mapping for photorealistic style transfer

In professional cinema, the intended artistic look of the movie informs the creation of a static 3D LUT that is applied on set, where further manual modifications to the image appearance are registered as 10-parameter transforms in a colour decision list (CDL). The original RAW footage and its corresponding LUT and CDL are passed-on to the post-production stage where the fine-tuning of the final look is performed during colour grading.

In many cases, the director wants to emulate the style and look present in a reference image, e.g. a still from an existing movie, or a photograph, or a painting, or even a frame from a previously shot sequence in the current movie. The manual creation of a LUT and CDL for this purpose may require a significant amount of work from very skilled artists and technicians, while the state of the art in the academic literature offers promising but partial solutions to the photorealistic style transfer problem, with limitations regarding artifacts, speed and manual interaction.

In this section we review the works by Zabaleta and Bertalmío presented in [15, 16], where we proposed a method that automatically transfers the style, in terms of luminance, colour palette and contrast, from a reference image to the source raw footage. It consists of three separable operations: global luminance matching (for which we use the TMO introduced in [4]), global colour transfer and local contrast matching. As it just takes into account the statistics of source and reference images, no training is required. The total transform is not static but adapts to the changes in the source footage. The computational complexity of the procedure is extremely low and allows for real-time implementation in-camera, for on-set monitoring. While the method is proposed as a substitute for the need to specify a LUT and a CDL, it's compatible with further refinements performed via LUTs, CDLs and grading, both on-set and in post-production. The results are free from artifacts and provide an excellent approximation to the intended look, bringing savings in pre-production, shooting and post-production time.

Let's detail in order each of the three stages of the algorithm, first for the case of style transfer from a reference image to a single frame in demosaicked RAW format. Later we will show how to extend this process to the whole video sequence.

Luminance transfer

In this step the demosaiced white balanced RAW image, which is linear RGB data captured by camera sensors, is non-linearly transformed to match the luminance of the reference image.

We use the global tone curve of the TM algorithm we proposed in [4], that calculates a variable exponent γ such that when applied to the original RAW image I it performs a constrained histogram equalisation creating a non-linear image I_{eq},

$$I_{eq}(x) = TM_I(I(x)) = I(x)^{\gamma(I(x))}, \tag{10.10}$$

where the subindex I in TM_I denotes that the tone curve is image-dependent.

Now let us recall the gamma correction formula of the Rec.709 standard, which we'll call gc:

$$gc(x) = \begin{cases} 4.5x, & 0 \le x < 0.018 \\ 1.099x^{0.45} - 0.099, & 0.018 \le x \le 1 \end{cases} \tag{10.11}$$

The inverse of this function allows to obtain values that are proportional to light intensity values in the scene:

$$gc^{-1}(y) = \begin{cases} \frac{y}{4.5}, & 0 \le x < 0.081 \\ \left(\frac{y+099}{1.099}\right)^{\frac{1}{0.45}}, & 0.081 \le x \le 1 \end{cases} \tag{10.12}$$

Let S_0 be the original source (demosaiced and white balance balanced) RAW image and R the reference image, which is a non-linear linear image (e.g. a JPEG image). If we assume, as a rough approximation, that the nonlinearity with which the reference image has been created is the gamma correction function of Eq. (10.11), and taking into account that the TM method of Eq. (10.10) performs an approximate histogram equalisation, we have that

$$TM_{S_0}(S_0) \simeq TM_R(gc^{-1}(R)) \tag{10.13}$$

We define the output S_1 of the luminance matching step, the first stage of our method, as

$$S_1 = gc(TM_R^{-1}(TM_{S_0}(S_0))), \tag{10.14}$$

where this transformation is applied independently to each channel of the RGB source image S_0.

Fig. 10.22 shows an example result of luminance transfer.

FIGURE 10.22 Luminance transfer example

From left to right: Demosaiced white balanced source RAW image S_0, luminance matched source image S_1, and reference image R. A non-linear transformation is applied to the source image so its luminance histogram approximates the reference luminance histogram. Video from RED Digital Cinema and image from pexels.com.

Colour transfer

We seek to transfer the colours of the reference image to the resulting non-linear image S_1 from the previous step. Statistical properties are often used to define a global mapping between two input images, especially when correspondences between images are not available. Following the same line as in [17–19] but with some key modifications outlined below, we transfer the 1st order statistics (mean and standard deviation) from R to S_1.

Principal Component Analysis (PCA) is applied to the RGB source and reference images to find decorrelated spaces for each of them. We have done experiments of PCA in other colour spaces, such as IPT [20], but results show that colour transfer by PCA is better in RGB. The colour transfer method can be summarised as a chain of transforms: a translation for matching the mean between images, a rotation for axes alignment and a scaling for matching the standard deviation along each axis. The matching matrix M for colour transfer is

$$M = T_R \cdot R_R \cdot V \cdot R_S^{-1} \cdot T_S, \tag{10.15}$$

where T_R and T_S are the translation matrices with the mean information of the reference and source images respectively, R_R and R_S are the rotation matrices composed of the eigenvectors of the PCA covariance matrices for reference and source images respectively, and V is the scaling matrix

$$V = \begin{bmatrix} 1 & 0 & 0 \\ 0 & \sqrt{\lambda_2^R/\lambda_2^S} & 0 \\ 0 & 0 & \sqrt{\lambda_3^R/\lambda_3^S} \end{bmatrix} \tag{10.16}$$

where λ_i^j are the eigenvalues of the PCA covariance matrices along the i axis, $i \in \{2, 3\}$, for reference and source images $j \in \{R, S\}$.

The resulting image S_2 from this colour transfer step will be

$$S_2(x) = M \cdot S_1(x) \tag{10.17}$$

where x is a pixel and M is the matrix defined in Eq. (10.15).

This type of technique has been used for colour transfer in some previous works [17,19], but we have introduced two modifications that are key for our purposes. First, we decided not to modify the luminance in this colour transfer stage; for this reason, and assuming as it's done in the literature that the first axis of the PCA corresponds to the luminance channel of the image [21], we make the scaling matrix V leave unmodified the standard deviation along the first axis. Second, we add a constraint to the rotation matrices to prevent large colour shifts. By doing PCA analysis and simply assigning the source axes to the reference axes, the geometry of the resulting mapping might not be as it was intended. The resulting image could have the colour proportions as expected, but locally the colours would be swapped. To avoid that, the method finds the rotation where the source axes are assigned to the closest reference axes.

Fig. 10.23 shows an example result of colour transfer.

FIGURE 10.23 Colour transfer example

From left to right: Luminance matched source image S_1 from the previous step, colour-matched image S_2 and reference image R.

Local contrast transfer

For modifying the contrast we will work on the first axis of the PCA, which we mentioned above contains the luminance and contrast information. The resulting image S_3 will be obtained by a local contrast transformation based on the second stage of the TMO of [4] and that we will apply to S_2:

$$S_3 = LC(S_2) \tag{10.18}$$

We seek to transfer the local contrast from the reference to the source image. If we define local contrast of an image as the difference between light intensity $I(x)$ and its local mean value $\mu(x)$, our approach transfers the standard deviation of the local contrast from the reference to the source image through the following formula:

$$S_3(x) = \mu(x) + \frac{\sigma_R}{\sigma_S}(S_2(x) - \mu(x)), \tag{10.19}$$

where x is a pixel, $\mu(x)$ is the local mean of S_2, σ_S is the standard deviation of the local contrast of the source image S_2 and σ_R is the standard deviation of the local contrast of the reference image R.

This modification is applied only to the luminance axis, therefore the second and third axes in the PCA decomposition of S_2 and S_3 remain equal.

Fig. 10.24 shows an example result of local contrast transfer, which is the final output of our method for still images.

FIGURE 10.24 Contrast transfer example

From left to right: colour-matched source image S_2 from the previous step, local contrast transferred image S_3 (final result) and reference image R.

Video style transfer

We extend now the problem of transferring the style from a reference image to a source image to the problem of transferring the style from a reference image to all the frames in a video. Applying an independent style transfer to each frame of a video sequence might result in strong texture flickering and temporal incoherence, even if neighbour frames are similar in content and colour. It is also computationally expensive to calculate the parameters for each frame without making profit from the common content between frames. In order for our method to have a low computational cost and to impose coherence between frames, we propose to apply the same style transfer transformation to all the frames in the video sequence.

First, the luminance, colour transfer and contrast transfer transformations (explained previously) are calculated for the first frame of the video, then those transformations are applied to each frame in the sequence.

Let f be the luminance transformation

$$S_1 = f(S_0), \tag{10.20}$$

where S_0 is the first frame of the video. Let g be the colour transfer transformation,

$$S_2 = g(S_1). \tag{10.21}$$

And let k be the quotient $\frac{\sigma_R}{\sigma_S}$ obtained when applying the local contrast transformation to S_2.

Then, the same transformation f calculated above is applied to each frame of the video, followed by g, and finally the local contrast of each frame is adjusted multiplying it by the constant k.

Temporal coherence is guaranteed by applying the same transformation to all the frames in the video. This approach has very low computational cost and produces temporal-coherent and flickering-free results.

Experiments and results

In this section we provide some results obtained by our style transfer method for image sequences, with example frames shown in Figs. 10.25 and 10.26.

We must say that quantitative evaluation of stylisation is very difficult since there is no ground truth and different artistic choices can be rated differently by users. However, the results are in general visually pleasing, and free of artifacts or flickering effects and show temporal and spatial consistency. The selection of the reference image is very important for the method. As we can see in Fig. 10.27 when the source and reference images are very different in content and composition, the results are not as good as expected. Generally, statistical colour-mapping methods suffer from the same limitation and some of them use segmentation or user interaction to solve this issue [17,19].

10.5 Tone mapping for dehazing

Our TM approach introduced in [4] and improved in Chapter 9 may also be used to perform image dehazing, as detailed in the work of Vazquez-Corral et al. [22] that we now proceed to summarise.

The image dehazing problem consists in removing from images the degradations introduced by atmospheric phenomena like fog or haze, that scatter light and reduce the visibility of objects. The most common strategy in the literature consists in estimating the scene radiance by inverting a light propagation model known as Kochsmieder law:

$$I(x) = t(x)J(x) + (1 - t(x))A, \tag{10.22}$$

where x represents a pixel location, $I(x)$ is the intensity captured by the camera sensor, $J(x)$ is the radiance in a hypothetical haze-free scene, $t(x)$ is the transmission of light in the atmosphere, inversely related to the scene's depth, and A is the airlight, a vector quantity describing the predominant colour of the atmosphere. Under good weather conditions, $t = 1$ and no scattering happens. Otherwise, the image I is a degraded version of J, with the amount of degradation observed on

FIGURE 10.25 Example results

Original images on top, reference images on the left and resulting images. Images from the authors, from pexels.com and from RED Digital Cinema.

an object being proportional to the object's distance (usually referred to as *scene depth*).

Dehazing approaches usually estimate the haze distribution across the scene, which should be inversely related to transmission $t(x)$, and also the airlight illuminant A. Once these estimates are available, the dehazed image is approximated by J, obtained by inverting Eq. (10.22):

$$J(x) \simeq \frac{I(x) - (1 - t(x))A}{t(x)}. \tag{10.23}$$

A more realistic model considers some amount of noise $\eta(x)$, due to sensor and scene sources, and the solution to the dehazing problems becomes

$$J(x) \simeq \frac{I(x) - (1 - t(x))A}{t(x)} - \frac{\eta(x)}{t(x)}. \tag{10.24}$$

FIGURE 10.26 Example results

Left: Selected frames from the original RAW video (gamma corrected for visualisation). Center: Resulting frames for reference (A). Right: Resulting frames for reference (B). Video from RED Digital Cinema and images from pexels.com.

FIGURE 10.27 Limitations of the method

From left to right: Original image (gamma corrected for visualisation), style transferred image and reference image. The dissimilarity between the source and the reference result in an incoherent style transferred image, with unreal skin colour. Images Video from RED Digital Cinema and image from pexels.com.

Since transmission $t(x)$ varies in [0, 1] and it is inversely related to depth in the scene, we can conclude that noise will be strongly amplified in far-away areas of the scene. This problem is exacerbated by the fact that $t(x)$ is a singe scalar quantity describing at the same time the propagation of light in the three chromatic components. This unavoidably leads to the appearance of colour artifacts and is the reason why most dehazing methods can't properly deal with sky regions.

These limitations often lead to a dehazing result whose image appearance is worse than the original, hazy input. Fig. 10.28 shows an example, comparing the results of two state-of-the-art methos with the approach we propose in [22].

FIGURE 10.28 Example of artifacts appearing in current image dehazing methods

Example of artifacts appearing in current image dehazing methods. From left to right: original image, result from [23], result from [24], result from proposed method. Figure from [22].

Our dehazing method works in the HSV colourspace and it is only applied to the saturation (S) and value (V) channels, the hue (H) channel is left untouched. The motivation for this can be seen in Fig. 10.29. Here we present an original haze-free image and the same image with a synthetic fog layer on top of it. The corresponding H, S, and V channels of both images are shown in the first two rows, and the histogram of their differences is displayed in the third row. From this figure, we can see that performing image dehazing amounts to increasing the saturation and decreasing the value components while leaving the hue component unaltered. For this reason, once the image is converted to the HSV space, our method only modifies the saturation and value of the input hazy image.

On the V channel we choose to apply a global tone mapping curve (like the one introduced in [4] and improved in Chapter 9) performing constrained histogram equalisation. The reason is that we saw in Fig. 10.29 that the V difference is negative, so the histogram of the V component of a hazy image is skewed towards 1, and therefore distributing these values uniformly across the available dynamic range should increase contrast and detail visibility in the image.

When we modify the V component the chroma of the image changes, because $Chroma_{HSV} = S \times V$. Based on psychophysical studies that state that observers perceive the same level of colourfulness independently of the amount of haze in the scene, we modify the S component so that the chroma remains constant.

Finally, we perform contrast enhancement on the modified V component by emulating the contrast adaptation process observed at retinal level, as discussed in Chapter 4.

The pipeline for our method is shown in Fig. 10.30.

Fig. 10.31 compares our proposed method with the state-of-the-art for a series of challenging images, where we can see that the competing algorithms produce explicit artifacts while our method doesn't.

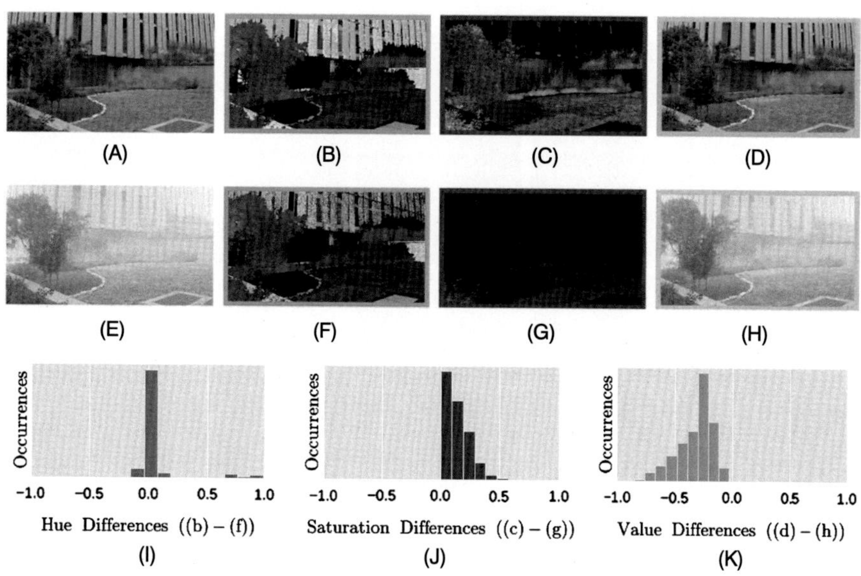

FIGURE 10.29 Comparison between a fog and a non-fog image in the HSV space.

First two rows, from left to right: Original image, hue, saturation and value. Last row: Histogram of differences between the fog and the non-fog image. We can see that hue stays the same, while saturation is always higher for the non-fog image, and value is always higher for the fog image. Figure from [22].

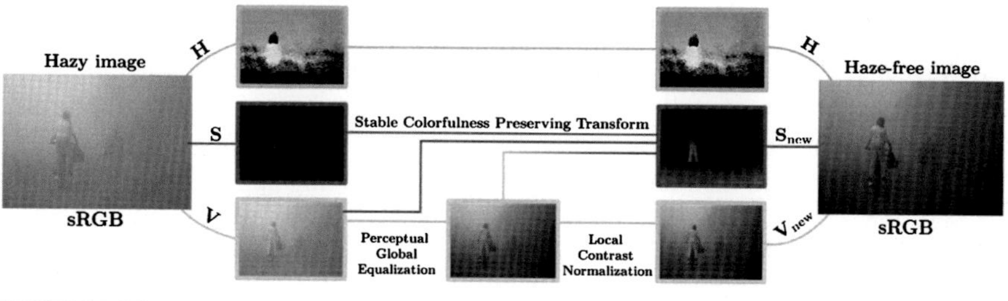

FIGURE 10.30

Pipeline of the method in [22].

We also performed a psychophysical evaluation, where observers were shown a hazy input and two dehazed versions of it and they were asked to select which of the three images they preferred. In 52% of the cases they selected the original, hazy image, which shows that there is still a long way to go regarding the development of new image dehazing methods capable of producing aesthetically pleasant results.

The psychophysical experiment also produced other interesting results: our method is the one preferred by observers, followed by the methods of [23] and [25],

FIGURE 10.31 Comparison of dehazing results

Comparison of dehazing results. From left to right: original, result from [25], result from [24], result from [26], result from proposed method. Figure from [22].

and this ranking is *inconsistent* with the one produced by the metrics commonly used for dehazing. This is the same situation we had for gamut mapping (Section 8.4.4) and tone mapping (Section 9.5), where we could not find any objective metric that had a good correlation with the observers' scores.

Finally we want to mention that, since the tone mapping approach we have used for dehazing is based on efficient representation, and we have seen in Chapters 4 and 7 the intimate connections between contrast adaptation, histogram equalisation and the Retinex theory of colour, it's not surprising that we can also propose state-of-the-art methods for dehazing based on local histogram equalisation, like the ones in [27–29].

Furthermore, and this was indeed quite an unexpected connection, we have shown in [30] that dehazing and colour estimation by Retinex can be seen as dual problems, i.e. any method to solve one of them can also solve the other.

10.6 Tone mapping for backlit images

In a backlit scene the contrast between foreground and background is too high to be captured properly in a SDR image, regardless of the exposure settings, and therefore the foreground appears too dark, with loss of visible detail.

In [31] Vazquez-Corral et al. propose a method to deal with backlit images that is based on the TM approach introduced in [4] and improved in Chapter 9. Our method is inspired by the observation that segmentation-based methods for backlit images (which linearly stretch the contrast in each segmented region) tend to fail when a dark object is present in the bright background, because the object is mis-labelled as belonging to the dim foreground. To alleviate this problem we propose a variational-based region split where both the original input values and the local contrast of the object are considered to create a set of weight maps. The local contrast will have a high value for a dark object in the bright region, while it will have a low value for any object of the same luminance found in the dim region, therefore allowing us to distinguish between the two cases. These weight maps are obtained by minimising an energy functional that is an extension of the local histogram equalisation method introduced in [32] and discussed in detail in Chapter 7, that is related to efficient representation, neural models and the Retinex theory of colour vision. See Fig. 10.32 for an example.

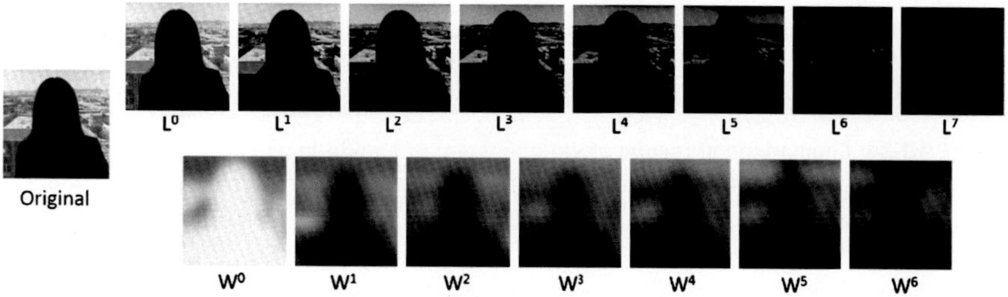

FIGURE 10.32 Weight maps

Original image (left). The iterates from the gradient descent (top) and the weight maps associated to these iterates (bottom). Figure from [31].

Each weight map defines a set of pixels, whose values are used to compute the parameters of the global tone curve of the TM method of [4]. We compute as many tone-curves as weight maps, we apply the tone-curves to the original image, and finally we merge the tone-mapped images. Figs. 10.33, 10.34 and 10.35 show some example results and comparisons with state-of-the-art methods.

10.7 Encoding curves and colour issues in HDR

To finish this chapter we present two common types of colour problems that might be encountered with HDR image creation and editing, where the encoding curve plays a role, and we propose methods to solve these issues.

Firstly, for HDR creation from a multiple-exposure sequence of low dynamic range (LDR) images, we will see that the gamma correction process is not constant and the gamma value changes from one exposure to the next, which violates the as-

FIGURE 10.33 Comparison of results of enhancement of backlit images

From left to right: original image, naturalness preserving method [33], Fu et al. fusion scheme [34], result from proposed method. Figure from [31].

FIGURE 10.34 Comparison of results of enhancement of backlit images

From left to right: original image, multi-scale Retinex approach [35], Fu et al. variational model [36], result from proposed method. Figure from [31].

FIGURE 10.35 Comparison of results of enhancement of backlit images

From left to right: original image, Dong et al. [37], Li and Wu [38], result from proposed method. Figure from [31].

sumptions of all HDR creation methods, introducing significant colour artifacts in the results.

Secondly, we will see how to match the colours of different views of the same scene captured with cameras using unknown encoding curves (gamma correction, logarithmic, PQ or HLG).

10.7.1 HDR imaging from multiple exposures

The dynamic range of light intensities in a natural scene may easily be above what digital cameras are able to represent. The net result is that standard cameras are only able to capture different intervals of the luminance range at different exposure times; in particular, bright areas are properly captured using short exposure times, while dark areas are better captured using longer exposure times.

To overcome this limitation Mann and Picard, in their seminal work [39], introduced the idea of creating a HDR picture of a static scene by combining a set of LDR images taken with different exposure times, proposing a parametric method to estimate the camera response function (CRF) that transforms the linear data into non-linear form. This was soon followed by other very influential approaches for the problem that differ in the way the constant CRF is estimated, e.g. Debevec and Malik [40], Mitsunaga and Nayar [41], Tsin et al. [42]. Later works tackled more general cases, like dynamic scenes with camera and/or object motion [43–49], or video [50–52], and it can be said that multi-exposure HDR creation is an ongoing research topic [47,53,50,54].

10.7.1.1 An important note about the intrinsic limitation of multi-exposure methods

An underlying assumption in multi-exposure HDR creation is that if the exposure times vary sufficiently, we can recover the full dynamic range of the scene and the resulting HDR image will be a *"radiance map"*, as Debevec and Malik put it [40], i.e. the HDR image values will be proportional to the scene radiance values.

The effect of scatter or glare in the context of HDR imaging was pointed out and extensively studied by McCann and Rizzi, see [55] and references therein, where they show through practical examples how it is not appropiate to refer to the outputs of multi-exposure HDR imaging techniques as *radiance maps*, since they are not proportional to the radiance values of the scene, and also it is wrong to assume that we can simply, without any further considerations, use multi-exposure HDR methods to recover the full dynamic range of any given scene, because veiling-glare limitations cannot be overcome by taking any number of different exposures. If we take a picture of the scene using a photo camera, the effects of reflections and scattering in the optic system can be approximately modeled by a convolution with a point spread function (PSF). This PSF has a shape like a blur kernel with a very slow decay, that implies that the light reaching each image point is effectively contributing a small fraction of its intensity to every other point in the image, regardless of how far apart they are: this is the *veiling glare*. Clearly, the brighter an image is, the higher the glare it will have. Given the linearity of the convolution operation and assuming a very long tail for the PSF, the image that is formed on the sensor can be roughly approximated as the original irradiance plus a homogeneous veiling glare component that can be estimated as a percentage c of the mean value of the image, and c is no smaller than 0.25% [56]. The net result is that *the DR of the sensor image is usually much lower than that of the original scene data, and the darkest areas in the sensor image will have values that will usually be way off from the actual values as measured on the original scene.* This is due to the optics, it has nothing to do with the dynamic range capabilities of the camera sensor, and it does not depend either on the exposure time that is actually used to capture the picture, so these problems cannot be solved by taking different exposures, regardless of the number of them.

We remark these implications because we feel that this very important information is usually missing from works on HDR imaging, and in particular from the multi-exposure literature.

10.7.1.2 The image formation model assumed by multi-exposure methods

Multiple-exposure approaches that use non-linear input pictures assume the following image formation model:

$$J(p) = f\left(E(p)\Delta t\right), \tag{10.25}$$

where Δt is the exposure time, p is a pixel location, $E(p)$ is the scene radiance value at p, f is a non-linear transform usually denoted as the CRF, and finally $J(p)$ is the resulting 8-bit image value, corresponding to one colour channel. Analogous expressions hold for each of the three colour channels, for which the function f might be different. In a static scene the values $E(p)$ remain constant, so taking a stack of N pictures by varying the exposure times gives us for each image

$$J_i(p) = f\left(E(p)\Delta t_i\right), \quad i = 1, \ldots, N, \tag{10.26}$$

where the subindex i denotes the different exposures *and it is also assumed that the function f remains constant as Δt_i changes*. Multiple-exposure methods estimate the inverse g of the CRF f, $g \equiv f^{-1}$, apply it to the image values $J_i(p)$ and then divide by the exposure time Δt_i so as to obtain one estimate of $E(p)$ for each image i in the stack:

$$\frac{g(J_i(p))}{\Delta t_i} = E(p), \quad g \equiv f^{-1}. \tag{10.27}$$

These N estimates of $E(p)$ are then averaged in order to provide the final output, the HDR value for pixel p.

We can see then how most multiple-exposure approaches share a set of building assumptions for the camera capture:

1. Different colour channels are independent.
2. The camera response remains constant while changing the exposure.

These assumptions made sense for film photography, but are not an accurate model of how digital cameras work. And as a result, HDR creation methods that implicitly or explicitly make the same assumptions, but work on images captured digitally, produce results that often show visible artifacts. In [62] by Gil Rodríguez et al. we study this problem and propose a method to solve it, and in this section we review those results.

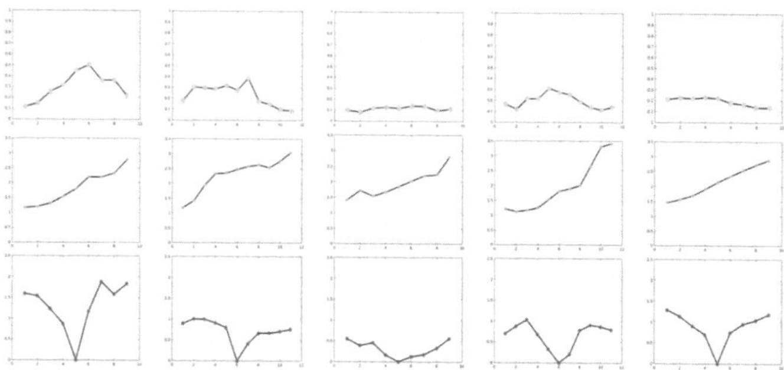

FIGURE 10.36 Gamma correction value and colour correction matrix change in multiple exposure sequences

Columns 1 to 4 correspond to different sequences taken with different cameras, shown in rows 1 to 4 of Fig. 10.37. The last column corresponds to the average over the HDR Survey database [57]. Top row: average value of non-diagonal elements of A for each image in the sequence. Middle row: gamma-correction value for each image in the sequence. Bottom row: difference between colour correction matrix of each image in the sequence with respect to colour correction matrix of reference image (middle-exposure).

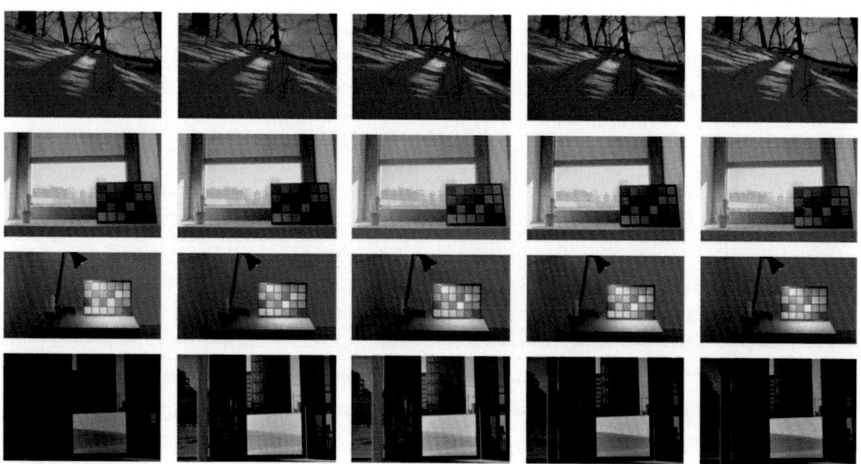

FIGURE 10.37 Incorrect assumptions produce visual artifacts in HDR from multiple exposures

From left to right: tone-mapped HDR result obtained from RAW pictures (i.e. ground truth) (first column), tone-mapped HDR result obtained with the multiple-exposure combination methods of Debevec and Malik [40] (column 2), Mitsunaga and Nayar [41] (column 3), Lee et al. [49] (column 4), and with our proposed approach (last column). Cameras used: Nikon (first row, from the HDR Survey [57]), Pentax (second row), Panasonic (third row), Canon (fourth row). Tone mapping methods used: Drago et al. [58] (row 1), Mai et al. [59] (row 2), Mantiuk et al. [60] (row 3), Ferradans et al. [61] (row 4).

Digital cameras follow a typical camera colour processing pipeline [63] that can be expressed as

$$\begin{bmatrix} R \\ G \\ B \end{bmatrix}_{out} = \left(A \cdot \begin{bmatrix} R \\ G \\ B \end{bmatrix}_{in} \right)^{\gamma}, \tag{10.28}$$

where $[R, G, B]^t_{in}$ is the sensor triplet (usually in 12 or 14 bits), $[R, G, B]^t_{out}$ is the pixel value at the end of the pipeline (in 8 bits per channel), A is a 3×3 matrix that combines the different colour channels taking into account white-balance, colour encoding, colour characterisation and a gain value, and γ is a value, typically between $1/1.8$ and $1/3$, performing gamma correction (notice that we omit demosaicing, denoising, compression, etc: for a complete explanation of these pipeline processes see [64]).

We claim that, contrary to what has been assumed in the multi-exposure literature,

1. The three channels R, G, B are not independent, because the matrix A is not diagonal as it incorporates colour processing steps like colour characterisation that involve all channels.
2. The CRF changes from one picture to the next in the multi-exposure scenario. The camera automatically modifies the γ value and the A matrix, implying that the non-linear transform f in Eq. (10.26) is not constant and therefore that *there is no CRF to speak of because the camera response has changed from image to image*.

To support this claim and highlight how generalised this camera behaviour is, we have performed tests on multiple-exposure sequences coming from four different camera models, where during capture only the exposure time changed, and with results recorded both in linear (RAW) and non-linear (JPEG) form. Having the same picture in these two versions allows us to estimate the values for γ and the matrix A with Eq. (10.28), using the RAW data for the $[R, G, B]^t_{in}$ values and the JPEG data for the $[R, G, B]^t_{out}$ values.

The first row of Fig. 10.36 plots, for different sequences taken with four different camera models, a value that measures how far the matrix A of each image in the sequence departs from being diagonal: we have chosen for this to compute the average of the absolute value of the non-diagonal elements of A normalised by its maximum value. The fact that these values are consistently above 0.1 shows that the three channels R, G, B are not independent. The second row of Fig. 10.36 plots the value of $1/\gamma$ for each image in the sequence, which is ordered from shortest to longest exposure time. We can see that, for all sequences, as the exposure time increases the value of $1/\gamma$ also increases, and the change is quite substantial. The third row of Fig. 10.36 plots the difference between the matrix A of each image in the sequence with respect to the matrix A of the middle-image in the stack (we compute this difference as the Frobenius norm $\|\cdot\|_F$ of the difference between the matrices). Again we see that the cameras are changing A from one exposure to the next.

In Fig. 10.37, each row corresponds to a camera model from a different camera maker. The columns show tone-mapped results of the HDR pictures obtained with different multi-exposure combination methods: from the RAW pictures, which would be the 'ground truth' or the best result we can aim for (first column), from the JPEG pictures using the multiple-exposure combination methods of Debevec and Malik [40] (column 2), Mitsunaga and Nayar [41] (column 3), Lee et al. [49] (column 4), and our proposed method (last column). We can see that the previous multiple-exposure methods that take the non-linear JPEG inputs produce results which have visible problems, like hue shifts and colour artifacts; to underline that these artifacts are not due to the particular tone-mapping method used, each row employs a different, state-of-the-art tone-mapping algorithm.

In the next section we will introduce a method that, considering all three channels simultaneously, removes the fluctuations in γ and A, effectively making the CRF constant for the whole sequence. Its results for the above sequences are shown in the last column of Fig. 10.37.

10.7.1.3 Colour-matching all exposures

The colour-matching process is key to our method, because it is the one that removes fluctuations in γ and A, and in practice turns the CRF constant for all images in the multi-exposure sequence. The basic idea was introduced in [65] for the particular problem of colour stabilisation among different shots of the same scene taken with different cameras: using the model of the camera colour processing pipeline of Eq. (10.28), find the parameter values so as to obtain an optimal match to a reference image. While there are works on multi-exposure HDR creation that match geometry and/or colour to a reference image in the stack (e.g. see [47] and references therein), the novelty of our approach lies in the use of a camera processing model, which allows our results to be more accurate, as we will see in Section 10.7.1.4.

Given the reference image I_{ref}, for each other image I_i in the sequence we do the following. Let p be a scene point appearing both in I_{ref} and I_i, which therefore produces in the sensor the same irradiance triplet $[R,G,B]^t_p$ in both pictures. Allowing for camera and/or object motion, point p may appear at different locations p_{ref} and p_i in both pictures, and the pixel values $[R,G,B]^t_{p_{ref}}$ and $[R,G,B]^t_{p_i}$ are also different:

$$\begin{bmatrix} R \\ G \\ B \end{bmatrix}_{p_{ref}} = \left(A_{ref} \begin{bmatrix} R \\ G \\ B \end{bmatrix}_p \right)^{\gamma_{ref}} ; \begin{bmatrix} R \\ G \\ B \end{bmatrix}_{p_i} = \left(A_i \begin{bmatrix} R \\ G \\ B \end{bmatrix}_p \right)^{\gamma_i}$$

These equalities can be combined into a single equation:

$$\left(\begin{bmatrix} R \\ G \\ B \end{bmatrix}_{p_{ref}} \right)^{1/\gamma_{ref}} - H_i \left(\begin{bmatrix} R \\ G \\ B \end{bmatrix}_{p_i} \right)^{1/\gamma_i} = \begin{bmatrix} 0 \\ 0 \\ 0 \end{bmatrix}, \qquad (10.29)$$

where H_i is the unknown 3×3 matrix $H_i = A_{ref} {A_i}^{-1}$. Thus, each pair p_{ref}, p_i of pixel correspondences gives us an equation of the form of (10.29), and we can find $H_i, \gamma_i, \gamma_{ref}$ by solving a system of equations. We only have 9 unknowns for H_i and two for γ_i, γ_{ref}, but we have many pixel correspondences: most of the image if there is motion, or the whole image if the scene is static. Accordingly the system is overdetermined, and the solution can be found by an optimisation procedure. This is in contrast with the approach taken both in [65] and in the preliminary version of our method [66], where the optimisation is done in two stages, first finding estimates for γ_i, γ_{ref}, and then for H_i. Our new single shot optimisation allows for a significant improvement in accuracy, as will be shown in Section 10.7.1.4.

Once H_i, γ_i have been found we can produce I_i', the linearised and colour-corrected (w.r.t. the reference image) version of I_i:

$$I_i' = H_i I_i^{1/\gamma_i}. \tag{10.30}$$

In Fig. 10.38 we show an example of this procedure.

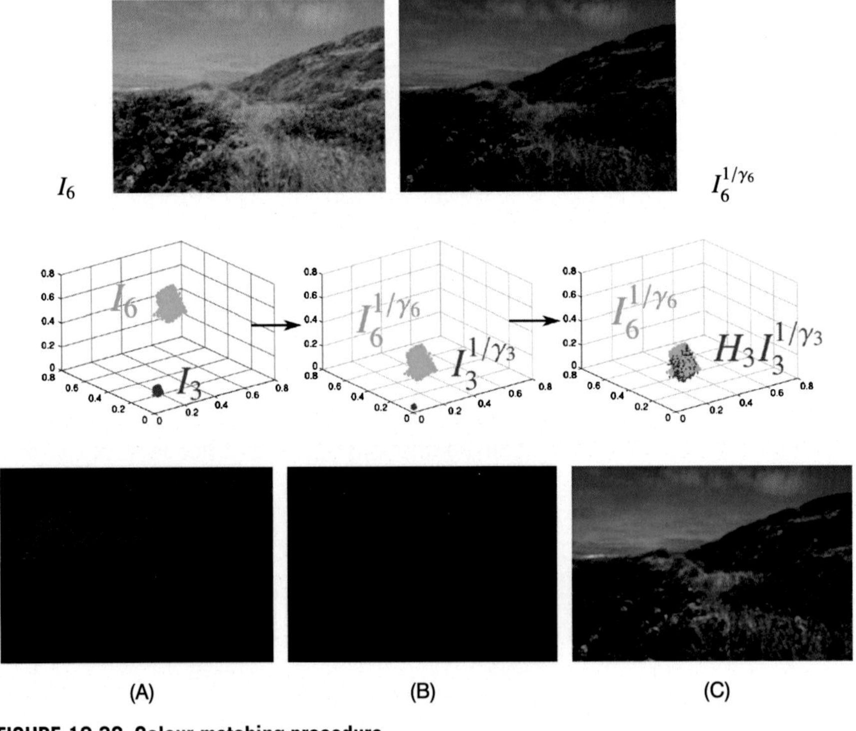

FIGURE 10.38 Colour matching procedure

Top row: the reference image I_6 on the left, and its linearised version I_6^{1/γ_6} on the right. Middle row: 3D point clouds representing the (R,G,B) values for I_6 in green and for I_3 in magenta. Bottom row: the corresponding images of I_3 for each plot in the middle row. (A) I_3; (B) I_3^{1/γ_3}; (C) $H_3 I_3^{1/\gamma_3}$.

10.7.1.4 Evaluation

We compare our approach against seven multiple-exposure HDR methods. Four of them operate only on static scenes: the classical methods of Debevec and Malik (DM) [40] and Mitsunaga and Nayar (MN) [41], and the recent methods by Lee et al.

FIGURE 10.39 Comparing results of methods for HDR from multiple exposures

From left to right: scenes *RITTiger*, *HancockKitchenInside*, *TheNarrows2* and *MasonLake1* from the HDR Survey. From top to bottom: results from DM [40], Lee13 [49], Sen [68] and finally our approach. All images tone-mapped with [60].

(Lee13) [49] and by Gil et al. [66]. The other three works also on dynamic scenes: Lee et al. (Lee14) [67], Sen et al. (Sen) [68], and Hu et al. (Hu) [47].

To show that the errors due to fluctuations in camera parameters can result in very visible artifacts, in Fig. 10.39, top to bottom, we compare the outputs of DM [40], Lee13 [49], Sen [68], and our approach, for the scenes *RITTiger*, *HancockKitchenInside*, *TheNarrows2* and *MasonLake1*, from left to right. All results have been tone-mapped with the method in [60]. For the scene *RITTiger* we can see that DM [40] (first row) presents a red cast in the image, therefore Sen [68] (third row) has the same colour cast since it uses DM as input. In *TheNarrows2*, Lee13 [49] (second row) shows very noticeable colour issues. Finally, for the *MasonLake1* scene, the method of Lee13 et al. presents a blue cast, while DM presents a reddish cast, and the method of Sen shows a banding artifact effect on the sky region.

To highlight that the visual problems described before are not due to a particular choice of tone-mapping algorithm, Fig. 10.40 shows the same HDR results but tone-mapped with two different methods, [60] for the first two columns and [58] for the last ones. The scenes are *AirBellowsGap* (columns 1 and 3) and *LabWindow* (columns 2 and 4), while the multiple-exposure HDR methods to compare are, from top to bottom: MN [41], Lee14 [67], Hu [47], and our approach. We can see how the previously existing methods produce colour artifacts in the sky and sun of the *Air-*

FIGURE 10.40 Comparing results of methods for HDR from multiple exposures

First and third column: *AirBellowsGap* scene. Second and fourth column: *LabWindow* scene. Columns 1 and 2: results tone-mapped with [60]. Columns 3 and 4: results tone-mapped with [58]. From top to bottom: results from MN [41], Lee14 [67], Hu [47] and finally our approach.

BellowsGap scene, and in the curtains, sky and background of the *LabWindow* scene, which are apparent for both of the tone-mapping methods used.

10.7.1.5 Dynamic scenes

It is worth emphasising that the proposed algorithm does not require image registration, only a set of pixel correspondences. Therefore, it can be used on dynamic scenes as well: in particular, Step 1 of our method can be employed as a pre-processing step to colour-stabilise the inputs of HDR methods operating on dynamic scenes, enhancing their performance. To illustrate this, we consider the algorithm of Sen et al. [68], which receives linearised images as input. Consequently, given a stack of non-linear images, we compare three linearising approaches: 1) CRF computed by Debevec and Malik [40], 2) radiometric calibration by Lee et al. [49], and 3) applying Step 1 of our method, using as a reference the image in the mid-point of the sequence and finding pixel correspondences with SIFT [69]. We conducted this experiment on a stack of five images from the dataset presented in [68], the *Skater* sequence (that comes in JPEG format). In Fig. 10.41 we present the HDR outputs obtained using the three different linearisation approaches. The zoomed-in details allow us to see how linearisation by [40] (left) produces artifacts in overexposed areas, whereas linearisation

FIGURE 10.41 HDR results on a dynamic scene

HDR results on a dynamic scene applying the HDR creation method of Sen et al. [68] with three different linearisation techniques: Debevec and Malik [40] (left), Lee et al. [49] (middle), and Step 1 of our proposed method (right), taking as reference the image in the mid-point of the sequence. HDR results tone-mapped with Mantiuk et al. [60]. Original image from [68].

with [49] produces results that, although free from artifacts, have lower contrast and less saturated colours than what can be obtained with our method.

10.7.2 Colour matching images with unknown non-linear encodings

Colour matching techniques aim to map the colours of one image, defined as source, to those of a second image, defined as reference. A particular case is colour stabilisation, where the two pictures are taken from the same scene and differ in terms of colour. These differences in colour may be caused either by the use of different camera models, which follow different internal procedures tailored by the manufacturer or by the use of the same camera but under different settings like white balance, exposure time, etc.

In the industry, solutions for bringing consistency across image shots usually involve very skilled manual work, done by colourists during colour grading in movie post-production and by technicians using camera control units (CCU) [70] in live TV broadcasts. They may also require a proper characterisation of the cameras used and their settings like with the ACES framework [71], or the presence of colour-charts in the shots.

Colour matching or colour stabilisation requires to linearise the images first, i.e. to apply the inverse of the encoding curve used by the camera to perform perceptual linearisation. We saw in Chapter 5 that the most common forms for these encoding curves are gamma correction,

$$I_{out} = (A \cdot I_{lin})^{1/\gamma}, \tag{10.31}$$

logarithmic encoding,

$$I_{out} = c \log_{10}(a \cdot A \cdot I_{lin} + b) + d, \tag{10.32}$$

and the PQ and HLG curves.

In this section we present an overview of our colour stabilisation framework from [72] with some recent updates. The framework takes as input an image pair encoded

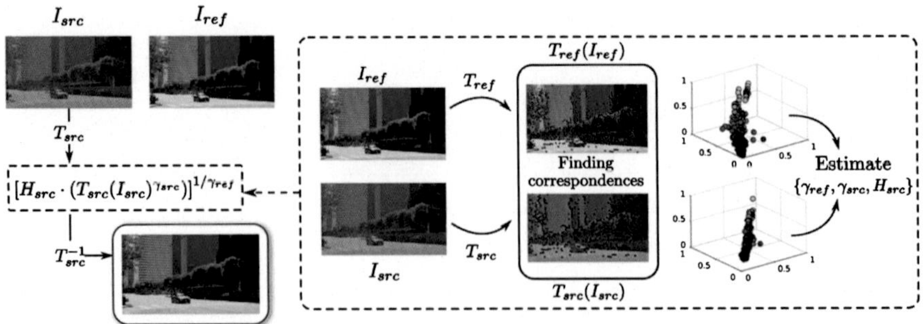

FIGURE 10.42 Flowchart of the proposed colour stabilisation method

Flowchart of the proposed colour stabilisation method. Given two non-linear encoded images, reference (I_{ref}) and source (I_{src}), we apply the transformations T_{ref} and T_{src} to the image pair. These transformations are defined as the power 10 function $10^{\times}$, in case of a given log, PQ or HLG-encoded image; and as the identity Id, in case of gamma corrected input. Then we compute a set of correspondences pts_{ref} and pts_{src}, using some standard feature descriptor (e.g. SIFT [69]). From this set of corresponding pixel locations, we estimate the parameters $\{\gamma_{ref}, \gamma_{src}, H_{src}\}$ in the pixel values correspondences. The computed values are applied to the $T_{src}(I_{src})$ image. Finally, T_{src}^{-1} function is applied to the colour matched image.

with unknown non-linear curves (gamma, log, PQ or HLG), and its main steps can be outlined as follows:

1. If source or reference are not gamma-corrected, we transform them into gamma-corrected.
2. We colour-stabilise the images by estimating a 4×4 matrix and two power law values.
3. Finally, if the original reference image was not gamma-corrected, we undo the transformation performed in the first step.

We refer the reader to the flowchart of the proposed model in Fig. 10.42.

10.7.2.1 From log, PQ or HLG to gamma-corrected images

Let us consider a log-encoded image as in Eq. (10.32),

$$I_{\log} = c \log_{10}(a \cdot A \cdot I_{lin} + b) + d. \tag{10.33}$$

If we apply a power 10 function to Eq. (10.33), we obtain the following expression

$$\begin{aligned} 10^{I_{\log}} &= 10^{\log_{10}((a \cdot A \cdot I_{lin} + b)^c)} \cdot 10^d \\ &= (a \cdot A \cdot I_{lin} + b)^c \cdot 10^d. \end{aligned} \tag{10.34}$$

In logarithmic encoding curves, the value of parameter b is usually small. Therefore, we can simplify Eq. (10.33) by neglecting b,

$$10^{I_{\log}} = (a \cdot A \cdot I_{lin})^c \cdot 10^d = (K \cdot I_{lin})^c, \tag{10.35}$$

where $K = a \cdot A \cdot 10^{d/c}$ is a matrix with the same size as A. Notice how Eq. (10.35) has the same form as Eq. (10.31). Therefore, by applying the power 10 function to a log-encoded image, it can now be treated as a regular gamma-corrected picture.

For images with PQ or HLG encodings, we treat them as if they were under a logarithmic encoding i.e. we apply a power 10 function and assume that the result now is a gamma-corrected image. This makes sense because, as explained in Chapter 5, the PQ and HLG encoding curves have a logarithmic behaviour over a significant expanse of their range.

10.7.2.2 Colour stabilisation

If I_{ref} or I_{src} (or both) are not gamma corrected, we transform them into gamma-corrected images I'_{ref} and I'_{src}, as explained in the previous section. Then we compute a set of correspondences pts_{ref} and pts_{src}; we use SIFT [69] for this purpose, although it can be replaced by any other method. It is important to note that we compute the correspondences between $I'_{ref} \leftrightarrow I'_{src}$, and $I'_{src} \leftrightarrow I'_{ref}$, and select those that appear in both directions. This allows us to discard some potentially incorrect correspondences. Let us now define the pixel values in the corresponding locations of I'_{ref} and I'_{src} as

$$\{(R'_{ref}, G'_{ref}, B'_{ref})^t\}_i, \text{ and } \{(R'_{src}, G'_{src}, B'_{src})^t\}_i, \tag{10.36}$$

where $i = 1, \ldots, N$ denotes the number of correspondences.

We follow the idea from the colour stabilisation model proposed in [65],

$$H_{src} \cdot I'^{\gamma_{src}}_{src} \sim I'^{\gamma_{ref}}_{ref}, \tag{10.37}$$

which we adapted in Section 10.7.1 to colour-match images in a multiple-exposure scenario. Recall that H_{src} was a 3×3 matrix that transforms colours from the source to match the ones of the reference, and γ_{ref}, γ_{src} are inverse gamma correction values.

We have extended the matrix H_{src} as a projective transformation with size 4×4 (inspired by colour homography [73,74]). By doing this, the model can deal not only with pixels in the core of the colour gamut, but also with those values that appear on the border, which are the most affected by gamut mapping and tone mapping. Then, from the set of correspondences, we can build a system of equations considering matrix size 4×4 and homogeneous coordinates,

$$H_{src} \cdot \begin{bmatrix} R'_{src} \\ G'_{src} \\ B'_{src} \\ 1 \end{bmatrix}^{\gamma_{src}} - \begin{bmatrix} R'_{ref} \\ G'_{ref} \\ B'_{ref} \\ 1 \end{bmatrix}^{\gamma_{ref}} = 0, \tag{10.38}$$

$$H_{ref} \cdot \begin{bmatrix} R'_{ref} \\ G'_{ref} \\ B'_{ref} \\ 1 \end{bmatrix}^{\gamma_{ref}} - \begin{bmatrix} R'_{src} \\ G'_{src} \\ B'_{src} \\ 1 \end{bmatrix}^{\gamma_{src}} = 0,$$

where $\{\gamma_{src}, \gamma_{ref}, H_{src}, H_{ref}\}$ are the unknowns. We perform a single optimisation process, where the only constraint is $H_{ref} \cdot H_{src} \sim Id$ (the identity). This constraint assures that the transformation H_{src} has an inverse, and that it's represented by H_{ref} (which corresponds to the matrix that would transfer the colours of the source into the reference). The objective function considers the differences on (R,G,B) points, plus the differences on Lab colour space, using 3×1 non-homogeneous coordinates. In this way, we bring the corresponding point clouds (colour matched and reference) as close as possible. The objective function is defined as

$$E(\mathcal{V}) = E_{RGB}(\mathcal{V}) + E_{Lab}(\mathcal{V}), \tag{10.39}$$

where $\mathcal{V} = \left\{\gamma_{ref}, \gamma_{src}, H_{ref}, H_{src}\right\}$ is the set of unknowns, and E_{RGB} and E_{Lab} are the errors in the RGB and Lab colour spaces, respectively. These terms are defined as

$$\begin{aligned} E_{RGB}(\mathcal{V}) = \sum_{\substack{RGB_{src} \\ RGB_{ref}}} & \left\| RGB_{ref} - g^1_{\mathcal{V}}(RGB_{src}) \right\|_2 \\ & + \left\| RGB_{src} - g^2_{\mathcal{V}}(RGB_{ref}) \right\|_2, \end{aligned} \tag{10.40}$$

$$\begin{aligned} E_{Lab}(\mathcal{V}) = \sum_{\substack{RGB_{src} \\ RGB_{ref}}} & \left\| Lab\left(RGB_{ref}\right) - Lab\left(g^1_{\mathcal{V}}(RGB_{src})\right) \right\|_2 \\ & + \left\| Lab(RGB_{src}) - Lab\left(g^2_{\mathcal{V}}(RGB_{ref})\right) \right\|_2, \end{aligned} \tag{10.41}$$

where RGB_{src}, RGB_{ref} are the (R, G, B) values of corresponding points pts_{src} and pts_{ref}, $Lab(\cdot)$ corresponds to the colour transformation from RGB to Lab, and the functions $g^1_{\mathcal{V}}(\cdot)$ and $g^2_{\mathcal{V}}(\cdot)$ are defined as

$$g^1_{\mathcal{V}}(RGB_{src}) = \left(H_{src} \cdot RGB_{src}^{\gamma_{src}}\right)^{1/\gamma_{ref}} \text{ and} \tag{10.42}$$

$$g^2_{\mathcal{V}}\left(RGB_{ref}\right) = \left(H_{ref} \cdot RGB_{ref}^{\gamma_{ref}}\right)^{1/\gamma_{src}}. \tag{10.43}$$

Finally, the matrices and non-linearities are applied to the entire images as in Eq. (10.37), and we obtain the colour matched image:

$$I''_{src} = \left[H_{src} \cdot I'^{\gamma_{src}}_{src}\right]^{1/\gamma_{ref}}. \tag{10.44}$$

10.7.2.3 Undo power 10 function

If I_{src} was not gamma-corrected, we apply a $\log_{10}$ function to the result of the previous step so as to undo the power 10 transform applied at the beginning.

10.7.2.4 Experiments with HDR encodings

		ΔE^*_{00}	PSNR L	CPSNR	CID	RMSE
Kotera	μ	3.344	32.505	30.567	0.110	0.045
	$\hat{\mu}$	4.117	30.379	29.061	0.114	0.052
Pitie	μ	1.022	40.047	40.134	0.035	0.021
	$\hat{\mu}$	0.582	43.069	42.568	0.004	0.006
Reinhard	μ	1.861	35.311	35.020	0.062	0.040
	$\hat{\mu}$	2.058	32.023	31.408	0.042	0.037
Xiao	μ	1.891	32.965	32.789	0.061	0.032
	$\hat{\mu}$	2.066	30.379	30.244	0.054	0.033
Ferradans	μ	4.820	24.692	24.624	0.183	0.073
	$\hat{\mu}$	3.865	25.364	25.485	0.145	0.052
Park	μ	1.624	38.250	36.795	0.044	0.029
	$\hat{\mu}$	1.418	37.730	37.142	0.018	0.017
Gil	μ	1.775	36.086	35.636	0.068	0.029
	$\hat{\mu}$	1.659	31.674	31.295	0.046	0.022
Ours	μ	**0.310**	**48.324**	**47.649**	**0.002**	**0.005**
	$\hat{\mu}$	**0.239**	**49.435**	**49.131**	**0.001**	**0.004**

FIGURE 10.43 Colour stabilisation errors

Colour stabilisation errors for different methods applied to content encoded with PQ, HLG and ARRI Log-C. Results show mean (μ) and median ($\hat{\mu}$) averages over 10 pairs. In the case of the PQ curve, we set up the absolute luminance of the display to 1000 cd/m^2.

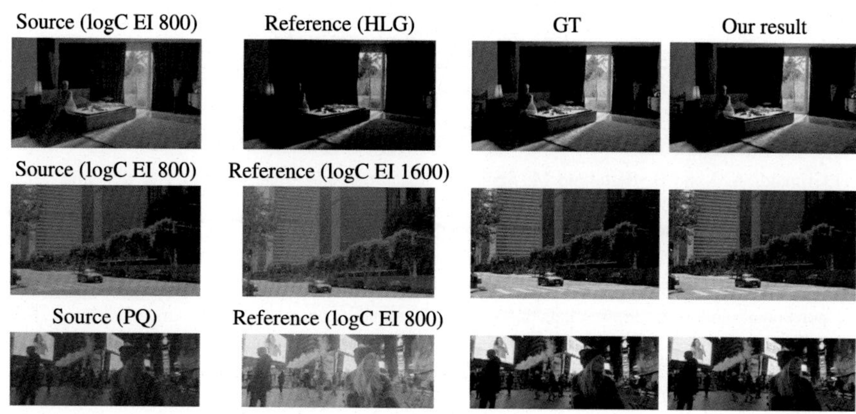

FIGURE 10.44 Colour matching examples with PQ and HLG

Examples with PQ and HLG. From left to right, source, reference, GT and our result. Each row represents a different comparison and scenario. The GTs and our results are tone mapped using [3]. Images from ARRI [75]. In the case of PQ curve, we set up the absolute luminance of the display to 1000 cd/m^2.

In this section, we colour match a pair of images encoded using different transfer functions: PQ, HLG and Log C ARRI curves.

10.7.2.5 Dataset

The dataset we use for experiments is provided by ARRI [75]. This data contains HDR videos. The linear RAW data is obtained by using *ARRIRAW Converter* [76]. We select three different scenes, and for each scene we set a reference image encoded with one of the 3 different options {PQ, HLG, Log C} (a random Log C ARRI curve). Then, we build the data by selecting all the possible combination pairs for each image reference (total of 9 pairs). We add an extra pair comparing two different Log C ARRI curves. Therefore, we have a total of 10 image pairs.

10.7.2.6 Results and comparisons

We compare our method with the algorithms: Reinhard et al. [18] (Reinhard), Kotera [17] (Kotera), Xiao and Ma [19] (Xiao), Pitié et al. [77] (Pitie), Ferradans et al. [78] (Ferradans), Park et al. [79] (Park), and Gil Rodríguez et al. [72] (Gil). The last method also considers the inputs as log-encoded images. In order to compute the quantitative results, we undo the non-linearity (since it is known) of the resulting colour matched image and the GT, and then apply a γ correction of $1/2.2$, as done in the previous experiments. From the data in Fig. 10.43, it is apparent that our method is accurate when working with real data and common situations.

Fig. 10.44 presents the image results. The reference and the source are presented without tone mapping, in order to appreciate the differences between applying the different curves, while the ground truth and our result have been tone mapped with the TMO from [3]. As it can be seen in the last column in Fig. 10.44, our method recovers the colours and appearance of the reference image, in different input situations.

References

[1] Grimaldi A, Kane D, Bertalmío M. Statistics of natural images as a function of dynamic range. Journal of Vision 2019;19(2).

[2] Sikudova E, Pouli T, Artusi A, Akyuz AO, Banterle F, Mazlumoglu ZM, et al. A gamut-mapping framework for color-accurate reproduction of hdr images. arXiv preprint. arXiv:1711.08925v1, 2017.

[3] Cyriac P, Kane D, Bertalmío M. Automatic, viewing-condition dependent contrast grading based on perceptual models. In: Proceedings of the SMPTE annual technical conference and exhibition; 2016. p. 1–11.

[4] Cyriac P, Kane D, Bertalmío M. Optimized tone curve for in-camera image processing. Electronic Imaging 2016;2016(13):1–7.

[5] Fairchild MD. The hdr photographic survey. In: Color and imaging conference, vol. 2007. Society for Imaging Science and Technology; 2007. p. 233–8.

[6] Schuck M, Ludé P. An analysis of system contrast in digital cinema auditoriums. SMPTE Motion Imaging Journal 2016;125(4):40–9.

[7] Tydtgat C, Maes D, Stojmenovik G, Grillet A. Modeling of achievable contrast and its impact on hdr projection in commercial cinema environments. SMPTE Motion Imaging Journal 2016;125(4):1–7.

[8] Froehlich J, Grandinetti S, Eberhardt B, Walter S, Schilling A, Brendel H. Creating cinematic wide gamut hdr-video for the evaluation of tone mapping operators and hdr-displays. In: Digital photography X, vol. 9023. International Society for Optics Photonics; 2014. p. 90230X.
[9] Bartleson C. Optimum image tone reproduction. Journal of the SMPTE 1975;84(8):613–8.
[10] Sugito Y, Cyriac P, Kane D, Bertalmío M. Improved high dynamic range video coding with a nonlinearity based on natural image statistics. International Journal of Signal Processing Systems 2017;5(3):100–5. https://doi.org/10.18178/ijsps.5.3.100-105.
[11] Ström J, Samuelsson J, Dovstam K. Luma adjustment for high dynamic range video. In: 2016 Data compression conference (DCC). IEEE; 2016. p. 319–28.
[12] Narwaria M, Da Silva MP, Le Callet P. Hdr-vqm: an objective quality measure for high dynamic range video. Signal Processing: Image Communication 2015;35:46–60.
[13] Bjontegaard G. Calculation of average psnr differences between rd-curves. VCEG-M33, 2001.
[14] Sugito Y, Bertalmío M. Practical use suggests a re-evaluation of hdr objective quality metrics. In: Eleventh international conference on quality of multimedia experience (QoMEX 2019); 2019.
[15] Zabaleta I, Bertalmío M. Photorealistic style transfer for cinema shoots. In: 2018 colour and visual computing symposium (CVCS). IEEE; 2018. p. 1–6.
[16] Zabaleta I, Bertalmío M. In-camera, photorealistic style transfer for on-set automatic grading. In: SMPTE 2018. SMPTE; 2018. p. 1–12.
[17] Kotera H. A scene-referred color transfer for pleasant imaging on display. In: IEEE international conference on image processing 2005, vol. 2. IEEE; 2005. p. II–5.
[18] Reinhard E, Adhikhmin M, Gooch B, Shirley P. Color transfer between images. IEEE Computer Graphics and Applications 2001;21(5):34–41.
[19] Xiao X, Ma L. Color transfer in correlated color space. In: Proceedings of the 2006 ACM international conference on virtual reality continuum and its applications. ACM; 2006. p. 305–9.
[20] Ebner F, Fairchild MD. Development and testing of a color space (ipt) with improved hue uniformity. In: Color and imaging conference, vol. 1998. Society for Imaging Science and Technology; 1998. p. 8–13.
[21] Brown M, Süsstrunk S. Multi-spectral sift for scene category recognition. In: CVPR 2011. IEEE; 2011. p. 177–84.
[22] Vazquez-Corral J, Galdran A, Cyriac P, Bertalmío M. A fast image dehazing method that does not introduce color artifacts. Journal of Real-Time Image Processing 2018:1–16.
[23] He K, Sun J, Tang X. Single image haze removal using dark channel prior. IEEE Transactions on Pattern Analysis and Machine Intelligence 2011;33(12):2341–53.
[24] Meng G, Wang Y, Duan J, Xiang S, Pan C. Efficient image dehazing with boundary constraint and contextual regularization. In: Proceedings of the IEEE international conference on computer vision; 2013. p. 617–24.
[25] Choi LK, You J, Bovik AC. Referenceless prediction of perceptual fog density and perceptual image defogging. IEEE Transactions on Image Processing 2015;24(11):3888–901.
[26] Wang S, Cho W, Jang J, Abidi MA, Paik J. Contrast-dependent saturation adjustment for outdoor image enhancement. JOSA A 2017;34(1):7–17.
[27] Galdran A, Vazquez-Corral J, Pardo D, Bertalmío M. A variational framework for single image dehazing. In: European conference on computer vision. Springer; 2014. p. 259–70.
[28] Galdran A, Vazquez-Corral J, Pardo D, Bertalmio M. Enhanced variational image dehazing. SIAM Journal on Imaging Sciences 2015;8(3):1519–46.
[29] Galdran A, Vazquez-Corral J, Pardo D, Bertalmio M. Fusion-based variational image dehazing. IEEE Signal Processing Letters 2016;24(2):151–5.
[30] Galdran A, Alvarez-Gila A, Bria A, Vazquez-Corral J, Bertalmío M. On the duality between retinex and image dehazing. In: Proceedings of the IEEE conference on computer vision and pattern recognition; 2018. p. 8212–21.
[31] Vazquez-Corral J, Cyriac P, Bertalmío M. Perceptually-based restoration of backlit images. In: Color and imaging conference, vol. 2018. Society for Imaging Science and Technology; 2018. p. 32–7.
[32] Bertalmío M, Caselles V, Provenzi E, Rizzi A. Perceptual color correction through variational techniques. IEEE Transactions on Image Processing 2007;16(4):1058–72.

[33] Wang S, Zheng J, Hu HM, Li B. Naturalness preserved enhancement algorithm for non-uniform illumination images. IEEE Transactions on Image Processing 2013;22(9):3538–48.
[34] Fu X, Zeng D, Huang Y, Liao Y, Ding X, Paisley J. A fusion-based enhancing method for weakly illuminated images. Signal Processing 2016;129:82–96.
[35] Petro AB, Sbert C, Morel JM. Multiscale retinex. Image Processing On Line 2014:71–88.
[36] Fu X, Zeng D, Huang Y, Zhang XP, Ding X. A weighted variational model for simultaneous reflectance and illumination estimation. In: Proceedings of the IEEE conference on computer vision and pattern recognition; 2016. p. 2782–90.
[37] Dong X, Wang G, Pang Y, Li W, Wen J, Meng W, et al. Fast efficient algorithm for enhancement of low lighting video. In: 2011 IEEE international conference on multimedia and expo. IEEE; 2011. p. 1–6.
[38] Li Z, Wu X. Learning-based restoration of backlit images. IEEE Transactions on Image Processing 2018;27(2):976–86. https://doi.org/10.1109/tip.2017.2771142.
[39] Mann S, Picard RW. On being 'undigital' with digital cameras: extending dynamic range by combining differently exposed pictures. In: Proceedings of IS&T; 1995. p. 442–8.
[40] Debevec PE, Malik J. Recovering high dynamic range radiance maps from photographs. In: Proceedings of the 24th annual conference on computer graphics and interactive techniques. New York, NY, USA: SIGGRAPH; 1997. p. 369–78.
[41] Mitsunaga T, Nayar S. Radiometric self calibration. In: IEEE conference on computer vision and pattern recognition (CVPR), vol. 1; 1999. p. 374–80.
[42] Tsin Y, Ramesh V, Kanade T. Statistical calibration of CCD imaging process. In: IEEE international conference on computer vision (ICCV), vol. 1; 2001. p. 480–7.
[43] Mann S. Comparametric equations with practical applications in quantigraphic image processing. IEEE Transactions on Image Processing (TIP) 2000;9(8):1389–406.
[44] Kang SB, Uyttendaele M, Winder S, Szeliski R. High dynamic range video. ACM Transactions on Graphics (TOG) 2003;22(3):319–25.
[45] Kim SJ, Pollefeys M. Robust radiometric calibration and vignetting correction. IEEE Transactions on Pattern Analysis and Machine Intelligence (TPAMI) 2008;30(4):562–76. https://doi.org/10.1109/TPAMI.2007.70732.
[46] Grossberg MD, Nayar SK. High dynamic range from multiple images: which exposures to combine. In: Proceedings international conference on computer vision workshop (ICCVW) on color and photometric methods in computer vision; 2003.
[47] Hu J, Gallo O, Pulli K, Sun X, Deghosting HDR. How to deal with saturation? In: IEEE conference on computer vision and pattern recognition (CVPR); 2013. p. 1163–70.
[48] Oh TH, Lee JY, Tai YW, Kweon IS. Robust high dynamic range imaging by rank minimization. IEEE Transactions on Pattern Analysis and Machine Intelligence (TPAMI) 2015;37(6):1219–32.
[49] Lee JY, Matsushita Y, Shi B, Kweon IS, Ikeuchi K. Radiometric calibration by rank minimization. IEEE Transactions on Pattern Analysis and Machine Intelligence (TPAMI) 2013;35(1):144–56.
[50] Tocci MD, Kiser C, Tocci N, Sen P. A versatile HDR video production system. ACM Transactions on Graphics (TOG) 2011;30(4):41.
[51] Kronander J, Gustavson S, Unger J. Real-time HDR video reconstruction for multi-sensor systems. In: ACM SIGGRAPH. New York, NY, USA: Posters; 2012.
[52] Guthier B, Kopf S, Wichtlhuber M, Effelsberg W. Parallel implementation of a real-time high dynamic range video system. Integrated Computer-Aided Engineering 2014;21(2):189–202. https://doi.org/10.3233/ICA-130461.
[53] Aguerrebere C, Delon J, Gousseau Y, Musé P. Best algorithms for HDR image generation. A study of performance bounds. SIAM Journal on Imaging Sciences 2014;7(1):1–34.
[54] Kronander J, Gustavson S, Bonnet G, Unger J. Unified HDR reconstruction from raw CFA data. In: IEEE international conference on computational photography; 2013. p. 1–9.
[55] McCann John J, Rizzi Alessandro. The Art and Science of HDR Imaging, vol. 26. John Wiley & Sons; 2011.
[56] Tomić Ivana, Karlović Igor, Jurič Ivana. Practical assessment of veiling glare in camera lens system. Journal of Graphic Engineering and Design 2014;5(2):23.

[57] Fairchild MD. The HDR photographic survey. Color and Imaging Conference 2007;2007(1):233–8.

[58] Drago F, Myszkowski K, Annen T, Chiba N. Adaptive logarithmic mapping for displaying high contrast scenes. In: Computer graphics forum. Wiley Online Library; 2003. p. 419–26.

[59] Mai Z, Mansour H, Mantiuk R, Nasiopoulos P, Ward RK, Heidrich W. Optimizing a tone curve for backward-compatible high dynamic range image and video compression. IEEE Transactions on Image Processing (TIP) 2011;20(6):1558–71.

[60] Mantiuk R, Daly S, Kerofsky L. Display adaptive tone mapping. In: ACM transactions on graphics (TOG), vol. 27. ACM; 2008. p. 68.

[61] Ferradans S, Bertalmío M, Provenzi E, Caselles V. An analysis of visual adaptation and contrast perception for tone mapping. IEEE Transactions on Pattern Analysis and Machine Intelligence (TPAMI) 2011;33(10):2002–12.

[62] Gil Rodríguez R, Vazquez-Corral J, Bertalmío M. Issues with common assumptions about the camera pipeline, and their impact in HDR imaging from multiple exposures. SIAM Journal on Imaging Sciences 2019.

[63] Bianco S, Bruna A, Naccari F, Schettini R. Color space transformations for digital photography exploiting information about the illuminant estimation process. Journal of the Optical Society of America (JOSA) 2012;29(3):374–84.

[64] Bertalmío M. Image processing for cinema. 1st ed. Chapman & Hall/CRC mathematical and computational imaging sciences series. CRC Press – Taylor & Francis; 2014.

[65] Vazquez-Corral J, Bertalmío M. Color stabilization along time and across shots of the same scene, for one or several cameras of unknown specifications. IEEE Transactions on Image Processing (TIP) 2014;23(10):4564–75.

[66] Gil Rodríguez R, Vazquez-Corral J, Bertalmío M. The intrinsic error of exposure fusion for hdr imaging, and a way to reduce it. In: Proceedings of the British machine vision conference (BMVC). BMVA Press; 2015. p. 126.1–126.12.

[67] Lee C, Li Y, Monga V. Ghost-free high dynamic range imaging via rank minimization. IEEE Signal Processing Letters 2014;21(9):1045–9.

[68] Sen P, Kalantari NK, Yaesoubi M, Darabi S, Goldman DB, Shechtman E. Robust patch-based HDR reconstruction of dynamic scenes. ACM Transactions on Graphics (TOG) 2012;31(6):203.

[69] Lowe DG. Object recognition from local scale-invariant features. In: IEEE international conference on computer vision (ICCV), vol. 2; 1999. p. 1150–7.

[70] MediaCollege. CCU (camera control unit) operations. https://www.mediacollege.com/video/production/camera-control/, 2012.

[71] Postma P, Chorley B. Colour grading with colour management. In: SMPTE15: persistence of vision – defining the future; 2015. p. 1–8.

[72] Gil Rodríguez R, Vazquez-Corral J, Bertalmío M. Color-matching shots from different cameras having unknown gamma or logarithmic encoding curves. In: SMPTE annual technical conference & exhibition; 2017. p. 1–15.

[73] Gong H, Finlayson G, Fisher R. Recoding color transfer as a color homography. In: Proceedings of the British machine vision conference (BMVC). BMVA Press; 2016. p. 17.1–1711..

[74] Gong H, Finlayson GD, Fisher RB, Fang F. 3D color homography model for photo-realistic color transfer re-coding. The Visual Computer 2017:1–11.

[75] Andriani S, Brendel H, Seybold T, Goldstone J. Beyond the Kodak image set: a new reference set of color image sequences. In: IEEE international conference on image processing (ICIP); 2013. p. 2289–93.

[76] ARRI. ARRIRAW converter. http://www.arri.com/camera/alexa/tools/ar-riraw_converter/, 2018. Accessed 2018.

[77] Pitié F, Kokaram AC, Dahyot R. Automated colour grading using colour distribution transfer. Computer Vision and Image Understanding 2007;107(1):123–37.

[78] Ferradans S, Papadakis N, Peyré G, Aujol J. Regularized discrete optimal transport. SIAM Journal on Imaging Sciences 2014;7(3):1853–82.

[79] Park J, Tai YW, Sinha SN, Kweon IS. Efficient and robust color consistency for community photo collections. In: IEEE conference on computer vision and pattern recognition (CVPR); 2016. p. 430–8.

CHAPTER

11 Open problems: an argument for new vision models rather than new algorithms

11.1 The linear receptive field is the foundation of vision models

We recall from Chapter 2 the definition of Receptive Field (RF): the RF of a neuron is the extent of the visual field where light influences the neuron's response.

The "standard model" of vision is grounded on the concept of a linear RF. From Carandini et al. [1]: *"At the basis of most current models of neurons in the early visual system is the concept of linear receptive field. The receptive field is commonly used to describe the properties of an image that modulates the responses of a visual neuron. More formally, the concept of a receptive field is captured in a model that includes a linear filter as its first stage. Filtering involves multiplying the intensities at each local region of an image (the value of each pixel) by the values of a filter and summing the weighted image intensities."*

And from Olshausen and Field [2]: *"And there has even emerged a fairly well-agreed-on 'standard model' for V1 in which simple cells compute a linearly weighted sum of the input over space and time (usually a Gabor-like function), which is then normalised by the responses of neighboring neurons and passed through a pointwise nonlinearity. Complex cells are similarly explained in terms of a summation over the outputs of a local pool of simple cells with similar tuning properties but different positions or phases. [...] The net result is often to think of V1 as a kind of 'Gabor filter bank.' There are numerous papers showing that this basic model fits much of the existing data well, and many scientists have come to accept this as a working model of V1 function."*

While there have been considerable improvements on and extensions to the standard model, the linear RF remains as the foundation of most vision models:

- In neuroscience, where models of single-neuron neurophysiological activity in the retina, the LGN and the cortex begin with a linear RF stage [1].
- In visual perception, where models of perceptual phenomena are based on convolving the visual input with a bank of linear filters [3].

Vision Models for High Dynamic Range and Wide Colour Gamut Imaging. https://doi.org/10.1016/B978-0-12-813894-6.00016-8

- In computational and mathematical neuroscience, very diverse approaches also assume a linear filtering of the signal [4–6].

Artificial Neural Networks (ANNs) are inspired by classical models of biological neural networks, and for this reason they are also based on the linear RF, which is their essential building block. For instance, in his classical treatise on ANNs [7], Haykin states that one of the three basic elements of the neural model is *"an adder for summing the input signals, weighted by the respective synaptic strengths of the neuron; the operations described here constitute a linear combiner."* The other two basic elements are the weights of the linear summation and a nonlinear activation function. Therefore, ANNs can be seen as a cascade of linear and nonlinear (L+NL) modules.

L+NL representations are very popular models in vision science [8]. In visual neuroscience, most modelling techniques for analysing spike trains consist of a cascade of L+NL stages [1], while in visual perception the most successful models are also in L+NL form [3].

But we saw in Chapter 3 how despite the enormous advances in the field, the most relevant questions about colour vision and its cortical representation remain open [9, 10]: which neurons encode colour, how does V1 transform the cone signals, how shape and form are perceptually bound, and how do these neural signals correspond to colour perception.

A key message is that the parameters of L+NL models *depend on the image stimulus*, as seen in Chapters 3 and 4, a topic we shall discuss in the next section. Probably not coincidentally, the effectiveness of these models decays considerably when they are tested on natural images. This has grave implications for our purposes, since in colour imaging most essential methodologies also assume a L+NL form: brightness perception models (a nonlinearity, possibly shifted by the output of a linear filter), contrast sensitivity functions (a linear model, possibly applied after a nonlinearity), colour spaces and colour appearance models (with linear filters implemented as matrix multiplications and signal differences, and nonlinearities that may take different forms such as power laws or Naka-Rushton equations) and image quality metrics (based on L+NL vision models or based on ANNs, as seen in Chapters 8 and 9).

We have seen in the book that there are many things which L+NL models can't satisfactorily explain. HDR/WCG imaging has further put to the test the capabilities of vision models and highlighted their limits. Many essential questions remain open, and solutions that were good enough for SDR and standard colour gamuts do not cut it now. To mention just a couple of key examples: there are no good models of brightness perception for HDR images, so for instance there can't be fully automated methods to re-master SDR content in HDR as the optimal level of diffuse white changes from shot to shot [11,12]; inter-observer differences could be ignored to define the colour matching functions (CMFs) for standard colour gamut monitors where the primaries were not too saturated, but with WCG displays inter-observer differences can be substantial, to the point where it may make sense to define CMFs for individuals [13].

As a result, both tone and gamut mapping remain open, challenging problems for which there are neither fully effective automatic solutions nor accurate vision models. But we must remark that *there are* solutions to these problems: they are manual solutions, performed by cinema professionals, who have the ability to modify images so that their appearance on screen matches what the real-world scene would look like to an observer in it [14]. Remarkably, artists and technicians with this ability are capable of achieving what neither state-of-the-art automated methods nor up-to-date vision models can. Put in other words, the manual techniques of cinema professionals seem to have a "built-in" vision model.

11.2 Inherent problems with using a linear filter as the basis of a vision model

The fundamental issue is the following: the responses of visual neurons, as well as visual perception phenomena in general, are highly nonlinear functions of the visual input.

The RF of a neuron is characterised by finding the linear filter that provides the best correlation between visual input (often, white noise) and neuron response. The problem is that model performance degrades quickly if any aspect of the stimulus, like the spatial frequency or the contrast, is changed, because **the resulting RF depends on the stimulus, since the visual system is nonlinear** [15,1,16].

This is an essential, inherent limitation of these models, and it's such a key point that we think it's best to let top vision scientists explain it themselves.

From Olshausen and Field [2]: *"Everyone knows that neurons are nonlinear, but few have acknowledged the implications for studying cortical function. Unlike linear systems, where there exist mathematically tractable textbook methods for system identification, nonlinear systems cannot be teased apart using some straightforward, structuralist approach. That is, there is no unique 'basis set' with which one can probe the system to characterize its behavior in general. Nevertheless, the structuralist approach has formed the bedrock of V1 physiology for the past four decades. Researchers have probed neurons with spots, edges, gratings, and a variety of mathematically elegant functions in the hope that the true behavior of neurons can be explained in terms of some simple function of these components. However, the evidence that this approach has been successful is lacking. We simply have no reason to believe that a population of interacting neurons can be reduced in this way."*

From Wandell [15] (Chapter 7): *"Multiresolution theory [uses] a neural representation that consists of a collection of component-images, each sensitive to a narrow band of spatial frequencies and orientations. This separation of the visual image information can be achieved by using a variety of convolution kernels, each of which emphasizes a different spatial frequency range in the image. This calculation might be implemented in the nervous system by creating neurons with a variety of receptive field properties. [...] There is a bewildering array of experimental methods – ranging from detection to pattern adaptation to masking – whose results are inconsistent with the central notions of multiresolution representations."*

From Olshausen and Field [2]: *"The Gabor function has been argued to provide a good model of cortical receptive fields. However, the methods used to measure the receptive field in the first place generally search for the best-fitting linear model. They are not tests of how well the receptive field model actually describes the response of the neuron. [...] The results demonstrate that these models often fail to adequately capture the actual behavior of neurons. [...] There is only one way to map a non-linear system with complete confidence: present the neuron with all possible stimuli. The scope of this task is truly breathtaking. Even an* 8×8 *pixel patch with 6 bits of gray level requires searching* $2^{384} > 10^{100}$ *possible combinations (a googol of combinations). If we allow for temporal sensitivity and include a sequence of 10 such patches, we are exceeding* 10^{1000}*. With the estimated number of particles in the universe estimated to be in the range of* 10^{80}*, it should be clear that this is far beyond what any experimental method could explore. The deeper question is whether one can predict the responses of neurons from some combinatorial rule of the responses derived from a reduced set of stimuli. The response of the system to any reduced set of stimuli cannot be guaranteed to provide the information needed to predict the response to an arbitrary combination of those stimuli. Of course, we will never know this until it is tested, and that is precisely the problem: the central assumption of the elementwise, reductionist approach has yet to be thoroughly tested."*

From Carandini et al. [1]: *"In the past few years, a number of laboratories have begun using natural scenes as stimuli when recording from neurons in the visual pathway. For example, David et al. (2004) have explored two different types of models [...]. These models can typically explain between 30 and 40 per cent of the response variance of V1 neurons. One could possibly obtain a better fit to the data by including additional terms [...] but it is still sobering to realize that the receptive field component per se, which is the bread and butter of the standard model, accounts for so little of the response variance. Moreover, the way in which these models fail does not leave one optimistic that the addition of modulatory terms or pointwise nonlinearities will fix matters. [...] Thus, there appears to be a qualitative mismatch in predicting the responses of cortical neurons to time-varying natural images that will require more than tweaking to resolve. What seems to be suggested by the data is that a more complex, network nonlinearity is at work here and that describing the behavior of any one neuron will require one to include the influence of other simultaneously recorded neurons."*

From Olshausen [17]: *"At the end of the day we are faced with this simple truth: No one has yet spelled out a detailed model of V1 that incorporates its true biophysical complexity and exploits this complexity to process visual information in a meaningful or useful way. The problem is not just that we lack the proper data, but that we don't even have the right conceptual framework for thinking about what is happening. In light of the strong nonlinearities and other complexities of neocortical circuits, one should view the existing evidence for filters or other simple forms of feature extraction in V1 with great skepticism. The vast majority of experiments that claim to measure and characterize 'receptive fields' were conducted assuming a linear systems identification framework. We are now discovering that for many V1 neurons these receptive field models perform poorly in predicting responses to complex, time-varying natural images. Some argue that with the right amount of tweaking and by including proper gain control mechanisms and other forms of contextual modulation that you can get these models to work. My own view is that the standard model is not just in need of revision, it is the wrong starting point and needs to be discarded altogether. What is needed in its place is a model that embraces the true biophysical complexity and structure of cortical micro-circuits, especially dendritic nonlinearities."*

11.3 Conclusion: vision-based methods are best, but we need new vision models

Now that we reach the very end of the book we hope that the reader may share our conclusions, based on the exposition and reported results in this and all previous chapters:

- Imaging techniques based on vision models are the ones that perform best for tone and gamut mapping and a number of other applications.
- The performance of these methods is still far below what cinema professionals can achieve.
- Vision models are lacking, most key problems in visual perception remain open.
- Rather than be improved or revisited, a change of paradigm seems to be needed for vision models.

Our proposal is to explore models based on local histogram equalisation, with fine-tuning by movie professionals. We have shown that this approach yields very promising outcomes in tone and gamut mapping, but results can still be improved, and new vision models developed. Local histogram equalisation is intrinsically nonlinear, and it's closely related with theories that advocate that spatial summation by neurons is nonlinear [18,19], and those using nonlinear time series analysis of oscillations in brain activity [20].

A change of paradigm in vision models, with intrinsically nonlinear frameworks developed by mimicking the techniques of cinema professionals, could clearly have a really wide impact, much wider than the HDR/WCG domain, given that, as mentioned above, the L+NL formulation is prevalent not only in vision science and imaging applications, it's the basis of ANNs as well.

References

[1] Carandini M, Demb JB, Mante V, Tolhurst DJ, Dan Y, Olshausen BA, et al. Do we know what the early visual system does? Journal of Neuroscience 2005;25(46):10577–97.

[2] Olshausen BA, Field DJ. How close are we to understanding v1? Neural Computation 2005;17(8):1665–99.
[3] Graham NV. Beyond multiple pattern analyzers modeled as linear filters (as classical v1 simple cells): useful additions of the last 25 years. Vision Research 2011;51(13):1397–430.
[4] Wilson HR, Cowan JD. Excitatory and inhibitory interactions in localized populations of model neurons. Biophysical Journal 1972;12(1):1–24.
[5] Atick JJ, Redlich AN. What does the retina know about natural scenes? Neural Computation 1992;4(2):196–210.
[6] Lindeberg T. A computational theory of visual receptive fields. Biological Cybernetics 2013;107(6):589–635.
[7] Haykin SS. Neural networks and learning machines, vol. 3. Pearson Education Upper Saddle River; 2009.
[8] Martinez-Garcia M, Cyriac P, Batard T, Bertalmio M, Malo J. Derivatives and inverse of cascaded linear+nonlinear neural models. PLoS ONE 2018;13(10):e0201326.
[9] Solomon SG, Lennie P. The machinery of colour vision. Nature Reviews. Neuroscience 2007;8(4):276.
[10] Conway BR, Chatterjee S, Field GD, Horwitz GD, Johnson EN, Koida K, et al. Advances in color science: from retina to behavior. Journal of Neuroscience 2010;30(45):14955–63.
[11] Boitard R, Smith M, Zink M, Damberg G, Ballestad A. Using high dynamic range home master statistics to predict dynamic range requirement for cinema. In: SMPTE 2018; 2018. p. 1–28.
[12] Ploumis S, Boitard R, Jacquemin J, Damberg G, Ballestad A, Nasiopoulos P. Quantitative evaluation and attribute of overall brightness in a high dynamic range world. In: SMPTE 2018; 2018. p. 1–16.
[13] Fairchild MD, Heckaman RL. Metameric observers: a Monte Carlo approach. Color and imaging conference, vol. 2013. Society for Imaging Science and Technology; 2013. p. 185–90.
[14] Van Hurkman A. Color correction handbook: professional techniques for video and cinema. Pearson Education; 2013.
[15] Wandell BA. Foundations of vision, vol. 8. Sinauer Associates Sunderland, MA; 1995.
[16] DeAngelis G, Anzai A. A modern view of the classical receptive field: linear and non-linear spatio-temporal processing by v1 neurons. The Visual Neurosciences 2004;1:704–19.
[17] Olshausen BA. 20 years of learning about vision: questions answered, questions unanswered, and questions not yet asked. In: 20 years of computational neuroscience. Springer; 2013. p. 243–70.
[18] Poirazi P, Brannon T, Mel BW. Pyramidal neuron as two-layer neural network. Neuron 2003;37(6):989–99.
[19] Polsky A, Mel BW, Schiller J. Computational subunits in thin dendrites of pyramidal cells. Nature Neuroscience 2004;7(6):621.
[20] Andrzejak RG, Rummel C, Mormann F, Schindler K. All together now: analogies between chimera state collapses and epileptic seizures. Scientific Reports 2016;6:23000.

Index

Printed in the United States
By Bookmasters